HANDBOOK OF ELECTRONIC FORMULAS, SYMBOLS AND DEFINITIONS

John R. Brand

D1447274

VNR VAN NOSTRAND REINHOLD COMPANY
NEW YORK CINCINNATI ATLANTA DALLAS SAN FRANCISCO
LONDON TORONTO MELBOURNE

9-21-81

Van Nostrand Reinhold Company Regional Offices:
New York Cincinnati Atlanta Dallas San Francisco

Van Nostrand Reinhold Company International Offices:
London Toronto Melbourne

Copyright © 1979 by Litton Educational Publishing, Inc.

Library of Congress Catalog Card Number: 78-26242
ISBN: 0-442-20999-1

All rights reserved. No part of this work covered by the copyright hereon may be reproduced or used in any form or by any means—graphic, electronic, or mechanical, including photocopying, recording, taping, or information storage and retrieval systems—without permission of the publisher.

Manufactured in the United States of America

Published by Van Nostrand Reinhold Company
135 West 50th Street, New York, N.Y. 10020

Published simultaneously in Canada by Van Nostrand Reinhold Ltd.

15 14 13 12 11 10 9 8 7 6 5 4 3 2 1

Library of Congress Cataloging in Publication Data

Brand, John R
Handbook of electronics formulas, symbols, and definitions.

Includes index.
1. Electronics—Handbooks, manuals, etc. I. Title.
TK7825.B7 621.381'02'02 78-26242
ISBN 0-442-20999-1

PREFACE

The *Handbook of Electronics Formulas, Symbols and Definitions* has been compiled for engineers, technicians, armed forces personnel, commercial operators, students, hobbyists, and all others who have some knowledge of electronic terms, symbols, and theory.

The author's intention has been to provide:

A small, light reference book that may be easily carried in an attaché case or kept in a desk drawer for easy access.

A source for the majority of all electronic formulas, symbols, and definitions needed or desired for today's passive and active analog circuit technology.

A format in which a desired formula may be located almost instantly without the use of an index, in the desired transposition, and in sufficiently parenthesized linear form for direct use with any scientific calculator.

Sufficient information, alternate methods, approximations, schematic diagrams, and/or footnotes in such a manner so that technicians and hobbyists may understand and use the majority of thc formulas, and that is acceptable and equally useful to engineers and others very knowledgeable in the field.

All manuscript material has been checked several times for accuracy; however, it is possible that a few errors remain. Readers are encouraged to report any errors found.

All comments, suggestions, criticisms, improvements, additional formulas, or additional material that may be useful in improving subsequent editions will also be welcomed. Readers who have the time, inclination, and knowledge to prepare a full page of additional material in the proper format for inclusion in subsequent editions will be credited at the bottom of any page used.

ACKNOWLEDGMENTS

Much of the material is this *Handbook* is based upon a small loose-leaf notebook containing formulas and other reference material compiled over many years. With the passage of time, the sources of this material have become unknown. It is impossible therefore to list and give the proper credit.

It is possible, however, to give richly deserved recognition to Juergen Wahl for his assistance during preparation of the manuscript. His suggestions, comments, proofreading, and checking of formulas for accuracy are all greatly appreciated.

Special thanks are due to my wife and family for their understanding and acceptance of long periods of neglect, without which this book would not have been possible.

INTRODUCTION

All formulas in this *Handbook* use only the basic units of all terms. It is especially easy in this age of scientific calculators to convert to and from basic units.

Formulas in all sections are listed alphabetically by symbol with the exception of applicable passive circuit symbols, where, for a given resultant, all series circuit formulas are listed first, followed by parallel and complex circuit formulas.

If the symbol for an electronic term is unknown, a liberally cross-referenced listing of electronic terms and their corresponding symbols may be found in the appendix.

Symbols of all reactive magnitude terms in formulas have been consistently given the signs conventionally associated with them to maintain capacitive or inductive identity. In rectangular quantities, this also allows identification of the complex number as representing a series equivalent impedance/voltage or a parallel equivalent admittance/current.

To prevent possible confusion, all symbols representing vector quantities in polar or rectangular form have been printed in boldface.

A number of formulas have the potential to develop a zero divisor. Conventional mathematics prohibits a division by zero, and calculators will overflow if this is attempted. However, formulas noted ⓓ allow the manual conversion of the reciprocal of zero to infinity and the reciprocal of infinity to zero. Division by zero in formulas noted ⊗ is prohibited.

Textbooks conventionally use italic (slanted) type for quantity symbols and roman (upright) type for unit symbols. However, this *Handbook* follows the example of almost all technical manuals, using roman type for both quantity and unit symbols.

CONTENTS

Preface iii
Introduction v

Section 1 Passive Circuits

1.1 English Letters 1
1.2 Greek Letters 185

Section 2 Transistors

2.1 Static Conditions 201
2.2 Small Signal Conditions 221

Section 3 Operational Amplifiers

3.1 Symbols and Definitions 253
3.2 Formulas and Circuits 277

Appendix

A Table of 5% Value Ratios 321
B Electronic Terms and their Corresponding Symbols 327

SECTION ONE

PASSIVE CIRCUITS

1.1 ENGLISH LETTERS

A — Ampere, Amplification etc.

A = Symbol for ampere.

A = Basic unit of electric current.

A = Coulombs per second.

A = $6.24 \cdot 10^{18}$ elementary charges per second (electrons or holes).

A = Unit often used with multiplier prefixes.
$pA = 10^{-12}$ A, $nA = 10^{-9}$ A, $\mu A = 10^{-6}$ A
$mA = 10^{-3}$ A, $kA = 10^{3}$ A, etc.

A = Symbol for area. Area is measured in various unit such as in^2, ft^2, cm^2, m^2 etc.

Ah = Symbol for ampere-hours. One ampere-hour equals 3600 coulomb (C).

At or A = Symbol for ampere turn, the SI unit of magneto-motive force.

A_i = Symbol for current amplification.
See—Active Circuits

A_v = Symbol for voltage amplification.
See—Active Circuits.

---

a = Symbol for atto. A multiplier prefix for 10^{-18}.

a = Substitute for greek letter alpha. (Not recommended)
See—α

a = Not recommended as a quantity symbol.

B Susceptance Definitions

B = Symbol for susceptance

B = The ease with which an alternating current of a given frequency at a given potential flows in a circuit containing only pure capacitive and/or inductive elements. The imaginary part of admittance. The reciprocal of reactance in any purely reactive circuit. The reciprocal of a pure reactance in parallel with other elements.

B = Magnitude of susceptance measured in mho (old) or siemens (new). Siemens (S) and mho (Ω^{-1}) are equal.

$B = |B| = B_{\text{absolute value}} = B_{\text{magnitude}}$

B = Complete description of susceptance

$\mathbf{B} = B\angle{\pm 90^\circ} = 0 \pm jB = 0 - (\pm B)\,j$

$\mathbf{B}_C$ = Capacitive susceptance

$\mathbf{B}_C = B\angle{+90^\circ} = 0 + jB = 0 - (-B)\,j$

$\mathbf{B}_L$ = Inductive susceptance

$\mathbf{B}_L = B\angle{-90^\circ} = 0 - jB = 0 - (+B)\,j$

B_C = B magnitude identified as capacitive

B_L = B magnitude identified as inductive

$-B$ = B magnitude "given" the sign usually associated with capacitive quantities. $-B = B_C$

$+B$ = B magnitude "given" the sign usually associated with inductive quantities. $+B = B_L$

$\pm B$ = Identification of B as capacitive or inductive in many formulas.

$\pm B$ = Identification of B as capacitive or inductive in the resultant of all formulas in this handbook.

Susceptance, Series Circuits **B**	Applicable Notes	Terms
$B_C = -X_C/(X_C^2 + R_s^2)$	① ② ③	R_s, X_C
$B_L = X_L/(X_L^2 + R_s^2)$	① ② ③	R_s, X_L
$\pm B = \pm X_s/(X_s^2 + R_s^2)$	① ② ③	$R_s, \pm X_s$
$B_C = -X_C/Z^2$	① ② ③	X_C, Z
$B_L = X_L/Z^2$	① ② ③	X_L, Z
$\pm B = \pm X_s/Z^2$	① ② ③	$\pm X_s, Z$
$\pm B = -Y[\sin(\pm\theta_Y)]$	① ② ③ ④ ⑬	$Y, \pm\theta_Y$
$\pm B = [\sin(\pm\theta_Z)]/Z$	① ② ③ ④ ⑬	$Z, \pm\theta_Z$

B *Notes:*

① B IS INTRINSICALLY A PARALLEL CIRCUIT QUANTITY. B DERIVED FROM A SERIES CIRCUIT IS THE EQUIVALENT PARALLEL CIRCUIT REACTANCE IN RECIPROCAL FORM.

② R_s = Series R, X_s = Series X.

③ B and X are magnitudes, however both B and X have been "given" the signs usually associated with capacitive and inductive quantities. B_C therefore "equals" –B, B_L "equals" +B, X_C "equals" –X and X_L "equals" +X. This allows direct identification of a reactive quantity derived from any formula in this handbook.

④ The form $(\pm\theta)$ is used as a reminder that the sign of the phase angle determines the sign of B and therefore the identity of B as either capacitive or inductive.

Susceptance, Parallel Circuits **B**	Applicable Notes	Terms
$(B_C)_t = (-B_C)_1 + (-B_C)_2 \cdots + (-B_C)_n$ $-B_t = (-B_1) + (-B_2) \cdots + (-B_n)$	③ ⑩	B_C $-B$
$(B_L)_t = (B_L)_1 + (B_L)_2 \cdots + (B_L)_n$ $+B_t = (+B_1) + (+B_2) \cdots + (+B_n)$	③ ⑩	B_L $+B$
$(B_C)_t = -\omega(C_1 + C_2 \cdots + C_n)$	③ ⑤ ⑩	C
$(B_L)_t = \omega^{-1}(L_1^{-1} + L_2^{-1} \cdots + L_n^{-1})$	③ ⑤ ⑥ ⑩	L
$(B_C)_t = (-X_C)_1^{-1} + (-X_C)_2^{-1} \cdots + (-X_C)_n^{-1}$ $-B_t = (-X_1)^{-1} + (-X_2)^{-1} \cdots + (-X_n)^{-1}$	③ ⑥ ⑩	X_C $-X$
$(B_L)_t = (+X_L)_1^{-1} + (+X_L)_2^{-1} \cdots + (+X_L)_n^{-1}$ $+B_t = (+X_1)^{-1} + (+X_2)^{-1} \cdots + (+X_n)^{-1}$	③ ⑥ ⑩	X_L $+X$
$\|B\|$ = The magnitude of the imaginary part of $\mathbf{Y}_{RECT}$ $\pm B$ = The imaginary part of $\mathbf{Y}_{RECT}$ multiplied by $-j$.	③ ⑦ ⑧ ⑨	**Y** RECT

B *Notes:*

⑤ $\omega = 2\pi f$ = angular velocity

⑥ $x^{-1} = 1/x$

⑦ $\mathbf{Y}_{RECT} = G \pm j|B| \quad = G - j(\pm B) \quad = G - (\pm B)\,j$

⑧ $|B|$ = magnitude of B without knowledge of vectorial direction. $|B|$ therefore cannot be identified as either capacitive or inductive.

⑨ $(-j) \cdot (-j) = +1, \quad (-j) \cdot (+j) = -1$

⑩ $(x)_t$ = total x = equivalent x

Susceptance, Parallel Circuits **B**	Applicable Notes	Terms
$\pm B_t = B_L - B_C$ $\pm B_t = (\pm B_1) + (\pm B_2) \cdots + (\pm B_n)$	③ ⑩	B_C, B_L $-B$, $+B$
$\pm B_t = (\omega L)^{-1} - (\omega C)$ $\pm B_t = (\omega L_1)^{-1} - (\omega C_1) + (\omega L_2)^{-1} - (\omega C_2) \cdots$	③ ⑤ ⑥ ⑩	C L
$\lvert B\rvert = \sqrt{Y^2 - G^2}$	⑧	G, Y
$\pm B = -G\left[\tan(\pm\theta_Y)\right]$	③ ④	G, θ_Y
$\lvert B\rvert = \sqrt{Z^{-2} - R^{-2}}$	⑧ ⑪ ⑫	R_p, Z
$\pm B = \left[\tan(\pm\theta_Z)\right]/R$	③ ④ ⑫	R_p, θ_Z
$\pm B_t = X_L^{-1} - X_C^{-1}$ $\pm B_t = (\pm X_1)^{-1} + (\pm X_2)^{-1} \cdots + (\pm X_n)^{-1}$	③ ⑥ ⑩	X_C, X_L $-X$ $+X$
$\pm B = -Y\left[\sin(\pm\theta_Y)\right]$	③ ④ ⑬	Y, θ_Y
$\pm B = \left[\sin(\pm\theta_Z)\right]/Z$	③ ④ ⑬	Z, θ_Z

B *Notes:*

⑪ $x^{-2} = 1/x^2$

⑫ R_p = parallel resistance

⑬ If the admittance (Y) or the impedance (Z) and the associated phase angle (θ_Y or θ_Z) are known, it is immaterial if the circuit configuration (i.e., series or parallel) is unknown.

B Magnetic Flux Density, Bandwidth

B = Symbol for bel. (Rarely used)

B = Ten decibels (dB)
See—dB

B = Symbol for magnetic flux density.

B = The magnetic flux per unit area perpendicular to the direction of flux. (also known as magnetic induction)

B = Magnetic flux density measured in telsa (T), gauss (G), maxwell (Mx) per square centimeter and lines of flux per square inch.

telsa (T) = weber (Wb) per square meter
gauss (G) = 10^{-4} telsa
maxwell (Mx) = lines of flux
weber (Wb) = 10^8 maxwell (Mx) = 10^8 lines of flux

$B = \phi/A$ where A = cross sectional area of magnetic path.
ϕ = Total magnetic flux in weber, maxwell or lines of flux.

$B = \mu H$ where μ = permeability
H = magnetic force

B = Symbol for bandwidth (not recommended)

$\overline{B}$ = Symbol for bandwidth (not recommended)

BW or $\overline{BW}$ is the preferred symbol for bandwidth.
See—BW

B_1 = Symbol for unity gain bandwidth. ($BW_{(Av=1)}$)
See—Active Circuits, Opamp

3 dB Down Bandwidth

BW

Bandwidth

BW = Symbol for bandwidth

Other symbols for or abbreviations of bandwidth include: B, $\bar{B}$, $(f_2 - f_1)$, B.W., $\overline{BW}$, $BW_{-3\,dB}$

BW = The difference between the two frequencies of a continuous frequency band where the output has fallen to one half power. (−3 dB is very close to one half power)

BW = Bandwidth expressed in hertz (Hz).

$BW = (f_2 - f_1)_{-3\,dB}$

$BW = f_r/Q$

$BW = (f_r R)/X_{L\,(@\,fr)}$

$BW = R/(2\pi L)$

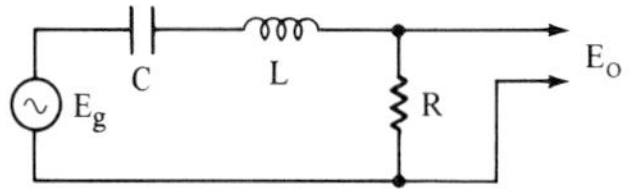

$f_r = \left(2\pi\sqrt{LC}\right)^{-1}$

$BW = (f_2 - f_1)_{-3\,dB}$

$BW = f_r/Q$

$BW = (f_r X_{C\,(@\,fr)})/R$

$BW = (2\pi RC)^{-1}$

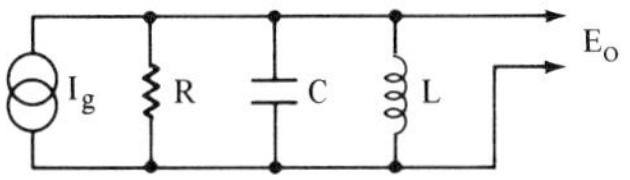

$f_r = \left(2\pi\sqrt{LC}\right)^{-1}$

$BW_{(Av=1)}$—See—Active Circuits, Opamp

$\overline{BW}$ = Average bandwidth. Effective noise bandwidth. See also—BW Active Circuits, Opamp

BW *Notes:*

See—Q for frequency to bandwidth ratio.
See—D for bandwidth to frequency ratio.
See also—d Active Circuits.

C

Capacitance etc.
Definitions

C = The symbol for capacitance.

C = 1. In a system of conductors and dielectric or in a capacitor, that property which permits the storage of electrical energy.

2. The property which determines the quantity of electric charge at a given potential.

3. In a system of conductors (plates) and dielectric (insulator) or in a capacitor, the ratio of the quantity of electric charge to the potential developed.

C = Capacitance (also known as capacity) measured in farad (F) units unless noted.

[This extremely large unit is very seldom used except in formulas. The resultant of all capacitance formulas should be converted to more practical units such as microfarads (μF) or picofarads (pF)]

C $\approx$ 7 pF per sq in of parallel plates separated by $\frac{1}{32}$ in of air.

C = The symbol for capacitor on part lists and schematics.

C = The symbol for coulomb (unit of quantity of charge) (Seldom used in electronics)

c = Obsolete symbol for cycles per second. [Use hertz (Hz)]

c = The symbol for the velocity of light or electromagnetic waves (physics). (Not recommended. Use v for velocity in electronics)

Capacitance, Series Circuits — C	Applicable Notes	Terms
$C_t = (C_1^{-1} + C_2^{-1} \cdots + C_n^{-1})^{-1}$	①	C
$C_x = (C_t^{-1} - C_1^{-1})^{-1}$	②	
$C_t = \omega^{-1}\left[(X_C)_1 + (X_C)_2 \cdots + (X_C)_n\right]^{-1}$	① ②	X_C
$C_t = -\omega^{-1}\left[(-X_1) + (-X_2) \cdots + (-X_n)\right]^{-1}$	③	$-X$
$C = D/(\omega R_s)$ Series reactive element must be capacitive	① ②	D R_s
$C = (\omega R_s Q)^{-1}$ Series reactive element must be capacitive.	① ②	Q R_s
$C = \left[-\omega R_s(\tan\theta_Z)\right]^{-1}$ θ_Z must be negative	① ②	R_s θ_Z
$C = \left[-\omega Z(\sin\theta_Z)\right]^{-1}$ θ_Z must be negative	① ②	Z θ_Z
Series to Parallel Conversion	① ②	C_s R_s
$C_p = \left[(\omega^2 R_s^2 C_s) + C_s^{-1}\right]^{-1}$	④	

C *Notes:*

① C = Capacitance, D = Dissipation Factor, Q = Quality Factor, R = Resistance, X_C and $-X$ = Capacitive Reactance, Z = Impedance, θ = Phase Angle, ω = Angular Velocity
Subscripts: C = capacitive, n = any number, p = parallel, s = series, t = total or equivalent, x = unknown

② $x^{-1} = 1/x$, $\omega = 2\pi f$

③ B and X are magnitudes, however both B and X are often "given" the signs usually associated with capacitive and inductive quantities. In all formulas in this handbook $-B = B_C$, $-X = X_C$, $+B = B_L$ and $+X = X_L$

④ Equivalent capacitance varies with frequency.

Capacitance, Parallel Circuits **C**	Applicable Notes	Terms
$C_t = \left[(B_C)_1 + (B_C)_2 \cdots + (B_C)_n\right]/\omega$	① ②	B_C
$C_t = \left[(-B_1) + (-B_2) \cdots + (-B_n)\right]/-\omega$	③ ⑤	$-B$
$C_t = C_1 + C_2 \cdots + C_n$	①	C
$C_t = \left[(X_C)_1^{-1} + (X_C)_2^{-1} \cdots + (X_C)_n^{-1}\right]/\omega$	① ②	X_C
$C_t = \left[(-X_1)^{-1} + (-X_2)^{-1} \cdots + (-X_n)^{-1}\right]/-\omega$	③	$-X$
$C_p = (\omega R_p D)^{-1}$ Parallel reactive element known to be capacitive	① ②	D, R_p
$C_p = \left[G(\tan\theta_Y)\right]/\omega$ θ_Y must be positive	① ② ⑤	G, θ_Y
$C_p = Q/(\omega R_p)$ Parallel reactive element known to be capacitive	① ②	Q R_p
$C_p = (\tan\theta_Z)/(-\omega R_p)$ θ_Z must be negative	① ②	R_p, θ_Z
$C_p = \left[Y(\sin\theta_Y)\right]/\omega$ θ_Y must be positive	① ② ⑤	Y, θ_Y
$C_p = (\sin\theta_Z)/(-\omega Z)$ θ_Z must be negative	① ②	Z, θ_Z
$C_p = \left[I_t(\sin\theta_I)\ /(\omega E)\right]$ θ_I must be positive	① ② ⑤	E I θ_I
Parallel to Series Conversion $C_s = (\omega^2 C_p R_p^2)^{-1} + C_p$	① ② ④	C_p R_p

C *Notes:* ⑤ B = Susceptance, E = rms Voltage, G = Conductance, I = rms Current, Y = Admittance

Capacitance Misc. Formulas	C	Applicable Notes	Terms
$C_r = (\omega^2 L)^{-1}$	C required for resonance. Series or parallel circuits	② ⑥	L
$C = Q/E$	C required for a charge of Q coulombs	⑥	E, Q
$C = Q^2/(2W)$	W = work equiv. stored energy in watt/sec Q = charge in coulombs	⑥	Q, W
$C = T/R$	C required for time constant T and resistor R	⑥	R, T
$C = (It)/E$	I = constant current E = voltage change after time t	⑥	C, E I

Capacitance of two parallel plates (conductors) separated by an insulator (dielectric)

$$C = (Ak)/(4.45d) \text{ approx. pF}$$

A = Useful area of each plate in square inches
d = Spacing or distance between plates in inches
k = Dielectric constant (Air = 1)

Capacitance of concentric cylinders (e.g., coaxial cable)

$$C = (7.354k)/\left[\log(D/d)\right] \text{ pF per foot length}$$

D = inside diameter of outside cylinder (inches)
d = outside diameter of inside cylinder (inches)
k = dielectric constant of material between cylinders (Air = 1)

C *Notes:*

⑥ C_r = Resonant Capacitance, E = dc Voltage, I = dc Current, L = Inductance, Q = Charge in coulombs, t = Time in sec., T = Time Constant, W = Work in joules

D

Dissipation Factor Definitions

D = The symbol for dissipation factor

D = 1. The ratio of energy dissipated to the energy stored in dielectric material, in certain electric elements, or in certain electric structures.

2. The inverse of the quality factor Q. (also known as the storage or merit factor)

3. In certain electric elements or structures, the absolute value of the cotangent of the phase angle of the alternating current with respect to the voltage, the voltage with respect to the alternating current, the impedance, or the admittance.

D = A factor which usually has a numerical value of from zero to one and is expressed in either decimal or percentage form.

D = A factor most commonly associated with capacitor specifications or measurements, however may be used in all Q factor applications.

$D \simeq$ Power factor when $D < .1$

D = A factor which is very useful for the calculation of equivalent series resistance. ($R_s = DX_C = D/(\omega C) = DX_L = D\omega L$)

D *Notes:*

① B = Susceptance, C = Capacitance, G = Conductance, L = Inductance, Q = Storage Factor, Quality Factor or Merit Factor, R = Resistance, X = Reactance, Y = Admittance, Z = Impedance, θ = Phase Angle, ω = Angular Velocity
Subscripts: p = parallel, s = series

② $\omega = 2\pi f$, $x^{-1} = 1/x$, $x^{-\frac{1}{2}} = 1/\sqrt{x}$, $x^{-2} = 1/x^2$

③ Not valid for LC circuits.

Dissipation Factor, Series Circuits **D**	Applicable Notes	Terms
$D = 1/Q$	①	Q
$D = \cot \theta$ Exception ③	① ⑥	θ
$D = \omega C_s R_s$	① ②	C_s R_s
$D = \sqrt{(Z\omega C_s)^2 - 1}$	① ② ⑧	C_s Z
$D = R_s/(\omega L_s)$	① ②	L_s R_s
$D = \sqrt{[Z/(\omega L_s)]^2 - 1}$	① ② ⑧	L_s Z
$D = R_s/X_s$	① ④	R_s X_s
$D = [(Z/R_s)^2 - 1]^{-\frac{1}{2}}$ Exception ③	① ② ⑧	R_s Z
$D = \sqrt{(Z/X_s)^2 - 1}$	① ④ ⑧	X_s Z
$D_r = \omega C_s R_s = R_s/(\omega L_s)$	① ② ⑦	C_s L_s R_s
$D_r = R_s/X_C = R_s/X_L$	① ② ⑦	X_C X_L R_s

D *Notes:*

④ X_s may be X_C or X_L but not $(X_L - X_C)$

⑤ B may be B_C or B_L but not $(B_L - B_C)$

⑥ cotan x = 1/(tan x)

⑦ D_r = Dissipation Factor at Resonance,
X_C = Capacitive Reactance,
X_L = Inductive Reactance

⑧ If the resultant under the radical sign is negative, a mistake has occurred.

Dissipation Factor, Parallel Circuits **D**	Applicable Notes	Terms
$D = 1/Q$	①	Q
$D = \cot\theta$ Exception ③	① ⑥	θ
$D = G/B$	① ⑤	B G
$D = \sqrt{(Y/B)^2 - 1}$	① ⑤ ⑧	B Y
$D = (R_p \omega C_p)^{-1}$	① ②	C_p R_p
$D = \sqrt{[Y/(\omega C_p)]^2 - 1}$	① ② ⑧	C_p Y
$D = \sqrt{(Z\omega C_p)^{-2} - 1}$	① ② ⑧	C_p Z
$D = \left[(Y/G)^2 - 1\right]^{-\frac{1}{2}}$ Exception ③	① ② ⑧	G Y
$D = (\omega L_p)/R_p$	① ②	L_p R_p
$D = \sqrt{(Y\omega L_p)^2 - 1}$	① ② ⑧	L_p Y
$D = \sqrt{[(\omega L_p)/Z]^2 - 1}$	① ② ⑧	L_p Z
$D = X_p/R_p$	① ④	R_p X_p
$D = \left[(R_p/Z)^2 - 1\right]^{-\frac{1}{2}}$ Exception ③	① ② ⑧	R_p Z
$D = \sqrt{(X_p/Z)^2 - 1}$	① ④ ⑧	X_p Z

dB

Decibel Definitions and Formulas

dB = The symbol for decibel

dB = 1. The standard logarithmic unit for expressing power gain or loss.

2. One tenth of a bel. (The basic bel unit is very seldom used)

3. A power ratio only—according to the original definition and to a few purists.

4. A commonly used convenient unit for expressing voltage and current ratios. See—dB Note 2

Formulas for Definitions 1, 2, & 3

$$dB = 10 \log (P_o/P_i)$$

$$dB = 20 \log (E_o/E_i) \quad \text{only when} \quad (Z_o\angle\theta_o) = (Z_i\angle\theta_i)$$

$$dB = 20 \log (I_o/I_i) \quad \text{only when} \quad (Z_o\angle\theta_o) = (Z_i\angle\theta_i)$$

$$dB = 20 \log \left[\left(E_o\sqrt{Z_i \cos\theta_i}\right)/\left(E_i\sqrt{Z_o \cos\theta_o}\right)\right]$$

$$dB = 20 \log \left[\left(I_o\sqrt{Z_o \cos\theta_o}\right)/\left(I_i\sqrt{Z_i \cos\theta_i}\right)\right]$$

Formulas for Definition 4

$$dB = 10 \log (P_o/P_i)$$
$$dB = 20 \log (E_o/E_i)$$
$$dB = 20 \log (I_o/I_i)$$

dB *Notes:*

① log = logarithm to the base 10, P = Power, E = rms Voltage, I = rms Current, θ = Phase Angle, Subscripts: i = Input, o = Output

② When using definition 4, it should be stated as dB voltage or current gain or loss, dB apparent power gain or loss, etc. ---, not as dB gain or loss or as dB power gain or loss.

③ See also—dBm notes, dB editorial—opamp

dBm

Power in dB Definitions and Formulas

dBm = Symbol for decibels referenced to one milliwatt.

dBm = Power level expressed in decibels above or below one milliwatt.

$dBm = L_{P(mW)}$

dBm = V.U. (volume units) (sinewave only)

$dBm = 10(\log P) + 30$

$dBm = 10[\log(1000\ P)]$

$dBm = 10\ [\log(E^2/R)]\ + 30$

$dBm = 10\ [\log(I^2 R)]\ + 30$

$dBm = 10\ [\log(EI \cos\theta)]\ + 30$

$dBm = 10\ [\log(I^2 Z \cos\theta)]\ + 30$

$dBm = 10\left(\log\left[(E^2 \cos\theta)/Z\right]\right) + 30$

$dBm = 10[\log(E^2 Y \cos\theta)]\ + 30$

dBm *Notes:*

① P = Power, E = dc or rms Voltage, I = dc or rms Current, Y = Admittance, Z = Impedance, θ = Phase Angle, cos = cosine, log = Logarithm to the base 10.

② When using a calculator to obtain the log of a number smaller than one, the value of both the characteristic and the mantissa are likely to be different than the value obtained from log tables. The calculator value will have both a negative characteristic and a negative mantissa. This is the correct value to use. (Log tables always have a positive mantissa)

E

Voltage Definitions

E = Symbol for electromotive force (emf) (emf is more commonly called voltage or potential)

E = The electric force which causes current to flow through a conductor.

E = Potential measured in volts (V)

$E = E_{dc}$ or $|E_{rms}|$

$\mathbf{E}$ = Complete description of voltage

$$\mathbf{E} = \mathbf{E}_{POLAR} = \mathbf{E}_{RECTANGULAR}$$

$$\mathbf{E} = E\angle\theta_E \quad = E_R + (\pm E_X)\,j$$

$$\mathbf{E}_R = E\angle 0^\circ \quad = E_R + 0j$$

$$\mathbf{E}_C = E\angle -90^\circ \quad = 0 + (-E_X)\,j \quad = 0 - jE_X$$

$$\mathbf{E}_L = E\angle +90^\circ \quad = 0 + (+E_X)\,j \quad = 0 + jE_X$$

$E_R = E_{magnitude}$ identified as resistive or real

$E_C = E_{magnitude}$ identified as capacitive

$E_L = E_{magnitude}$ identified as inductive

$-E_X = E_C$ "given" the sign associated with capacitive quantities.

$+E_X = E_L$ "given" the sign associated with inductive quantities.

$\pm E_X$ = Identification of E_X as capacitive or inductive in the resultant of many formulas.

e = The instantaneous value of voltage

Note: The symbol V is also used for voltage and predominates in active circuits. See–V, Active Circuits

Voltage, DC Circuits **E**	Applicable Notes	Terms	Circuit Type
$E_t = (\pm E_1) + (\pm E_2) \cdots + (\pm E_n)$	① ②	E	Series Circuits
$E_t = P_t/I$ $E_t = (P_1 + P_2 \cdots + P_n)/I$	① ②	I P	Series Circuits
$E_t = IR_t$ $E_t = I(R_1 + R_2 \cdots + R_n)$	① ②	I R	Series Circuits
$E_t = \sqrt{P_t R_t}$ $E_t = \sqrt{(P_1 + P_2 \cdots + P_n)(R_1 + R_2 \cdots + R_n)}$	① ②	P R	Series Circuits
$E = I_1/G_1 = I_t/G_t$	① ②	G I	Parallel Circuits
$E = \sqrt{P_1/G_1} = \sqrt{P_t/G_t}$	① ②	G P	Parallel Circuits
$E = P_1/I_1 = P_t/I_t$	① ②	I P	Parallel Circuits
$E = I_1 R_1 = I_t R_t$	① ②	I R	Parallel Circuits
$E = \sqrt{R_1 P_1} = \sqrt{R_t P_t}$	① ②	R P	Parallel Circuits
See–R, complex circuits See–R, delta to Y conversion See also–I and P if necessary Simplify circuit and use above formulas			Complex Circuits

Transient Voltages, Voltage Ratios

e E

$e_C = E\left[1 - (\epsilon^{-1})^{\frac{t}{RC}}\right]$ (E = Applied Voltage) $e_C = .6321\ E$ when $t = RC$ (1 time constant) $e_C = (It)/C$ (I = constant current)	Capacitor Voltage During Charge thru Resistor
$e_C = E/\epsilon^{\frac{t}{RC}}$ (E = Initial Cap. Voltage) $e_C = .3679\ E$ when $t = RC$ (1 time constant) $e_C = E - \left[(It)/C\right]$ (I = constant current)	Capacitor Voltage During Discharge thru Resistor
$e_L = E/\epsilon^{\frac{Rt}{L}}$ (E = Applied Voltage) $e_L = .3679\ E$ when $t = RC$ (1 time constant)	Inductor Voltage During Energization thru Resistor
$e_L = -L(di/dt)$ [(di/dt) = rate of current change in (ampere/seconds)]	Inductor Voltage Developed By Current Change
$E = Q/C$ (Q = Charge in coulombs)	Voltage Developed by Electric Charge

Voltage Ratios

$$E_{av} = \left[(2\sqrt{2})/\pi\right] E_{rms} = .9003\ E_{rms}$$
$$E_{av} = (2/\pi)\ E_{peak} = .6366\ E_{peak}$$
$$E_{peak} = (\sqrt{2})\ E_{rms} = 1.414\ E_{rms}$$
$$E_{p\text{-}p} = (2\sqrt{2})\ E_{rms} = 2.828\ E_{rms}$$
$$E_{rms} = \left[\pi/(2\sqrt{2})\right] E_{av} = 1.111\ E_{av}$$
$$E_{rms} = E_{eff}$$
$$E_{rms} = (1/\sqrt{2})\ E_{peak} = .7071\ E_{peak}$$
$$E_{rms} = \left[1/(2\sqrt{2})\right] E_{p\text{-}p} = .3535\ E_{p\text{-}p}$$

E *Notes:*

① General

B = Susceptance ⑥, C = Capacitance, e = Instantaneous Voltage, E = Voltage Magnitude or DC Voltage ⑥, **E** = Magnitude and Phase Angle of Voltage, f = Frequency, G = Conductance, I = Current, j = Imaginary Number ③, L = Inductance, P = Power, Q = Quantity of Electrical Charge, R = Resistance, X = Reactance ⑥, Y = Admittance, Z = Impedance, ϵ = Base of Natural Logarithms ③, π = Ratio of Circumference to diameter of a circle ③, θ = Phase Angle ⑥, ω = Angular Velocity ③

② Subscripts

C = capacitive, E = voltage, I = current, L = inductive, n = any number, o = output, p = parallel circuit, r = (of or at) resonance, s = series circuit, t = total or equivalent, X = reactive, Y = admittance, Z = impedance

③ Constants

$j = i \quad j = \sqrt{-1} \quad j = 90°$ multiplier, $\epsilon = 2.718+ \quad \epsilon^{-1} = .36788-$, $\pi = 3.1416- \quad 2\pi = 6.2832-$, $\omega = 2\pi f \quad \omega = 6.2832f$

④ Algebra

$x^{-1} = 1/x$, $x^{-2} = 1/x^2$, $x^{\frac{1}{2}} = \sqrt{x}$, $x^{-\frac{1}{2}} = 1/\sqrt{x}$, $x^{\frac{-1}{2}} = 1/\sqrt{x}$, $|x|$ = absolute value or magnitude of x

⑤ Trigonometry

sin = sine, cos = cosine, tan = tangent, $\tan^{-1}$ = arc tangent

⑥ Reminders

$\pm\theta$ --- use the sign of the phase angle

$\pm X$ --- $-X$ identifies X as capacitive (X_C)
$+X$ identifies X as inductive (X_L)

$\pm B$ --- $-B$ identifies B as capacitive (B_C)
$+B$ identifies B as inductive (B_L)

$\pm E_X$ --- $-E_X$ identifies E_X as capacitive (E_C)
$+E_X$ identifies E_X as inductive (E_L)

Voltage, Series Circuits **E**	Applicable Notes	Terms
$(E_C)_t = (E_C)_1 + (E_C)_2 \cdots + (E_C)_n$ $(E_L)_t = (E_L)_1 + (E_L)_2 \cdots + (E_L)_n$ $(E_R)_t = (E_R)_1 + (E_R)_2 \cdots + (E_R)_n$	① ②	E_C E_L E_R
$(\pm E_X)_t = (E_L)_1 - (E_C)_1 + (E_L)_2 - (E_C)_2$	① ②	E_C E_L
$(\pm E_X)_t = (\pm E_1) + (\pm E_2) \cdots + (\pm E_n)$	⑥	$-E_X$ $+E_X$
$E_t = \sqrt{E_R^2 + E_C^2}$ $E_t = \sqrt{E_R^2 + E_L^2}$	① ②	E_C E_R E_L E_R
$(E_C)_t = I\omega^{-1}(C_1^{-1} + C_2^{-1} \cdots + C_n^{-1})$	① ② ③ ④	I C
$(E_L)_t = I\omega(L_1 + L_2 \cdots + L_n)$	① ② ③	I L
$(E_R)_t = I(R_1 + R_2 \cdots + R_n)$	① ②	I R
$(E_C)_t = I\left[(X_C)_1 + (X_C)_2 \cdots + (X_C)_n\right]$	① ②	I X_C
$(-E_X)_t = I\left[(-X_1) + (-X_2) \cdots + (-X_n)\right]$	⑥	I −X
$(E_L)_t = I\left[(X_L)_1 + (X_L)_2 \cdots + (X_L)_n\right]$	① ②	I X_L
$(+E_X)_t = I\left[(+X_1) + (+X_2) \cdots + (+X_n)\right]$	⑥	I +X
$E = IZ$	①	I Z

Additional E magnitude formulas are included in **E** formulas starting on page 27.

Voltage, Parallel Circuits **E**	Applicable Notes	Terms
$E = I_t / [(B_C)_1 + (B_C)_2 \cdots + (B_C)_n]$	① ②	I_t B_C
$E = \lvert I_t / [(-B_1) + (-B_2) \cdots + (-B_n)] \rvert$	④ ⑥	I_t $-B$
$E = I_t / [(B_L)_1 + (B_L)_2 \cdots + (B_L)_n]$	① ②	I_t B_L
$E = I_t / [(+B_1) + (+B_2) \cdots + (+B_n)]$	⑥	I_t $+B$
$E = \lvert I_t / [(\pm B_1) + (\pm B_2) \cdots + (\pm B_n)] \rvert$	① ② ④ ⑥	I_t $\pm B$
$E = I_t / [\omega(C_1 + C_2 \cdots + C_n)]$	① ② ③	I_t C_p
$E = I_t / (G_1 + G_2 \cdots + G_n)$	① ②	I_t G
$E = I_t \omega [(L_p)_1^{-1} + (L_p)_2^{-1} \cdots + (L_p)_n^{-1}]^{-1}$	① ② ③ ④	I_t L_p
$E = I_t [(R_p)_1^{-1} + (R_p)_2^{-1} \cdots + (R_p)_n^{-1}]^{-1}$	① ② ④	I_t R_p
$E = I_t [(X_C)_1^{-1} + (X_C)_2^{-1} \cdots + (X_C)_n^{-1}]^{-1}$	① ②	I_t X_C
$E = \lvert I_t [(-X_p)_1^{-1} + (-X_p)_2^{-1} \cdots + (-X_p)_n^{-1}]^{-1} \rvert$	④ ⑥	I_t $-X_p$
$E = I_t [(X_L)_1^{-1} + (X_L)_2^{-1} \cdots + (X_L)_n^{-1}]^{-1}$	① ②	I_t X_L
$E = I_t [(+X_p)_1^{-1} + (+X_p)_2^{-1} \cdots + (+X_p)_n^{-1}]^{-1}$	④ ⑥	I_t $+X_p$
$E = \lvert I_t [(\pm X_p)_1^{-1} + (\pm X_p)_2^{-1} \cdots + (\pm X_p)_n^{-1}]^{-1} \rvert$	① ② ④ ⑥	I_t $\pm X_p$
$E = I/Y$	①	I Y
$E = IZ$	①	I Z

E Complex Voltages, Series & Differential

$$E_t = \left\{ \left([E_1 \cos\theta_1] + [E_2 \cos\theta_2] \cdots + [E_n \cos\theta_n] \right)^2 + \left([E_1 \sin(\pm\theta_1)] + [E_2 \sin(\pm\theta_2)] \cdots + [E_n \sin(\pm\theta_n)] \right)^2 \right\}^{1/2}$$

$$\theta_t = \tan^{-1}\left[\frac{\left([E_1 \sin(\pm\theta_1)] + [E_2 \sin(\pm\theta_2)] \cdots + [E_n \sin(\pm\theta_n)] \right)}{\left([E_1 \cos\theta_1] + [E_2 \cos\theta_2] \cdots + [E_n \cos\theta_n] \right)} \right]$$

Series Sum of $E_1\underline{/\theta_1}$, $E_2\underline{/\theta_2}$, --- $E_n\underline{/\theta_n}$

$$E_t = \sqrt{\left([E_1 \cos\theta_1] - [E_2 \cos\theta_2] \right)^2 + \left([E_1 \sin(\pm\theta_1)] - [E_2 \sin(\pm\theta_2)] \right)^2}$$

$$\theta_t = \tan^{-1}\left[\left([E_1 \sin(\pm\theta_1)] - [E_2 \sin(\pm\theta_2)] \right) \Big/ \left([E_1 \cos\theta_1] - [E_2 \cos\theta_2] \right) \right]$$

$E_1\underline{/\theta_1}$, $E_2\underline{/\theta_2}$ Differential

E Voltage & Phase Important Notes

1. It should be understood by the reader that the phase angle of voltage and current is the same one and only phase angle of a circuit or of a circuit element. The fact that current leads the voltage while the voltage lags the current in an inductive circuit means only that the signs of the voltage and current phase angles are different.
2. In a given circuit, the phase angle of voltage, current, impedance and admittance is the same one and only phase angle. The signs of the angle is the only difference. $\pm\theta_E = -(\pm\theta_I) = \pm\theta_Z = -(\pm\theta_Y)$.
3. The voltage phase angle uses the current phase angle as a reference (0°) while the current phase angle uses the voltage phase angle as a reference (0°). Due to this fact, if the voltage phase angle is expressed, the current phase angle is 0° and if the current phase angle is expressed, the voltage phase angle is 0°. It should be obvious that the voltage and current phase angles cannot be used at the same time.
4. The same applies to rectangular form voltage and current. Rectangular form current cannot have an imaginary (reactive) component when the rectangular form voltage has an imaginary (reactive) component. The reverse, obviously, is also true.
5. Due to this confusing situation and the high probability of error, the author DOES NOT RECOMMEND THE USE OF POLAR OR RECTANGULAR FORM VOLTAGE OR CURRENT WHERE EACH USES THE OTHER AS A REFERENCE. THE USE OF THE GENERATOR AS THE PHASE REFERENCE IS RECOMMENDED.
6. The following polar and rectangular form voltage formulas are listed for reference only. Proceed to the E_o and vector algebra E_o formulas.

E = The magnitude and phase angle of the voltage developed by current through a series circuit. ($\theta_I = 0°$)
See also—θ

$\mathbf{E}_{POLAR} = E \angle \pm\theta_E$

$\mathbf{E}_{RECT}$ = 1. The 0° and ±90° voltages which have a resultant equal to $\mathbf{E}_{POLAR}$.
2. The voltages developed by current through series resistance and net reactance.

$\mathbf{E}_{RECT} = E_R + (\pm E_X)\,j$

$\mathbf{E}_{RECT} = (E \cos\theta_E) + \left[E \sin(\pm\theta_E)\right] j$

	Applicable Notes	Terms
$\mathbf{E}_{POLAR} = \sqrt{E_R^2 + E_C^2}\ \angle \tan^{-1}(-E_C/E_R)$ $\mathbf{E}_{RECT} = E_R - jE_C$	① ② ③ ⑤	E_R E_C
$\mathbf{E}_{POLAR} = \sqrt{E_R^2 + E_L^2}\ \angle \tan^{-1}(E_L/E_R)$ $\mathbf{E}_{RECT} = E_R + jE_L$	① ② ③ ⑤	E_R E_L
$\mathbf{E}_{POLAR} = \sqrt{E_R^2 + E_{Xc}^2}\ \angle \tan^{-1}(-E_X/E_R)$ $\mathbf{E}_{RECT} = E_R + (-E_X)\,j$	① ② ③ ⑤ ⑥	E_R E_{Xc}
$\mathbf{E}_{POLAR} = \sqrt{E_R^2 + E_{XL}^2}\ \angle \tan^{-1}(+E_X/E_R)$ $\mathbf{E}_{RECT} = E_R + (+E_X)\,j$	① ② ③ ⑤ ⑥	E_R E_{XL}
$\mathbf{E}_{POLAR} = \sqrt{E_R^2 + (E_L - E_C)^2}\ \angle \tan^{-1}\left[(E_L - E_C)/E_R\right]$ $\mathbf{E}_{RECT} = E_R + (E_L - E_C)\,j$ $\mathbf{E}_{RECT} = E_R + (\pm E_X)\,j$	① ② ③ ⑤ ⑥	E_R E_C E_L

Voltage and Phase, Series Circuits **E**	Applicable Notes	Terms
$E = I\sqrt{R^2 + (\omega C)^{-2}}$ $\theta_E = \tan^{-1}(\omega CR)^{-1}$	① ② ③ ④ ⑤	I C R
$E = I\sqrt{R^2 + (\omega L)^2}$ $\theta_E = \tan^{-1}[(\omega L)/R]$	① ② ③ ⑤	I L R
$E = P/(I \cos \theta_Z)$ $\theta_E = \pm\theta_Z = -(\pm\theta_I)$	① ② ⑤ ⑥	I P ±θ
$E = (IR)/(\cos \theta)$ $\theta_E = \pm\theta_Z = -(\pm\theta_I)$	① ② ⑤ ⑥	I R ±θ
$E = \left\|(IX)/(\sin \theta)\right\|$ $\theta_E = \pm\theta_Z = -(\pm\theta_I)$	① ② ④ ⑤ ⑥	I |X| ±θ
$E = IZ$ $\theta_E = \pm\theta_Z = -(\pm\theta_I)$	① ② ⑥	I Z ±θ
$E = I\sqrt{R^2 + [(\omega L) - (\omega C)^{-1}]^2}$ $\theta_E = \tan^{-1}\left([(\omega L) - (\omega C)^{-1}]/R\right)$	① ② ③ ⑤	I C L R
$E = I\sqrt{R^2 + (X_L - X_C)^2}$ $\theta_E = \tan^{-1}[(X_L - X_C)/R]$	① ② ⑤	I R X_C X_L

See previous page for definitions, E_{POLAR} and E_{RECT}

E Voltage and Phase When a parallel circuit is driven by a current source

E = The magnitude and phase angle of the voltage developed by the total current through a parallel circuit. ($\theta_{IR} = 0°$) See also—θ

$\mathbf{E}_{POLAR} = E\,\angle{\pm\theta_E}$

$\mathbf{E}_{RECT}$ = 1. The 0° and ±90° voltages which have a resultant equal to $\mathbf{E}_{POLAR}$.

2. The series equivalent voltages of a parallel circuit.

3. The voltages developed by current through the series equivalent of a parallel circuit.

$\mathbf{E}_{RECT} = (E\cos\theta_E) + [E\sin(\pm\theta_E)]\,j$

$\mathbf{E}_{RECT} = (E_R)_s + [(\pm E_X)_s]\,j$

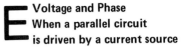

	Applicable Notes	Terms
$E = I_t / \sqrt{G^2 + (B_L - B_C)^2}$ $\theta_E = \tan^{-1}[(B_L - B_C)/G]$	① ② ⑤ ⑥	$I_t \pm B\ G$
$E = \lvert (I_t \sin\theta)/(B_L - B_C) \rvert$ $\theta_E = -(\pm\theta_I) = -(\pm\theta_Y)$	① ② ⑤ ⑥ ⑦	$I_t \pm B \pm\theta$
$E = I_t / \sqrt{R^{-2} + [(\omega L)^{-1} - (\omega C)]^2}$ $\theta_E = \tan^{-1}(R[(\omega L)^{-1} - (\omega C)])$	① ② ③ ④ ⑤	$I_t CL\ R$
$E = \lvert (I_t \sin\theta) / [(\omega L)^{-1} - (\omega C)] \rvert$ $\theta_E = \pm\theta_Z = -(\pm\theta_I) = -(\pm\theta_Y)$	① ② ③ ④ ⑥	$I_t CL \pm\theta$
$E = \lvert (I_t \cos\theta)/G \rvert$ $\theta_E = -(\pm\theta_I) = -(\pm\theta_Y) = \pm\theta_Z$	① ② ④ ⑤ ⑥	$I_t G \pm\theta$

Voltage and Phase, Parallel Circuits With Current Source **E**	Applicable Notes	Terms
$E = Z\sqrt{I_R^2 + (I_{X_L} - I_{X_C})^2}$ $\theta_E = \tan^{-1}\left[(I_{X_L} - I_{X_C})/I_R\right]$	① ② ⑤ ⑥	I_R $\pm I_X$ Z
$E = I_t\sqrt{R^{-2} + (X_L^{-1} - X_C^{-1})^2}$ $\theta_E = \tan^{-1}\left[R(X_L^{-1} - X_C^{-1})\right]$	① ② ④ ⑤ ⑥	I_t R $\pm X$
$E = (I_t R)\cos\theta$ $\theta_E = \pm\theta_Z = -(\pm\theta_I) = -(\pm\theta_Y)$	① ② ⑤ ⑥	I_t R $\pm\theta$
$E = \left\lvert I_t \sin\theta/(X_L^{-1} - X_C^{-1})\right\rvert$ $\theta_E = \pm\theta_Z = -(\pm\theta_I) = -(\pm\theta_Y)$	① ② ③ ④ ⑤ ⑥ ⑦	I_t $\pm X$ $\pm\theta$
$E = I_t/Y$ $\theta_E = -(\pm\theta_Y)$	① ② ⑥	I Y $\pm\theta_Y$
$E = I_t Z$ $\theta_E = \pm\theta_Z$	① ② ⑥	I Z $\pm\theta_Z$
$E = P/(I_t \cos\theta)$ $\theta_E = -(\pm\theta_I) = \pm\theta_Z$	① ② ⑤ ⑥	P I_t $\pm\theta$
$E = \sqrt{(PZ)/(\cos\theta)}$ $\theta_E = \pm\theta_Z = -(\pm\theta_I)$	① ② ⑤ ⑥	P Z $\pm\theta$

See previous page for definitions, E_{POLAR}, and E_{RECT}.

Complex Voltages, Series & Differential **E**	**Terms**
$E_t = \sqrt{(E_R)_t^2 + (E_X)_t^2}$ $\theta_t = \tan^{-1}[(\pm E_X)_t/(E_R)_t]$ $(E_R)_t = (E_1 \cos\theta_1) + (E_2 \cos\theta_2) \cdots + (E_n \cos\theta_n)$ $(\pm E_X)_t = [E_1 \sin(\pm\theta_1)] + [E_2 \sin(\pm\theta_2)] \cdots + [E_n \sin(\pm\theta_n)]$	$E_1\angle\theta_1$ $E_2\angle\theta_2$ $E_n\angle\theta_n$
$E_t = \sqrt{(E_R)_t^2 + (E_X)_t^2}$ $\theta_t = \tan^{-1}[(\pm E_X)_t/(E_R)_t]$ $(E_R)_t = (E_1 \cos\theta_1) - (E_2 \cos\theta_2)$ $(\pm E_X)_t = [E_1 \sin(\pm\theta_1)] - [E_2 \sin(\pm\theta_2)]$	$E_1\angle\theta_1$ $E_2\angle\theta_2$ Differential
$(\mathbf{E}_{RECT})_t = (\mathbf{E}_{RECT})_1 + (\mathbf{E}_{RECT})_2 \cdots + (\mathbf{E}_{RECT})_n$ $\mathbf{E}_{RECT} = E_R \pm jE_X = E_R + (\pm E_X)\,j$ $(\mathbf{E}_{RECT})_t = [(E_R)_1 + (E_R)_2 \cdots + (E_R)_n] + [(\pm E_X)_1 + (\pm E_X)_2 \cdots + (\pm E_X)_n]\,j$	$(\mathbf{E}_{RECT})_1$ $(\mathbf{E}_{RECT})_2$ $(\mathbf{E}_{RECT})_n$
$(\mathbf{E}_{RECT})_t = (\mathbf{E}_{RECT})_1 - (\mathbf{E}_{RECT})_2$ $\mathbf{E}_{RECT} = E_R \pm jE_X = E_R + (\pm E_X)\,j$ $(\mathbf{E}_{RECT})_t = [(E_R)_1 - (E_R)_2] + [(\pm E_X)_1 - (\pm E_X)_2]\,j$	$(\mathbf{E}_{RECT})_1$ $(\mathbf{E}_{RECT})_2$ Differential

E *Notes:*

① $E\angle\theta_E = \mathbf{E}_{POLAR}$
② $E_R = E_{0°} = +E$ = "Real" numbers ($-E_R = E_{180°}$)
③ $E = |E| = E_{polar\ magnitude}$, $\pm E_X = E_{\pm 90°}$
④ $+E_X = E_{+90°} = E_L = E \sin(+\theta)$
⑤ $-E_X = E_{-90°} = E_C = E \sin(-\theta)$

E_o Output Voltage & Phase

$E_o = E_g\left[(R_2/R_1) + 1\right]^{-1}$

$E_o = (E_g R_2)/(R_1 + R_2)$

$\theta_{Eo} = \theta_{Eg} = 0^\circ$

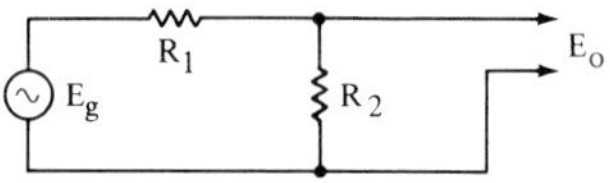

$E_o = (E_g R)/\sqrt{R^2 + X_C^2}$

$E_o = E_g(\cos\theta_{Zi})$

$\theta_{Eo} = -(-\theta_{Zi}) = \tan^{-1}(X_C/R)$

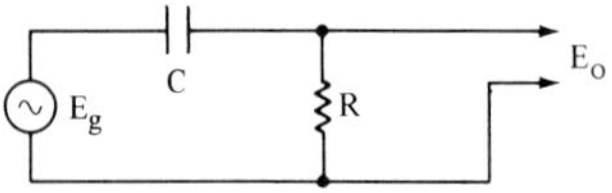

(E_o Leads E_g)

$E_o = (E_g R)/\sqrt{R^2 + X_L^2}$

$E_o = E_g(\cos\theta_{Zi})$

$\theta_{Eo} = -(+\theta_{Zi}) = \tan^{-1} - (X_L/R)$

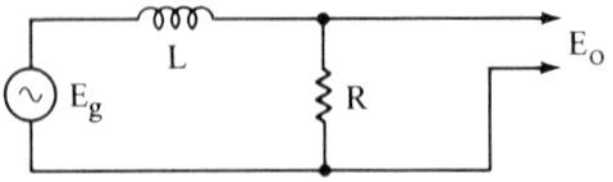

(E_o Lags E_g)

$E_o = (E_g X_C)/\sqrt{R^2 + X_C^2}$

$E_o = \left|E_g(\sin\theta_{Zi})\right|$

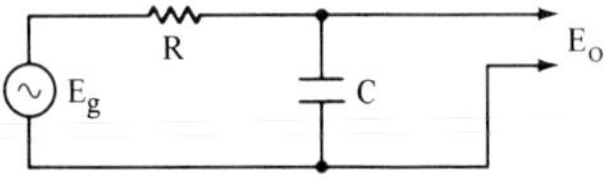

$\theta_{Eo} = -(-\theta_{Zi}) - 90^\circ = \left[\tan^{-1}(X_C/R)\right] - 90^\circ$ (E_o Lags)

E_o *Notes:*

① E_g = Generator voltage, Z_i = Input impedance, θ_{Eo} = Phase angle of output voltage

② $X_L = \omega L$, $X_C = (\omega C)^{-1}$, $\omega = 2\pi f$

③ $x^{-1} = 1/x$, $x^{-2} = 1/x^2$, $x^{\frac{1}{2}} = \sqrt{x}$, $x^{-\frac{1}{2}} = 1/\sqrt{x}$

④ $\tan^{-1}$ = arc tangent, sin = sine, cos = cosine

E_o Output Voltage & Phase

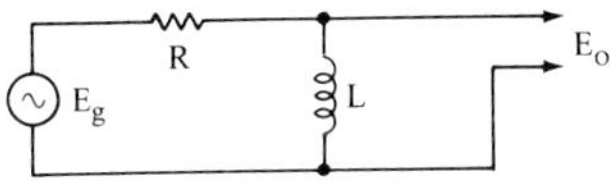

$$E_o = (E_g X_L)/\sqrt{R^2 + X_L^2}$$

$$E_o = \left|E_g(\sin\theta_{Zi})\right|$$

$$\theta_{Eo} = -(+\theta_{Zi}) + 90^\circ = 90^\circ - \left[\tan^{-1}(X_L/R)\right]$$

θ_{Eo} Leads θ_{Eg}

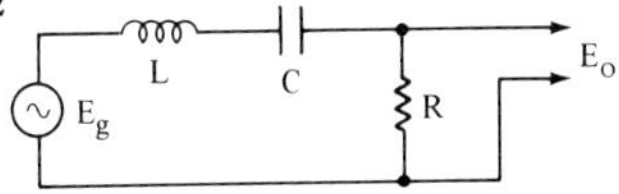

$$E_o = (E_g R)/\sqrt{R^2 + (X_L - X_C)^2}$$

$$E_o = E_g(\cos\theta_{Zi})$$

$$\theta_{Eo} = -(\pm\theta_{Zi}) = \tan^{-1}\left[(X_C - X_L)/R\right] = 0^\circ \text{ @ } f_r$$

$\theta_{Eo} = 0^\circ$ @ f_r, near $+90^\circ$ @ vlf, near -90° @ vhf

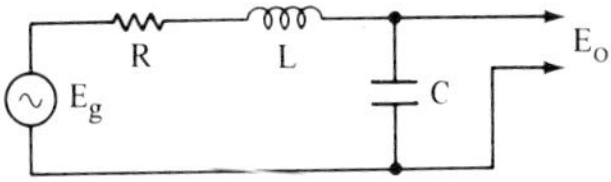

$$E_o = (E_g X_C)/\sqrt{R^2 + (X_L - X_C)^2}$$

$$\theta_{Eo} = -(\pm\theta_{Zi}) - 90^\circ = \left(\tan^{-1}\left[(X_C - X_L)/R\right]\right) - 90^\circ$$

$\theta_{Eo} = -90^\circ$ @ f_r, near 0° @ vlf, near -180° @ vhf

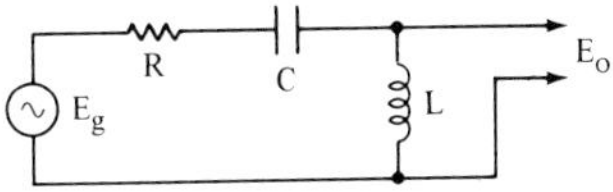

$$E_o = (E_g X_L)/\sqrt{R^2 + (X_L - X_C)^2}$$

$$\theta_{Eo} = +90^\circ - (\pm\theta_{Zi}) = 90^\circ - \left(\tan^{-1}\left[(X_L - X_C)/R\right]\right)$$

$\theta_{Eo} = +90^\circ$ @ f_r, near 0° @ vhf, near 180° @ vlf

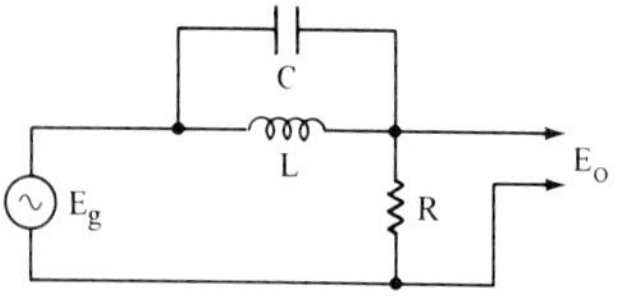

$E_o = (E_g R)/Z_i$

$E_o = E_g(\cos\theta_{Zi})$

$\theta_{Eo} = -(\pm\theta_{Zi})$

where $Z_i = \sqrt{R^2 + (X_L^{-1} - X_C^{-1})^{-2}}$ ⓓ

$\theta_{Zi} = \tan^{-1}\left[R(X_C^{-1} - X_L^{-1})\right]^{-1}$ ⓓ

$\theta_{Eo} = 0^\circ$ @ f_r, Lags θ_{Eg} below f_r, Leads θ_{Eg} above f_r

---

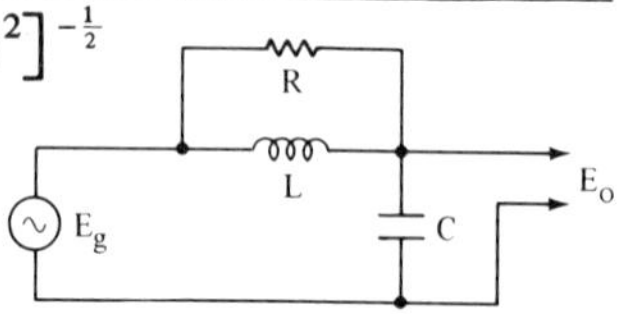

$E_o = (E_g X_C)\left[R_s^2 + (X_{Ls} - X_C)^2\right]^{-\frac{1}{2}}$

$\theta_{Eo} = \theta_{Xc} - \theta_{Zi}$

$\theta_{Eo} = (-90^\circ) - (\pm\theta_{Zi})$

$\theta_{Eo} = (-90^\circ) - \left(\tan^{-1}\left[(X_{Ls} - X_C)/R_s\right]\right)$

where $R_s = \left[(R/X_L^2) + R^{-1}\right]^{-1}$

$X_{Ls} = \left[(X_L/R^2) + X_L^{-1}\right]^{-1}$

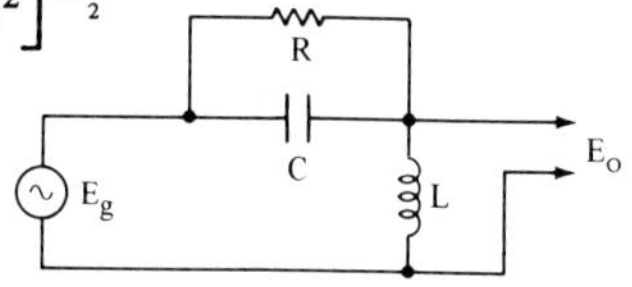

$E_o = (E_g X_L)\left[R_s^2 + (X_L - X_{Cs})^2\right]^{-\frac{1}{2}}$

$\theta_{Eo} = \theta_{X_L} - \theta_{Zi}$

$\theta_{Eo} = +90^\circ - (\pm\theta_{Zi})$

$\theta_{Eo} = +90^\circ - \left(\tan^{-1}\left[(X_L - X_{Cs})/R_s\right]\right)$

where $R_s = \left[(R/X_C^2) + R^{-1}\right]^{-1}$

$X_{Cs} = \left[(X_C/R^2) + X_C^{-1}\right]^{-1}$

$$E_o = E_g\left[Z_i(X_L^{-1} - X_C^{-1})\right]^{-1} \quad ⓓ$$

$$E_o = \left|E_g(\sin\theta_{Zi})^{-1}\right| \quad ⓓ$$

$$\theta_{Eo} = \pm 90^\circ - (\pm\theta_{Zi})$$

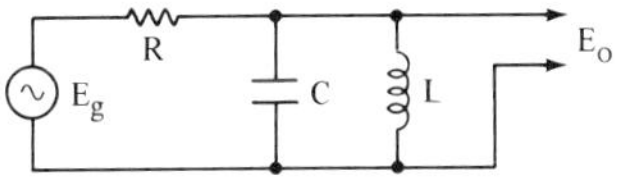

where $Z_i = \sqrt{R^2 + (X_L^{-1} - X_C^{-1})^{-2}}$ ⓓ

$\theta_{Zi} = \tan^{-1}\left[R(X_L^{-1} - X_C^{-1})\right]^{-1}$ ⓓ

$\theta_{Eo} = 0^\circ$ @ f_r, Leads θ_{Eg} below f_r, Lags θ_{Eg} above f_r

---

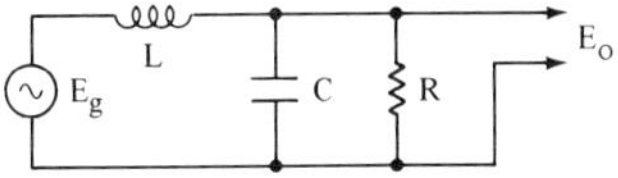

$$E_o = (E_g Z_2)/Z_i$$

$$\theta_{Eo} = \theta_{Z2} - \theta_{Zi}$$

$Z_i = \left[R_s^2 + (X_L - X_{Cs})^2\right]^{\frac{1}{2}}, \theta_{Zi} = \tan^{-1}\left[(X_L - X_{Cs})/R_s\right]$

$Z_2 = (R^{-2} + X_C^{-2})^{-\frac{1}{2}}, \theta_{Z2} = \tan^{-1}(R/X_C)$

where $R_s = \left[(R/X_C^2) + R^{-1}\right]^{-1}$

$X_{Cs} = \left[(X_C/R^2) + X_C^{-1}\right]^{-1}$

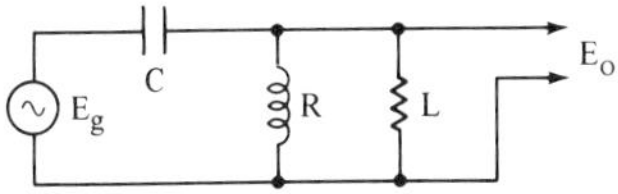

$$E_o = (E_g Z_2)/Z_i$$

$$\theta_{Eo} = \theta_{Z2} - \theta_{Zi}$$

$Z_i = \left[R_s^2 + (X_{Ls} - X_C)^2\right]^{\frac{1}{2}}, \theta_{Zi} = \tan^{-1}\left[(X_{Ls} - X_C)/R_s\right]$

$Z_2 = (R^{-2} + X_L^{-2})^{-\frac{1}{2}}, \theta_{Z2} = \tan^{-1}(R/X_L)$

where $R_s = \left[(R/X_L^2) + R^{-1}\right]^{-1}$

$X_{Ls} = \left[(X_L/R^2) + X_L^{-1}\right]^{-1}$

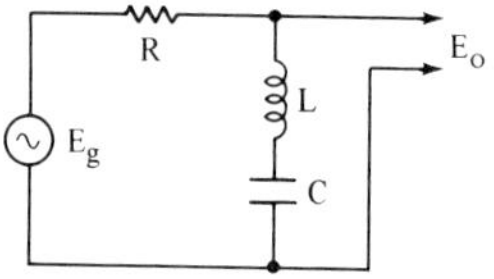

$$E_o = \left[E_g(X_L - X_C)\right]/Z_i$$

$$E_o = \left|E_g(\sin\theta_{Zi})\right|$$

$$\theta_{Eo} = \pm 90^\circ - (\pm\theta_{Zi})$$

where $Z_i = \sqrt{R^2 + (X_L - X_C)^2}$

$\theta_{Zi} = \tan^{-1}\left[(X_L - X_C)/R\right]$

$\theta_{Eo} = 0^\circ$ @ f_r, Lags θ_{Eg} below f_r, Leads θ_{Eg} above f_r

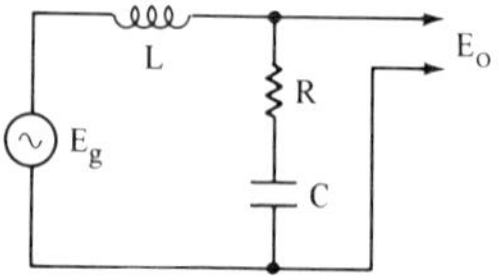

$$E_o = (E_g Z_2)/Z_i$$

$$\theta_{Eo} = \theta_{Z2} - \theta_{Zi}$$

$$\theta_{Eo} = (-\theta_{Z2}) - (\pm\theta_{Zi})$$

where $Z_i = \sqrt{R^2 + (X_L - X_C)^2}$

$\theta_{Zi} = \tan^{-1}\left[(X_L - X_C)/R\right]$

$Z_2 = \sqrt{R^2 + X_C^2}$, $\theta_{Z2} = \tan^{-1}(-X_C/R)$

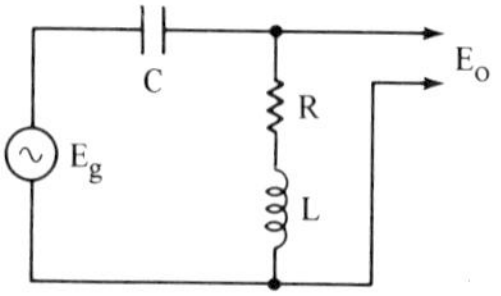

$$E_o = (E_g Z_2)/Z_i$$

$$\theta_{Eo} = \theta_{Z2} - \theta_{Zi}$$

$$\theta_{Eo} = (+\theta_{Z2}) - (\pm\theta_{Zi})$$

where $Z_i = \sqrt{R^2 + (X_L - X_C)^2}$

$\theta_{Zi} = \tan^{-1}\left[(X_L - X_C)/R\right]$

$Z_2 = \sqrt{R^2 + X_L^2}$, $\theta_{Z2} = \tan^{-1}(X_L/R)$

E_o Output Voltage & Phase

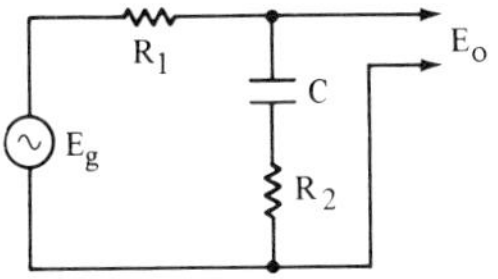

$$E_o = (E_g Z_2)/Z_i$$

$$\theta_{Eo} = \theta_{Z2} - \theta_{Zi}$$

$$\theta_{Eo} = (-\theta_{Z2}) - (-\theta_{Zi})$$

where $Z_i = \sqrt{(R_1 + R_2)^2 + X_C^2}$

$$\theta_{Zi} = \tan^{-1}[-X_C/(R_1 + R_2)]$$

$$Z_2 = \sqrt{R_2^2 + X_C^2}, \ \theta_{Z2} = \tan^{-1}(-X_C/R_2)$$

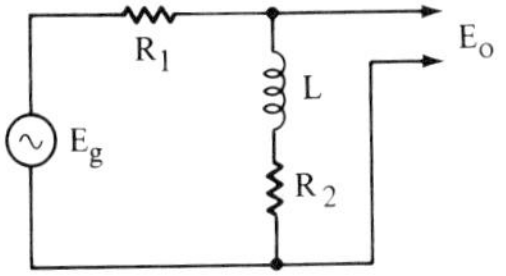

$$E_o = (E_g Z_2)/Z_i$$

$$\theta_{Eo} = \theta_{Z2} - \theta_{Zi}$$

$$\theta_{Eo} = (+\theta_{Z2}) - (+\theta_{Zi})$$

where $Z_i = \sqrt{(R_1 + R_2)^2 + X_L^2}$

$$\theta_{Zi} = \tan^{-1}\left[X_L/(R_1 + R_2)\right]$$

$$Z_2 = \sqrt{R_2^2 + X_L^2}, \ \theta_{Z2} = \tan^{-1}(X_L/R_2)$$

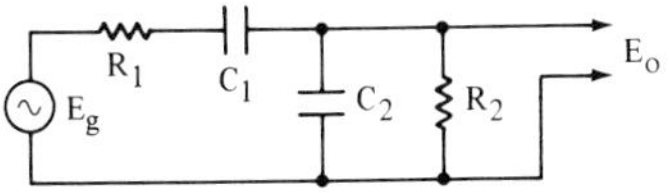

$$E_o = (E_g Z_2)/Z_i$$

$$\theta_{Eo} = \theta_{Z2} - \theta_{Zi}$$

$$E_o = \left[E_g(R_2^{-2} + X_{C2}^{-2})^{-\frac{1}{2}}\right] / \left[(R_1 + R_{2s})^2 + (X_{C1}^{-1} + X_{C2s}^{-1})^{-2}\right]^{\frac{1}{2}}$$

$$\theta_{Eo} = [\tan^{-1}(R_2/-X_{C2})] - \left(\tan^{-1}\left[(R_1 + R_{2s})/-(X_{C1}^{-1} + X_{C2s}^{-1})\right]\right)$$

where $R_{2s} = \left[(R_2/X_{C2}^2) + R_2^{-1}\right]^{-1}$

$$X_{C2s} = \left[(X_{C2}/R_2^2) + X_{C2}^{-1}\right]^{-1}$$

$E_o = I_g R$

$\theta_{Eo} = 0°$

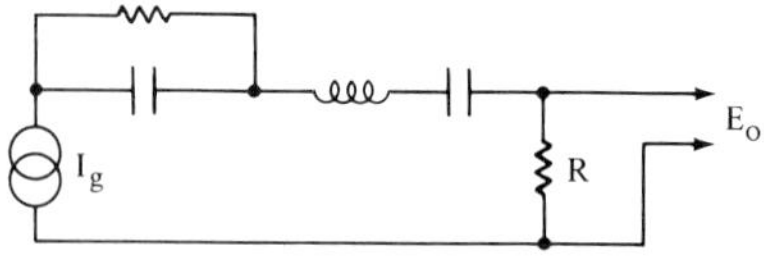

$E_o = I_g X_C$

$\theta_{Eo} = -90°(\theta_{Eo}$ Lags $\theta_{Ig})$

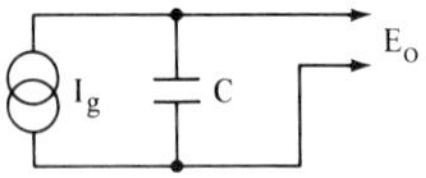

$E_o = I_g X_L$

$\theta_{Eo} = +90°$ $(\theta_{Eo}$ Leads $\theta_{Ig})$

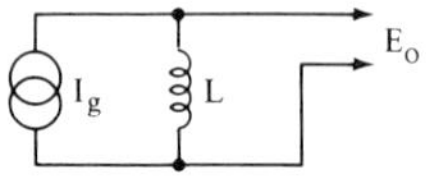

$E_o = I_g Z$

$E_o = I_g \sqrt{R^2 + X_C^2}$

$\theta_{Eo} = (-\theta_Z) = \tan^{-1}(-X_C/R)$

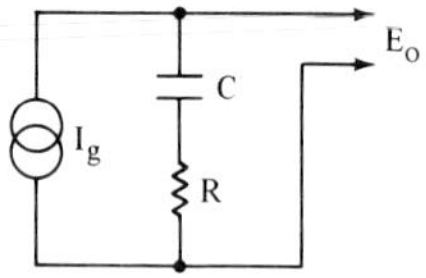

θ_{Eo} Lags θ_{Ig}

$E_o = I_g Z$

$E_o = I_g \sqrt{R^2 + X_L^2}$

$\theta_{Eo} = (+\theta_Z) = \tan^{-1}(X_L/R)$

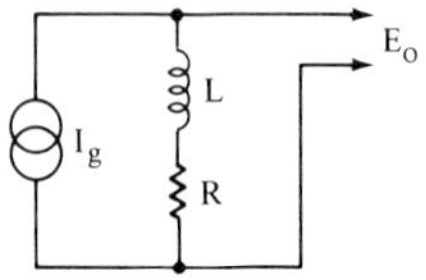

θ_{Eo} Leads θ_{Ig}

Note: —ꝏ— = Infinite impedance alternating current source

Networks Driven By Current Source | E_o | Output Voltage & Phase

$$E_o = I_g Z$$

$$E_o = I_g (R^{-2} + X_C^{-2})^{-\frac{1}{2}}$$

$$\theta_{Eo} = +(-\theta_Z) = \tan^{-1}(R/-X_C)$$

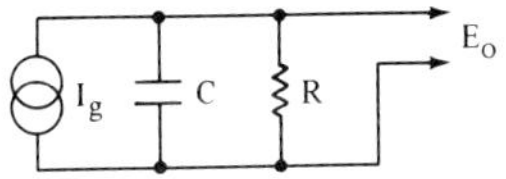

θ_{Eo} Lags θ_{Ig}

$$E_o = I_g Z$$

$$E_o = I_g (R^{-2} + X_L^{-2})^{-\frac{1}{2}}$$

$$\theta_{Eo} = +(+\theta_Z) = \tan^{-1}(R/X_L)$$

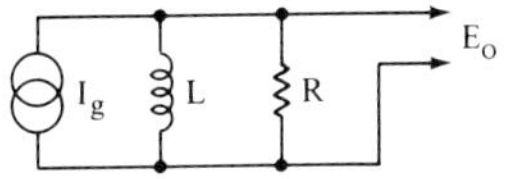

θ_{Eo} Leads θ_{Ig}

$$E_o = I_g Z$$

$$\theta_{Eo} = +(\pm\theta_Z)$$

$$E_o = I_g \left[R^{-2} + (X_L^{-1} - X_C^{-1})^2\right]^{-\frac{1}{2}} = I_g R \ @ \ f_r$$

$$\theta_{Eo} = \tan^{-1}\left[R(X_L^{-1} - X_C^{-1})\right]$$

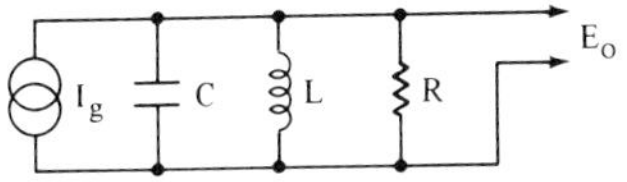

$$E_o = I_g Z$$

$$\theta_{Eo} = +(\pm\theta_Z)$$

$$E_o = I_g \sqrt{R^2 + (X_L - X_C)^2}$$

$$\theta_{Eo} = \tan^{-1}\left[(X_L - X_C)/R\right]$$

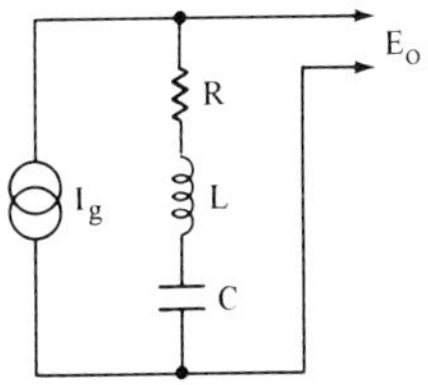

$$E_o = I_g Z$$

$$\theta_{Eo} = \pm\theta_Z$$

$$E_o = I_g\left[R^{-2} + (X_L - X_C)^{-2}\right]^{-\frac{1}{2}}$$

$$\theta_{Eo} = \tan^{-1}\left[R/(X_L - X_C)\right]$$

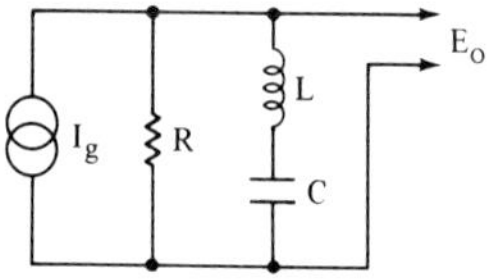

$$E_o = I_g Z$$

$$\theta_{Eo} = \pm\theta_Z = \pm 90°$$

$$E_o = \left|I_g\left[X_{C1}^{-1} + (X_L - X_{C2})^{-1}\right]^{-1}\right|$$

ⓓ

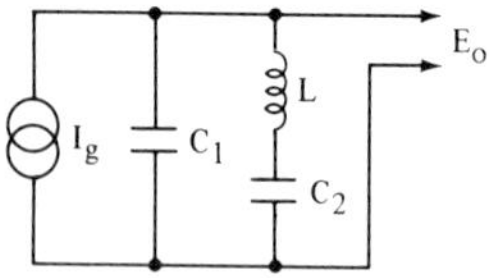

$$E_o = I_g Z$$

$$\theta_{Eo} = \pm\theta_Z = \pm 90°$$

$$E_o = \left|I_g\left[X_{L1}^{-1} + (X_{L2} - X_C)^{-1}\right]^{-1}\right|$$

ⓓ

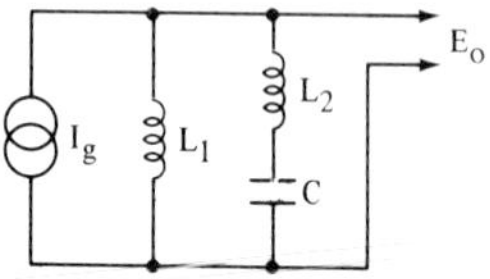

$$E_o = (I_g Z_i X_{C2})/(X_L - X_{C2})$$

$$\theta_{Eo} = (\pm\theta_{Zi}) + (-90°) - (\pm 90°)$$

$$E_o = I_g/\left[X_{C1}^{-1} + X_{C2}^{-1} - (X_L/X_{C1}X_{C2})\right]$$

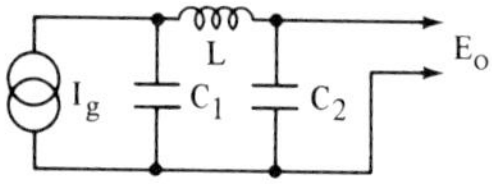

See also—**Z** complex circuits, **Y** complex circuits

Note ⓓ

If the reciprocal of zero is presented, $E_O = \infty$.

E Voltage Vector Algebra

Vector Algebra AC Ohms Law

$\mathbf{E}_g = E_g \underline{/0^\circ}$ or $\mathbf{I}_g = I_g \underline{/0^\circ}$

$\mathbf{E} = \mathbf{I}_g \mathbf{Z} = I_g Z \underline{/0^\circ + \theta_Z = \pm\theta_Z}$

$\mathbf{I} = \mathbf{E}_g / \mathbf{Z} = E_g / Z \underline{/0^\circ - \theta_Z = -(\pm\theta_Z)}$

$\mathbf{Z} = \mathbf{E}_g / \mathbf{I} = E_g / I \underline{/0^\circ - \theta_I = -(\pm\theta_I)}$

$\mathbf{Z} = \mathbf{E} / \mathbf{I}_g = E / I_g \underline{/\theta_E - 0^\circ = \pm\theta_E}$

Addition and Subtraction of Rect. Quantities

$$\mathbf{E}_1 + \mathbf{E}_2 = \mathbf{E}_{1\,(\text{RECT.})} + \mathbf{E}_{2\,(\text{RECT.})}$$

$$= \left[E_R + (\pm E_X)\,j\right]_1 + \left[E_R + (\pm E_X)\,j\right]_2$$

$$= \left[(E_R)_1 + (E_R)_2\right] + \left[(\pm E_X)_1 + (\pm E_X)_2\right] j$$

$$\mathbf{E}_1 - \mathbf{E}_2 = \left[(E_R)_1 - (E_R)_2\right] + \left[(\pm E_X)_1 - (\pm E_X)_2\right] j$$

$$\left|+E_X\right| = E_L \qquad \left|-E_X\right| = E_C$$

$$\mathbf{I}_1 + \mathbf{I}_2 = \mathbf{I}_{1(\text{RECT})} + \mathbf{I}_{2\,(\text{RECT.})}$$

$$= \left[I_R - (\pm I_X)\,j\right]_1 + \left[I_R - (\pm I_X)\,j\right]_2$$

$$= \left[(I_R)_1 + (I_R)_2\right] - \left[(\pm I_X)_1 + (\pm I_X)_2\right] j$$

$$\mathbf{I}_1 - \mathbf{I}_2 = \left[(I_R)_1 - (I_R)_2\right] - \left[(\pm I_X)_1 - (\pm I_X)_2\right] j$$

$$\left|+I_X\right| = I_L \qquad \left|-I_X\right| = I_C$$

Note: The rectangular current of a series circuit represents current through an equivalent parallel circuit.

Note: See $\mathbf{Z}_{\text{RECT}}$ for addition and subtraction of impedance

E_o

Output Voltage
Vector Algebra

$\mathbf{I}_g = I_g \angle 0°$

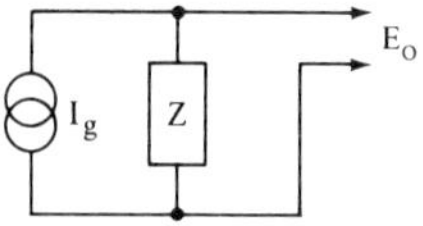

$\mathbf{Z}_i = \mathbf{Z}$

$\mathbf{Z}_o = \mathbf{Z}$

$\mathbf{E}_o = \mathbf{I}_g \mathbf{Z}$

$\mathbf{E}_g = E_g \angle 0°$

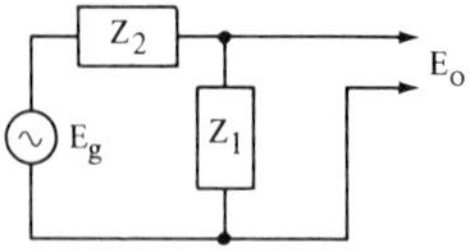

$\mathbf{Z}_i = \mathbf{Z}_1 + \mathbf{Z}_2$

$\mathbf{Z}_o = [\mathbf{Z}_1^{-1} + \mathbf{Z}_2^{-1}]^{-1}$

$\mathbf{Y}_o = \mathbf{Y}_1 + \mathbf{Y}_2$

$\mathbf{E}_o = (\mathbf{E}_g \mathbf{Z}_1)/\mathbf{Z}_i$

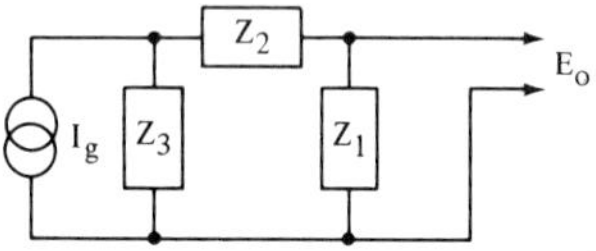

$\mathbf{I}_g = I_g \angle 0°$ $\quad \mathbf{E}_g = \mathbf{I}_g \mathbf{Z}_i$

$\mathbf{Z}_i = \left[\mathbf{Z}_3^{-1} + (\mathbf{Z}_2 + \mathbf{Z}_1)^{-1}\right]^{-1}$

$\mathbf{Z}_o = \left[\mathbf{Z}_1^{-1} + (\mathbf{Z}_2 + \mathbf{Z}_3)^{-1}\right]^{-1}$

$\mathbf{Y}_o = \mathbf{Y}_1 + (\mathbf{Y}_2^{-1} + \mathbf{Y}_3^{-1})^{-1}$

$\mathbf{E}_o = \mathbf{I}_g \mathbf{Z}_1 \left[1 - (\mathbf{Z}_i/\mathbf{Z}_3)\right]$

$\mathbf{E}_o$ Output Voltage Vector Algebra

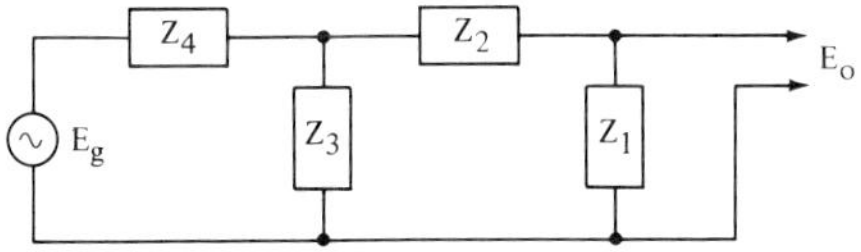

$\mathbf{E}_g = E_g \angle 0^\circ \quad \mathbf{I}_g = \mathbf{E}_g/\mathbf{Z}_i$

$\mathbf{Z}_i = \mathbf{Z}_4 + \left[\mathbf{Z}_3^{-1} + (\mathbf{Z}_2 + \mathbf{Z}_1)^{-1}\right]^{-1}$

$\mathbf{Z}_o = \left(\mathbf{Z}_1^{-1} + \left[\mathbf{Z}_2 + (\mathbf{Z}_3^{-1} + \mathbf{Z}_4^{-1})^{-1}\right]^{-1}\right)^{-1}$

$\mathbf{Y}_o = \mathbf{Y}_1 + \left[\mathbf{Y}_2^{-1} + (\mathbf{Y}_3 + \mathbf{Y}_4)^{-1}\right]^{-1}$

$\mathbf{E}_o = \mathbf{E}_g\left[1 - (\mathbf{Z}_4/\mathbf{Z}_i)\right] / \left[(\mathbf{Z}_2/\mathbf{Z}_1) + 1\right]$

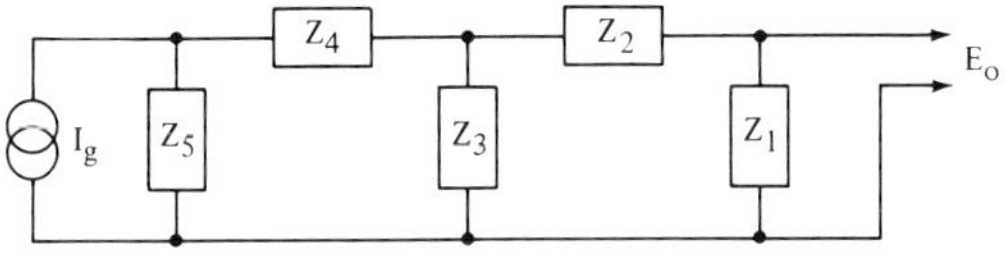

$\mathbf{I}_g = I_g \angle 0^\circ \quad \mathbf{E}_g = \mathbf{I}_g \mathbf{Z}_i$

$\mathbf{Z}_i = \left[\mathbf{Z}_5^{-1} + \left(\mathbf{Z}_4 + \left[\mathbf{Z}_3^{-1} + (\mathbf{Z}_2 + \mathbf{Z}_1)^{-1}\right]^{-1}\right)^{-1}\right]^{-1}$

$\mathbf{Z}_o = \left[\mathbf{Z}_1^{-1} + \left(\mathbf{Z}_2 + \left[\mathbf{Z}_3^{-1} + (\mathbf{Z}_2 + \mathbf{Z}_1)^{-1}\right]^{-1}\right)^{-1}\right]^{-1}$

$\mathbf{Y}_o = \mathbf{Y}_1 + \left(\mathbf{Y}_2^{-1} + \left[\mathbf{Y}_3 + (\mathbf{Y}_2^{-1} + \mathbf{Y}_1^{-1})^{-1}\right]^{-1}\right)^{-1}$

$\mathbf{E}_o = \left[\mathbf{I}_g(\mathbf{Z}_i - \mathbf{Z}_4)\right] / \left[(\mathbf{Z}_2/\mathbf{Z}_1) + 1\right]$

Current Source to Voltage Source Conversion

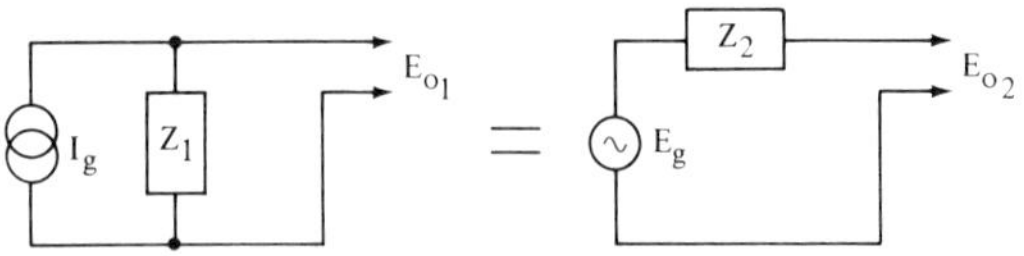

$E_g = I_g Z_1$

$\theta_{Eg} = \theta_{Ig} = 0^\circ$

$\mathbf{Z}_2 = \mathbf{Z}_1$ at all frequencies

$\mathbf{E}_{o2} = \mathbf{E}_{o1}$ at all frequencies

Voltage Source to Current Source Conversion

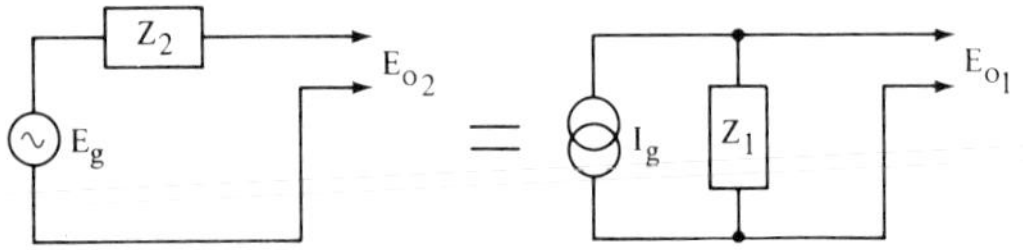

$I_g = E_g/Z_2$

$\theta_{Ig} = \theta_{Eg} = 0^\circ$

$\mathbf{Z}_1 = \mathbf{Z}_2$ at all frequencies

$\mathbf{E}_{o1} = \mathbf{E}_{o2}$ at all frequencies

Note: $\mathbf{E}_o$ may be loaded in any manner and the two outputs although changed will remain equal to each other.

e_N E_N

Noise Voltage

$e_{N(th)}$ = Thermal noise (white noise) voltage of resistance. (Other symbols of thermal noise voltage include E_N, E_{TH}, $E_{N(TH)}$, e_n, $e_{N(TH)}$, $e_{N(\sqrt{\sim})}$, $e_{N(\sqrt{Hz})}$, V_N, $V_{N(TH)}$ etc)

Note: Thermal noise voltage is always rms voltage regardless of symbol used.

$e_{N(th)} = \sqrt{4kT_K R\, BW}$

k = Boltzmann constant ($1.38 \cdot 10^{-23}$ J/°K)

T_K = Temperature in Kelvin. (°C + 273.15)

BW = Noise bandwidth in hertz. (Noise measured with infinite attenuation of frequencies outside of bandwidth)

$e_{N(\sqrt{Hz})}$ = Thermal noise per hertz. (per root hertz)

$e_{N(\sqrt{Hz})} = 1.283 \cdot 10^{-10}\sqrt{R}$ @ 25°C and 1 Hz bandwidth

$E_{N(EX)}$ = The noise (1/f noise) voltage (rms) of a resistor in excess of thermal noise.

$E_{N(EX)}$ = Resistor excess noise voltage (rms) in microvolts per volt of dc voltage drop per decade of frequency.

$E_{N(EX)} = 10^{-6}\, E_{dc}\left[\log^{-1}(\overline{NI}/20)\right]$

NI = Noise Index in dB (a specification)

NI = +10 to −20 dB (carbon composition)

NI = −10 to −25 dB (carbon film)

NI = −15 to −40 dB (metal film or wirewound)

E_N Note: $\log^{-1}$ = antilog$_{10}$

f

f = Symbol for femto.

f = A multiplier prefix meaning 10^{-15} unit.

f = Symbol for frequency.

f = The number of complete cycles per second of alternating current, sound, electromagnetic radiation, vibrations or certain other periodic events.

f = Frequency measured in hertz (Hz). (old cps)

f_c = Crossover or cutoff frequency. (3 dB down)

f_o = Oscillation, output or reference frequency.

f_r = Frequency of resonance.

$f = 1/t$ (t = time of one cycle)

$f = v/\lambda$

Sound in Air

$f \approx 1136/\lambda$ (λ in feet, @25°C)

$f \approx 346.3/\lambda$ (λ in meters, @25°C)

Electromagnetic waves including radio frequency and light in air or vacuum.

$f \approx (9.83 \cdot 10^8)/\lambda$ (λ in feet)

$f \approx (3 \cdot 10^8)/\lambda$ (λ in meters)

$f = 1/(2\pi X_C C)$

$f = X_L/(2\pi L)$

Notes: X = reactance, v = velocity, λ = wavelength, C = Capacitance, L = Inductance

f_c Crossover Frequency

$f_c = (2\pi CR)^{-1}$

$f_c = \frac{R}{2\pi L}$

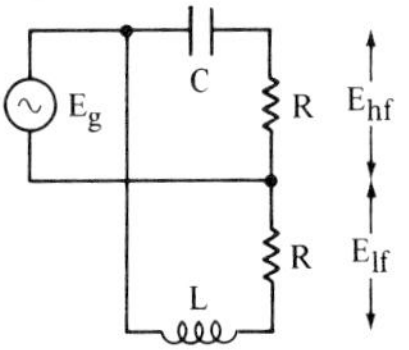

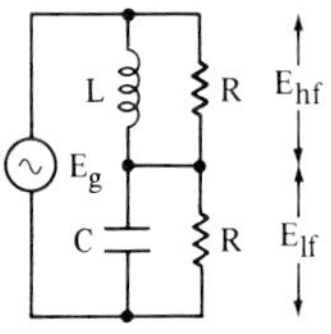

$X_L = X_C = R$ when $f = f_c$

$\mathbf{Z} = R$, $(\mathbf{E}_{lf} + \mathbf{E}_{hf}) = \mathbf{E}_g$ when $f = 0$ to ∞

$L = R/(2\pi f_c)$, $C = (2\pi f_c R)^{-1}$

$E_{C(max)} = E_{L(max)} = E_g$

$I_{L(max)} = I_{C(max)} = (E_g/R)$

$f_c = (2\sqrt{2}\pi CR)^{-1}$

$f_c = R/(\sqrt{2}\pi L)$

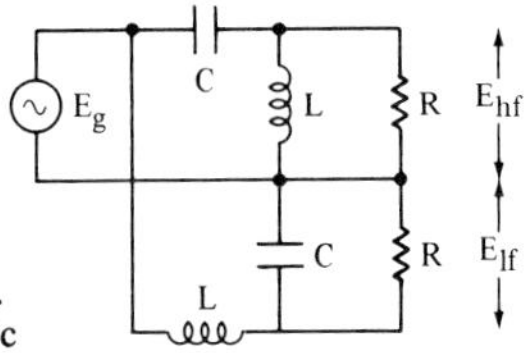

$X_L = X_C = \sqrt{2}\,R$ when $f = f_c$

$\mathbf{Z} = R$, $(\mathbf{E}_{lf} + \mathbf{E}_{hf}) = \mathbf{E}_g$ when $f = 0$ to ∞

$L = R/(2\pi f_c\sqrt{2})$, $C = (2\pi f_c R\sqrt{2})^{-1}$

$E_{C(max)} = E_{L(max)} = 1.272\ E_g$

$I_{L(max)} = I_{C(max)} = 1.029\ (E_g/R)$

f_c *Notes:*

① C = Capacitance, E = rms Voltage Magnitude, **E** = Polar or Rectangular form of Voltage, I = rms Current Magnitude, L = Inductance, R = Resistance, X = Reactance, **Z** = Polar or Rectangular form of Impedance

f_c Crossover Frequency

$$f_c = \left(\sqrt{2}\pi CR\right)^{-1}$$

$$f_c = \frac{R}{2\pi L\sqrt{2}}$$

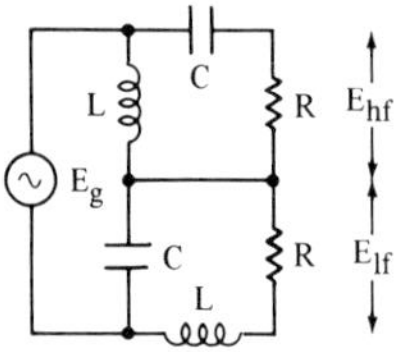

$$X_L = X_C = R/\sqrt{2} \quad \text{when} \quad f = f_c$$

$$\mathbf{Z} = R, \quad (\mathbf{E}_{lf} + \mathbf{E}_{hf}) = \mathbf{E}_g \quad \text{when} \quad f = 0 \text{ to } \infty$$

$$L = R/\left(2\pi f_c\sqrt{2}\right), \quad C = \left(\sqrt{2}\pi f_c R\right)^{-1}$$

$$E_{C(max)} = E_{L(max)} = 1.029\ E_g$$

$$I_{L(max)} = I_{C(max)} = 1.272\ (E_g/R)$$

3 dB Down Frequency — f_c — Cutoff Frequency

$$f_c = (2\pi CR)^{-1}$$

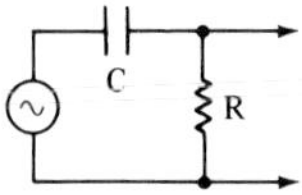

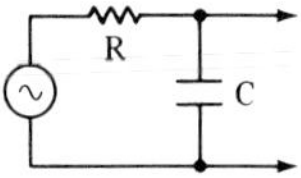

$$f_c = R/(2\pi L)$$

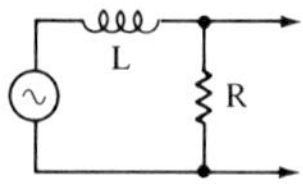

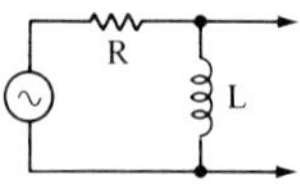

f_c *Notes:*

② E_C = Capacitor voltage, E_g = Generator voltage, E_L = Inductor voltage, E_{hf} = High freq. voltage, E_{lf} = Low freq. voltage, I_C = Capacitor current, I_L = Inductor current

③ $x^{-1} = 1/x$

④ $\pi = 3.1416$, $\sqrt{2} = 1.414$, $2\pi = 6.2832$, $\sqrt{2}\pi = 4.443$

f' f_{FC}

f_{FC} = Symbol for flare cutoff frequency.

f_{FC} = In an exponential horn of infinite length, the frequency below which no energy is coupled through the horn.

f_{FC} = .5 to .8 of the lowest frequency of interest in the usual exponential horn.

$f_{FC} = v/(18.13\,\ell_{2A})$

$f_{FC} = v/\left(18.13\sqrt{\ell_{2d}}\right)$

$f_{FC} = v/\left(18.13\sqrt{\ell_{2r}}\right)$

$f_{FC} = (mv)/(4\pi)$

f_{FC} *Notes:*

ℓ_{2A}, ℓ_{2d}, ℓ_{2r} = Length between points on the horn center line of double cross sectional area, double diameter, and double radius respectively.

m = Flare constant = $.6931/\ell_{2A} = .6931/\sqrt{\ell_{2d}}$

v = Velocity of sound ≈ 13,630 in./sec, 1136 ft/sec, 346.3 meters/sec, 34,630 cm/sec at 25°C

f' = The lowest frequency of "satisfactory" horn loading due to area of horn mouth.

f' = Frequency at which:

1. Mouth diameter equals $\frac{1}{4}$ wavelength
2. Mouth circumference equals one wavelength
3. Mouth diameter equals $\frac{1}{3}$ wavelength
4. Mouth diameter equals $\frac{1}{2}$ wavelength
5. Mouth diameter equals $\frac{2}{3}$ wavelength

Low frequency horns are almost always compromised to use criteria 1, 2 or 3. Wavelength–See λ

f_o Frequency of Oscillation or Output

$f_o \approx (2\pi\sqrt{LC})^{-1}$ (LC Oscillator)

$f_o \approx (2\pi RC\sqrt{6})^{-1}$ (Phase Shift Oscillator with three equal RC stages. May be phase lead or phase lag type)

$f_o \approx (2\pi R_1 C_1)^{-1}$ (Wein Bridge Oscillator) (when $R_1C_1 = R_2C_2$)

See–Active Circuits

Output frequency of a electromechanical generator

$f_o = N_{pp}(_{rps})$ (Number of pairs of poles times rev./sec)

f_r Resonant Frequency Definitions

f_r = Symbol for frequency of resonance.

f_r = 1. The frequency at which the circuit acts as a pure resistance. In a series circuit, the frequency at which the impedance is lowest. In a parallel circuit, the frequency at which the impedance is highest.

2. The frequency at which the inductive reactance equals the capacitive reactance.

Note: Definition 2 is commonly used due to simpler mathematics, but in many low Q circuits, it is a poor approximation.

When Q is high, the difference between the definitions is negligible.

Series Resonant Frequency

f_r

Resonant Frequency, Series Resonance

$f_r = \left(2\pi\sqrt{LC}\right)^{-1}$ Def. 1 & 2

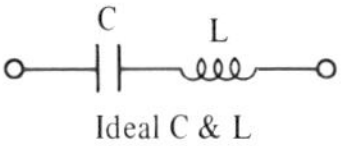

@ f_r $Z = 0$

$X_L = X_C$, $Q = \infty$

$f_r = \left(2\pi\sqrt{LC}\right)^{-1}$ Def. 1 & 2

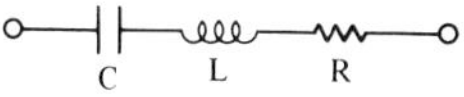

@ f_r $Z = R$

$\theta_Z = 0°$, $X_L = X_C$, $Q = X_L/R$

$f_r = \left[(LC) - (L/R)^2\right]^{-\frac{1}{2}}/(2\pi)$ Def. 1
$(LC > L^2/R^2)$

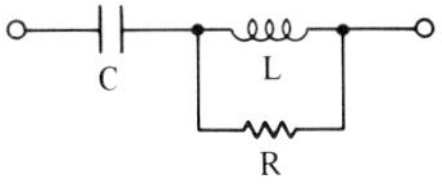

$f_r \approx \left(2\pi\sqrt{LC}\right)^{-1}$

@ f_r Def. 1 — @ f_r

$Z = \left[(R/X_L^2) + R^{-1}\right]^{-1}$ — $Z \approx X_L^2/R$

$\theta_Z = 0°$

$X_C = \left[(X_L/R^2) + X_L^{-1}\right]^{-1}$ — $X_C \approx X_L$

$Q = X_C\left[(R/X_L^2) + R^{-1}\right]$ — $Q \approx R/X_L$

f_r *Notes:*

① C = Capacitance, L = Inductance, Q = "Q" Factor, R = Resistance, R_C = Resistance in capacitive circuit, R_L = Resistance in inductive circuit, X_C = Capacitive reactance, X_L = Inductive reactance, Z = Impedance, θ_Z = Phase angle of impedance

Series Resonant Frequency	f_r	Resonant Frequency, Series Resonance

$$f_r = \sqrt{(LC)^{-1} - (CR)^{-2}}/(2\pi) \quad \text{Def. 1}$$

exception = $\sqrt{-x}$

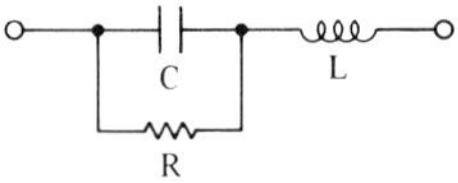

$$f_r \approx \left(2\pi\sqrt{LC}\right)^{-1}$$

@ f_r Definition 1

$$Z = \left[(R/X_C^2) + R^{-1}\right]^{-1} \qquad Z \approx X_C^2/R$$

$$\theta_Z = 0°$$

$$X_L = \left[(X_C/R^2) + X_C^{-1}\right]^{-1} \qquad X_L \approx X_C$$

$$Q = X_L\left[(R/X_C^2) + R^{-1}\right] \qquad Q \approx R/X_C$$

$$f_r = \sqrt{\left[(R_C^2C)^{-1} - L^{-1}\right]/\left[(L/R_L^2) - C\right]}\,/(2\pi) \quad \text{Def. 1}$$

exception = $\sqrt{-x}$

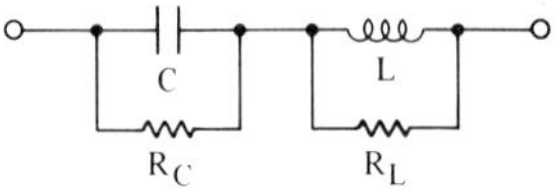

$$f_r \approx \left(2\pi\sqrt{LC}\right)^{-1}$$

@ f_r (Definition 1)

$$Z = \left[(R_C/X_C^2) + R_C^{-1}\right]^{-1} + \left[(R_L/X_L^2) + R_L^{-1}\right]^{-1}$$

$$\theta_Z = 0°$$

$$\left[(X_L/R_L^2) + X_L^{-1}\right] = \left[(X_C/R_C^2) + X_C^{-1}\right]$$

$$Q \approx \left[X_L(R_L^{-1} + R_C^{-1})\right]^{-1}$$

Parallel Resonant Frequency

f_r

Resonant Frequency, Parallel Resonance

$f_r = \left(2\pi\sqrt{LC}\right)^{-1}$ Def. 1 & 2

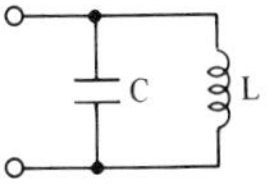

@ f_r, $Z = \infty$, $\theta_Z = 0°$

$X_C = X_L$, $Q = \infty$

$f_r = \left(2\pi\sqrt{LC}\right)^{-1}$ Def. 1 & 2

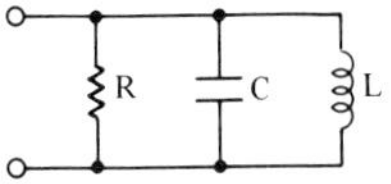

@ f_r, $Z = R$, $\theta_Z = 0°$

$X_C = X_L$, $Q = R/X_L$

$f_r = \sqrt{(LC)^{-1} - (R/L)^2}/(2\pi)$ Def. 1

exception $= \sqrt{-x}$

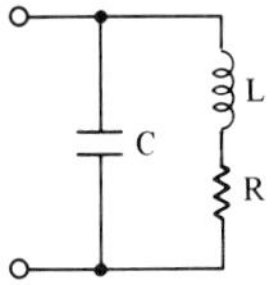

$f_r \approx \left(2\pi\sqrt{LC}\right)^{-1}$

@ f_r $\theta_Z = 0°$

$Z = (X_L^2/R) + R$ $\quad Z \approx X_L^2/R$

$X_C = (R^2/X_L) + X_L$ $\quad X_C \approx X_L$

$Q = \left[(X_L^2/R) + R\right]/X_C$ $\quad Q \approx X_L/R$

f_r *Notes:*

② $x^{-1} = 1/x$, $x^{\frac{1}{2}} = \sqrt{x}$, $x^{-\frac{1}{2}} = 1/\sqrt{x}$, $x^{-2} = 1/x^2$

③ Def. 1 = f_r Definition 1 (max or min Z plus $\theta_Z = 0°$)

④ ω_r = Resonant angular velocity = $2\pi f_r$

⑤ L_p, R_{Cp}, R_{Lp} = Parallel equivalent quantities of series quantities

See also—Q, Z, Y

Parallel Resonant Frequency

f_r

Resonant Frequency, Parallel Resonance

$f_r = \left[(LC) - (CR)^2\right]^{-\frac{1}{2}}/(2\pi)$ Def. 1 $(-x)^{-\frac{1}{2}}$ exception

$f_r \approx \left(2\pi\sqrt{LC}\right)^{-1}$

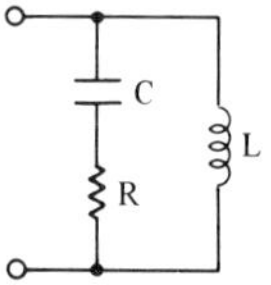

@ f_r Definition 1

$\theta_Z = 0°$

$Z = (X_C^2/R) + R$ $Z \approx X_C^2/R$

$X_L = (R^2/X_C) + X_C$ $X_L \approx X_C$

$Q = \left[(X_C^2/R) + R\right]/X_L$ $Q \approx X_C/R$

$f_r = \sqrt{\left[C^{-1} - (R_L^2/L)\right]/\left[L - (R_C^2 C)\right]}/(2\pi)$ Def. 1

exception $= \sqrt{-x}$

$f_r \approx \left(2\pi\sqrt{LC}\right)^{-1}$

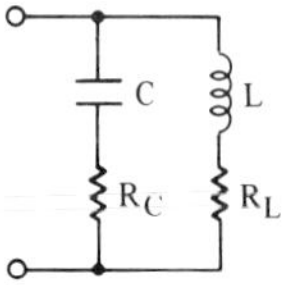

@ f_r Definition 1

$\theta_Z = 0°$

$Z = \left(\left[(X_C^2/R_C) + R_C\right]^{-1} + \left[(X_L^2/R_L) + R_L\right]^{-1}\right)^{-1}$

$\left[(R_C^2/X_C) + X_C\right] = \left[(R_L^2/X_L) + X_L\right]$ $X_C \approx X_L$

$Q \approx \left[\omega_r L_p(R_{Lp}^{-1} + R_{Cp}^{-1})\right]^{-1}$

$Q \approx \sqrt{L/\left[C(R_L + R_C)^2\right]}/(2\pi)$

$Q \approx X_C/(R_L + R_C)$

Frequency of Acoustical Resonance f_r Pipes and Tubes	Applicable Notes	
$f_r = v/(2\ell + 1.6d)$	①	Open Pipe
$\ell = (v/2f_r) - (d/1.25)$	②	
$d = (v/1.6f_r) - (1.25\ell)$		
$f_r = v/\left(2\ell + 1.8\sqrt{A}\right)$	③	
$\ell = (v/2f_r) - \left(\sqrt{A}/1.11\right)$		
$A = \sqrt{(v/1.8f_r) - (1.11\ell)}$	⑤	
$f_r = v/(4\ell + 1.6d)$	①	Stopped Pipe (one end closed)
$\ell = (v/4f_r) - (d/2.5)$	②	
$d = (v/1.6f_r) - (2.5\ell)$		
$f_r = v/\left(4\ell + 1.8\sqrt{A}\right)$	④	
$\ell = (v/4f_r) - \left(\sqrt{A}/2.22\right)$		
$d = \sqrt{(v/1.8f_r) - (2.22\ell)}$	⑤	

Pipe *Notes:*

① A = Cross sectional inside area of pipe
d = Inside diameter of pipe
ℓ = Length of pipe
v = Velocity of sound in air

② $v \simeq 13\,630$ inches per second @ 25°C
$v \simeq 1136$ feet per second @ 25°C
$v \simeq 346.3$ meters per second @ 25°C

③ Also has secondary resonances @ $2f_r$, $3f_r$, $4f_r$, $5f_r$, etc.

④ Also has secondary resonances @ $3f_r$, $5f_r$, $7f_r$, etc.

⑤ A, d, ℓ and v must all use the same unit of linear measure

Frequency of Acoustical Resonance	f_r	Applicable Notes	
$f_r = 2070[A/V^2]^{\frac{1}{4}}$ $f_r = 1948.7\sqrt{d/V}$ $V = d[1948.7/f_r]^2$ $d = V[f_r/1948.7]^2$		① ⑤ ⑥	Helmholtz Resonator (ported hollow sphere)
$f_r \approx 1424\sqrt{d/V}$ $V \approx d[1424/f_r]^2$ $d \approx V[f_r/1424]^2$	(Assuming speaker resonance is much lower than box resonance)	② ⑥	Closed Box Speaker Cabinet
$f_r \approx 2070\left[(.285A_1 + A_2)/V^2\right]^{\frac{1}{4}}$ $V \approx \left[2070^2\sqrt{.285A_1 + A_2}\right]/f_r^2$ $A_2 \approx \left[V^2(f_r/2070)^4\right] - .285A_1$		③ ⑤ ⑥	Ported Speaker Cabinet (Bass Reflex)
$f_r \approx 1713/\sqrt{(.85d_2 + \ell)\left[(V_1/d_2^2) - (.25\pi\ell)\right]}$ $\ell \approx \left[(1913d_2)^2/(f_r^2V_2)\right] - .85d_2$		④ ⑥	Ducted Port Speaker Cabinet

Cabinet *Notes:*

① A = Area of opening (port), d = Diameter of opening (port), V = Internal volume of sphere.

② d = Diameter of speaker opening, V = Internal volume of cabinet (neglect speaker volume)

③ A_1 = Area of speaker opening, A_2 = Area of port, V = Internal volume of cabinet (neglect speaker volume)

④ d_2 = Diameter of speaker opening and duct opening, V_1 = Internal volume of cabinet including duct volume. V_2 = Internal cabinet volume excluding duct, ℓ = Duct length.

⑤ $x^{\frac{1}{4}} = \sqrt{\sqrt{x}}$, $x^4 = (x^2)^2$

⑥ A, d, ℓ and V must all use the same unit of linear measure.

F

Farad, Force etc

F = Symbol for farad.

F = Basic unit of capacitance.

F = Capacitance required to store one coulomb of charge at one volt potential.

F = Extremely large unit. Seldom used without a prefix symbol.

$F = \mu F \cdot 10^6$ (Typewriter–use uF)

$F = nF \cdot 10^9$ (just coming into usage in USA)

$F = pF \cdot 10^{12}$ ($\mu\mu F$ is not recommended)

Note: The prefix symbol m (milli) should not be used with F due to long time previous use of m with F to indicate microfarads.

F = Symbol for magnetic, electrostatic and mechanical force.

F = Magnetomotive force when units are in gilberts or ampere turns [gilbert = 1.257 ampere turns (At)]

$F = \phi \mathcal{R}$ where ϕ = total flux and $\mathcal{R}$ = reluctance

---

Repulsive Electrostatic Force

$F = 9 \cdot 10^9 [Q_1 Q_2 / d^2]$ dynes

Q_1, Q_2 = charge in coulombs on two bodies

d = distance in cm separating two bodies

°F = Symbol for degrees.

°F = Unit of temperature. (USA)

F, F_n–See–NF (Noise Figure)

F_p–See–pf (Power Factor)

---

g G Conductance Definitions and DC Formulas, Mutual Conductance

G = Symbol for conductance.

G = The ease with which direct current flows in a circuit at a given potential. The ease with which alternating current at a given potential flows in a purely resistive circuit. The reciprocal of resistance in any purely resistive circuit. The reciprocal of a pure resistance in parallel with other elements. The real part of admittance. The reciprocal of the equivalent parallel circuit resistance in a series circuit.

G = Conductance in units of siemens (S).
[old unit mho (Ω^{-1} or ℧) is still common usage in USA]

G = A parallel circuit quantity which may be used as easily in parallel circuits as resistance is used in series circuits.

$G = R_P^{-1} \angle 0^\circ$ in terms of polar impedance.

Definitions

$G = 1/R$

$G = I/E$

$G = P/E^2$

$G = I^2/P$

$G_t = (R_1 + R_2 \cdots + R_n)^{-1}$ Series Circuits

$G_t = G_1 + G_2 \cdots + G_n$ Parallel Circuits

$G_t = R_1^{-1} + R_2^{-1} \cdots + R_n^{-1}$ Parallel Circuits

DC Formulas

g_m = Symbol for mutual conductance or transconductance. See–Active Circuits

$g_m = \Delta I_p / \Delta E_g$ (Vacuum Tubes)

Conductance, Series Circuits **G**	Applicable Notes	Terms
$G = (R_1 + R_2 \cdots + R_n)^{-1}$	① ②	R
$G = R/\left[R^2 + (\omega C)^{-2}\right]$	① ② ③	C R
$G = (\omega C Z^2)^{-1}$	① ② ③	C Z
$G = \omega C(\sin\theta)^2$	① ③ ④ ⑦	C θ
$G = I/E_R$	① ⑤	E_R I
$G = P/E_R^2$	① ⑤	E_R P
$G = I^2/P$	①	I P
$G = R/\left[R^2 + (\omega L)^2\right]$	① ③	L R
$G = (\omega L)/Z^2$	① ③	L Z
$G = (\sin\theta)^2/(\omega L)$	① ③ ④ ⑦	L θ

G *Notes:*

① G IS INTRINSICALLY A PARALLEL CIRCUIT QUANTITY. G DERIVED FROM A SERIES CIRCUIT IS THE EQUIVALENT PARALLEL CIRCUIT RESISTANCE IN RECIPROCAL FORM.
② $x^{-1} = 1/x$, $x^{-2} = 1/x^2$
③ $\omega = 2\pi f = 6.283f$ = angular velocity
④ sin, cos, tan = abbr. for sine, cosine and tangent
⑤ E_R = Voltage developed by a resistance
⑥ $|x|$ = Absolute value of x = Magnitude of x
⑦ θ may be θ_E, θ_I, θ_Y or θ_Z, B may be B_C or B_L, X may be X_C or X_L

Conductance, Series Circuits **G**	Applicable Notes	Terms
$G = R/[R^2 + X_C^2]$	①	X_C R
$G = X_C/Z^2$	①	X_C Z
$G = (\sin \theta_Z)^2/X_C$	① ④ ⑦	X_C θ
$G = R/[R^2 + X_L^2]$	①	X_L R
$G = X_L/Z^2$	①	X_L Z
$G = (\sin \theta)^2/X_L$	① ④ ⑦	X_L θ
$G = R\Big/\left(R^2 + \left[(\omega L) - (\omega C)^{-1}\right]^2\right)$	① ② ③	C L R
$G = \left\| \left[(\omega L) - (\omega C)^{-1}\right]/Z^2 \right\|$	① ② ③ ⑥	C L Z
$G = \left\|(\sin \theta)^2/\left[(\omega L) - (\omega C)^{-1}\right]\right\|$	① ② ③ ④ ⑥ ⑦	C L θ
$G = R/\left[R^2 + (X_L - X_C)^2\right]$	①	X_C X_L R
$G = \left\|(X_L - X_C)/Z^2\right\|$	① ⑥	X_C X_L Z
$G = \left\|(\sin \theta)^2/(X_L - X_C)\right\|$	① ④ ⑥ ⑦	X_C X_L θ

Conductance, Parallel Circuits G	Applicable Notes	Terms
$G = G_1 + G_2 \cdots + G_n$	⑧	G
$G = R_1^{-1} + R_2^{-1} \cdots + R_n^{-1}$	② ⑧	R
$G = \sqrt{Y^2 - B^2}$	⑨	B Y
$G = \sqrt{Z^{-2} - B^2}$	② ⑨	B Z
$G = \lvert B/(\tan\theta) \rvert$	④ ⑥ ⑦	B θ
$G = \sqrt{Y^2 - (\omega C)^2}$	③ ⑨	C Y
$G = \sqrt{Z^{-2} - (\omega C)^2}$	② ③ ⑨	C Z
$G = \lvert (\omega C)/(\tan\theta) \rvert$	③ ④ ⑥ ⑦	C θ
$G = P/E^2$		E P
$G = \sqrt{Y^2 - (\omega L)^{-2}}$	② ③ ⑨	L Y
$G = \sqrt{Z^{-2} - (\omega L)^{-2}}$	② ③ ⑨	L Z
$G = \lvert [(\omega L)(\tan\theta)]^{-1} \rvert$	③ ④ ⑥ ⑦	L θ
$G = \sqrt{Y^2 - X^{-2}}$	② ⑦ ⑨	X Y
$G = \sqrt{Z^{-2} - X^{-2}}$	② ⑦ ⑨	X Z

Conductance, Parallel Circuits **G**	Applicable Notes	Terms
$G = \left\| [X(\tan\theta)]^{-1} \right\|$	② ④ ⑦	X θ
$G = Y(\cos\theta)$	④ ⑦	Y θ
$G = (\cos\theta)/Z$	④ ⑦	Z θ
$G = \sqrt{Y^2 - (B_L - B_C)^2}$	⑨	B_C B_L Y
$G = \sqrt{Z^{-2} - (B_L - B_C)^2}$	② ⑨	B_C B_L Z
$G = \left\| (B_L - B_C)/(\tan\theta) \right\|$	④ ⑥ ⑦	B_C B_L θ
$G = \sqrt{Y^2 - [(\omega L)^{-1} - (\omega C)]^2}$	② ③ ⑨	C L Y
$G = \sqrt{Z^{-2} - [(\omega L)^{-1} - (\omega C)]^2}$	② ③ ⑨	C L Z
$G = \left\| [(\omega L)^{-1} - (\omega C)]/(\tan\theta) \right\|$	② ③ ④ ⑥ ⑦	C L θ
$G = \sqrt{Y^2 - (X_L^{-1} - X_C^{-1})^2}$	② ⑨	X_C X_L Y
$G = \sqrt{Z^{-2} - (X_L^{-1} - X_C^{-1})^2}$	② ⑨	X_C X_L Z
$G = \left\| (X_L^{-1} - X_C^{-1})/(\tan\theta) \right\|$	② ④ ⑥ ⑦	X_C X_L θ

G *Notes:*

⑧ In a purely parallel circuit, the values of parallel reactances are not relevant to the value of G.

⑨ A negative resultant under the radical sign indicates an error.

H

Henry Unit, Magnetic Field Strength

H = Symbol for henry.

H = Basic unit of inductance.

H = The inductance which develops one volt from current changing at the rate of one ampere per second.

$H = mH \cdot 10^3$

$H = \mu H \cdot 10^6$

H = Symbol for magnetic field strength.

H = Magnetomotive force per unit length.
Magnetizing force.
Magnetic intensity.

H = Gilberts per centimeter (CGS Oersteds).

H = Ampere turns per meter (SI A/m).

$H = F\ell$ where F = magnetomotive force
ℓ = length of magnetic path

$H = B/\mu$ where B = magnetic flux density
μ = permeability

H = B when magnetic path is air

H = B when permeability (μ) = 1

Ampere turns per inch = .495 Oersteds
Oersteds = 2.02 Ampere turns per inch

h

Hybrid Parameter, Height, Hour

h = Symbol for hybrid parameter.
See—Active Circuits

h = Symbol for height.

h = Symbol for hour.

h = Symbol for planck's constant.

Hz

Hertz

Hz = Symbol for hertz.

Hz = The basic unit of frequency equal to one cycle per second.

Hz = Unit often used with multiplier prefixes.

kHz = 10^3 Hertz (kilohertz)
MHz = 10^6 Hertz (megahertz)
GHz = 10^9 Hertz (gigahertz)

Hz = cps = c/s

Hz = 360° per second

Hz = 2π radians per second

Hz = Vectorial revolutions per second.

I = Symbol for electrical current.

I = 1. The movement of electrons through a conductor.
2. The rate of flow of electric charge.

I = Current in amperes (A). (Coulombs per sec.)

$I = \pm I_{dc}$ or $I_{ac(effective)}$

$I_{eff} = I_{rms}$

$I_{ac} = |I| = I_{absolute\ value} = I_{magnitude}$

θ_I = Phase angle of alternating current.

$\mathbf{I}$ = Complete description of alternating current.

$\mathbf{I} = \mathbf{I}_{POLAR}$ or $\mathbf{I}_{RECTANGULAR}$ ($\mathbf{I}_{POLAR} = \mathbf{I}_{RECT}$)

$\mathbf{I}_{POLAR} = I\underline{/\theta_I}$ = Vectorial current

$\mathbf{I}_{RECT} = (\pm I_R \pm I_X\, j)$ = Complex number current
where $\pm I_R$ = Current through a real or an equiv. parallel circuit resistance and
where $\pm I_X$ = Current through a real or an equivalent parallel circuit reactance.

$\mathbf{I}_{RECT}$ = Complex number form of current which expresses the 0° or 180° and the +90° or −90° vectors which have a resultant vector equal to $\mathbf{I}_{POLAR}$.

$\mathbf{I}_{RECT} = I_R - (\pm I_X)j$ in this handbook (one exception) whereby $+I_X$ identifies I_X as inductive and $-I_X$ identifies I_X as capacitive.

$\mathbf{I}_{RECT}$ = Mathematical equivalent of resistive and reactive currents in parallel regardless of actual circuit configuration.

---

i = Instantaneous value of current.
(exception: i_N = rms noise current)

I Direct Current Formulas

Formula	
$I = EG$ $I = P/E$ $I = E/R$ $I = \sqrt{P/R}$	General
$I = P_1/E_1 = P_2/E_2 = P_n/E_n$ $I = E_1/R_1 = E_2/R_2 = E_n/R_n$ $I = (E_1 + E_2 \cdots + E_n)/(R_1 + R_2 \cdots + R_n)$ $I = \sqrt{P_1/R_1} = \sqrt{P_2/R_2} = \sqrt{P_n/R_n}$ $I = \sqrt{(P_1 + P_2 \cdots + P_n)/(R_1 + R_2 \cdots + R_n)}$	Series Circuits
$I_t = I_1 + I_2 \cdots + I_n$ $I_t = EG_t = E(G_1 + G_2 \cdots + G_n)$ $I_t = P_t/E = (P_1 + P_2 \cdots + P_n)/E$ $I_t = E(R_1^{-1} + R_2^{-1} \cdots + R_n^{-1})$ $I_t = \sqrt{P_t G_t}$ $I_t = \sqrt{P_t(R_1^{-1} + R_2^{-1} \cdots + R_n^{-1})}$	Parallel Circuits

I *Notes:*

① General

B = Susceptance, C = Capacitance, e = Instantaneous Voltage, E = Voltage (dc or rms), f = Frequency, G = Conductance, i = Instantaneous Current, I = Current (dc or rms), L = Inductance, P = Power, Q = Quantity of Electric Charge, Q = Quality or Q Factor, R = Resistance, t = Time, T = Time Constant, X = Reactance, Y = Admittance, Z = Impedance, ϵ = Base of Natural Logarithms, θ = Phase Angle, –Continued on page 67

$I = Q/t$ (I produced by charge Q for t sec.)	
$i = (E/R)(\epsilon^{\frac{-t}{RC}})$ (E = Applied voltage) $i = .36788\,(E/R)$ @ $t = RC$ (one time constant) $I = (e_C C)/t$ (constant current)	Capacitor Charge
$i = (E/R)(\epsilon^{\frac{-t}{RC}})$ (E = Initial voltage) $i = .36788\,(E/R)$ @ $t = RC$ (one time constant) $I = (E - e_C)(C/t)$ (constant current)	Capacitor Discharge
$i = (E/R)(1 - \epsilon^{\frac{-Rt}{L}})$ (E = Applied Voltage) $i = .6321\,(E/R)$ @ $t = R/L$ (one time constant)	Inductor Energization
$I_{p\text{-}p} = (2\sqrt{2})\,I_{rms} = 2.828\,I_{rms}$ $I_{peak} = (\sqrt{2})\,I_{rms} = 1.414\,I_{rms}$ $I_{av} = \left[(2\sqrt{2})/\pi\right] I_{rms} = .9003\,I_{rms}$ $I_{av} = (2/\pi)\,I_{peak} = .6366\,I_{rms}$ $I_{rms} = \left[\pi/(2\sqrt{2})\right] I_{av} = 1.111\,I_{av}$ I_{rms} = effective current = dc equiv. current $I_{rms} = (1/\sqrt{2})\,I_{peak} = .707\,I_{peak}$ $I_{rms} = \left[1/(2\sqrt{2})\right] I_{p\text{-}p} = .3535\,I_{p\text{-}p}$	Current Ratios

I *Notes:*

① Continued
π = Circumference to Diameter Ratio, ω = Angular Velocity or Angular Frequency.

Series Circuit Current	Applicable Notes	Terms
$I = E_C\, \omega C$	① ② ③	E_C C
$I = E_L/(\omega L)$	① ② ③	E_L L
$I = E_R/R$	① ②	E_R R
$I = E_C/X_C$	① ②	E_C X_C
$I = E_L/X_L$	① ②	E_L X_L
$I = EY$	①	E Y
$I = E/Z$	①	E Z
$I = \sqrt{P/R}$	①	P R
$I = P/(E \cos \theta)$	① ⑤	E P θ
$I = (E \cos \theta)/R$	① ⑤	E R θ
$I = E\Big/\sqrt{R^2 + \left[(\omega L) - (\omega C)^{-1}\right]^2}$	① ③	E CL R
$I = E/\sqrt{R^2 + (X_L - X_C)^2}$	① ②	E $X_C X_L$ R
$I = \sqrt{P/(Z \cos \theta)}$	① ⑤	P Z θ

I *Notes:*

② Subscripts
C = capacitive, E = voltage, g = generator, I = current, L = inductive, n = any number, o = output, p = parallel, R = resistive, s = series, t = total or equivalent, X = reactive, Y = admittance, Z = impedance

③ Constants
$j = \sqrt{-1}$, = 90° multiplier, = mathematical i, $\epsilon = 2.718$, $\epsilon^{-1} = .36788$, $\pi = 3.1416$, $2\pi = 6.2832$, $\omega = 2\pi f$

④ Algebra
$x^{-1} = 1/x$, $x^{-2} = 1/x^2$, $x^{\frac{1}{2}} = \sqrt{x}$, $x^{-\frac{1}{2}} = 1/\sqrt{x}$, $x^{(-y/z)} = 1/x^{(y/z)}$, $|x|$ = absolute value or magnitude of x

Current, Parallel Circuits	Applicable Notes	Terms
$I_t = E(B_{C1} + B_{C2} \cdots + B_{Cn})$	① ②	E B_C
$I_t = E(B_{L1} + B_{L2} \cdots + B_{Ln})$	① ②	E B_L
$I_t = E\omega(C_1 + C_2 \cdots + C_n)$	① ② ③	E C
$I_t = E(G_1 + G_2 \cdots + G_n)$	① ②	E G
$I_t = \left[E(L_1^{-1} + L_2^{-1} \cdots + L_n^{-1})\right]/\omega$	① ② ③ ④	E L
$I_t = E(R_1^{-1} + R_2^{-1} \cdots + R_n^{-1})$	① ② ④	E R
$I_t = E(X_{C1}^{-1} + X_{C2}^{-1} \cdots + X_{Cn}^{-1})$	① ② ④	E X_C
$I_t = E(X_{L1}^{-1} + X_{L2}^{-1} \cdots + X_{Ln}^{-1})$	① ② ④	E X_L
$I = EY$	①	E Y
$I = E/Z$	①	E Z
$I = E(B_L - B_C)$	① ②	E B_C B_L
$I = E(X_L^{-1} - X_C^{-1})$	① ② ④	E X_C X_L
$I = E\sqrt{G^2 + (B_L - B_C)^2}$	① ②	E B_C B_L G
$I = E\sqrt{R^{-2} + (X_L^{-1} - X_C^{-1})^2}$	① ② ④	E X_C X_L R

I *Notes:*

⑤ Trigonometry
sin = sine, cos = cosine, tan = tangent, $\tan^{-1}$ = arc tangent

⑥ Reminders
$\pm\theta$ --- use the sign of the phase angle
$\pm X$, $\pm B$, $\pm I_X$ ---identifies X, B and I_X as capacitive or inductive
$-X$, $-B$, $-I_X$ are capacitive
$+X$, $+B$, $+I_X$ are inductive

⊗ Division by zero is prohibited

1. It should be understood that the phase angle of voltage, current, impedance and admittance is the same, one and only, phase angle in a given circuit. The fact that the sign of the voltage or impedance phase angle differs from the sign of the current or admittance phase angle means only that if the current leads the voltage, the voltage must lag the current by the same angle.
2. $\pm\theta_I = -(\pm\theta_E) = -(\pm\theta_Z) = \pm\theta_Y$
3. When using the phase angle of impedance (or admittance), a phase angle always exists when the circuit is reactive. The phase angle of voltage and current however can only exist for one of the two at the same time. When the voltage phase angle exists, the current phase angle must be 0° and when the current phase angle exists, the voltage phase angle must be 0°. This is explained by the fact the voltage uses the current as a reference and the current uses the voltage as a reference.
4. The same situation exists with voltage and current in rectangular form. When "imaginary" current exists, the voltage must be $E\angle 0°$ or $E + j0$ and when "imaginary" voltage exists, the current must be $I\angle 0°$ or $I + j0$.
5. For practical problems, the best method of minimizing confusion and errors is to use the phase angle of the generator as the 0° reference. If the generator is a current source, the phase angle of the total current is always 0° and if the generator is a voltage source, the phase angle of the total voltage is always 0°.

Note: The rectangular current of a series circuit driven by a voltage source represents the currents through an equivalent parallel circuit. The rectangular voltage of a parallel circuit driven by a current source represents the voltages across elements of the equivalent series circuit.

I = The magnitude and phase angle of the current developed by a voltage applied to a circuit. ($\theta_{E_d} = 0°$) See also—θ

$I_{POLAR} = I\underline{/\pm\theta_I} = I\underline{/-(\pm\theta_Z)}$

I_{RECT} = 1. The 0° and ±90° currents which produce a resultant equal to I_{POLAR}.

2. The current through resistance and net reactance in parallel.

3. The current through the parallel equivalent resistance and net reactance of a series circuit.

(*Note:* Only one current is possible in a series circuit.)

$I_{RECT} = I_R - (\pm I_X)\,j$

$I_{RECT} = \left[I \cos\theta_I\right] - \left[-I \sin(\pm\theta_I)\right] j$

$I_{RECT} = \left[I \cos\theta_Z\right] - \left[I \sin(\pm\theta_Z)\right] j$

Note: The above rectangular form is strongly recommended for most uses. The negative sign will always identify the complex quantity as current or admittance and as a parallel equivalent quantity. The use of θ_Z eliminates the double change of signs often needed and maintains the identity of the reactive quantity at all times.

Note: The rectangular form of current has been used by some as a substitute for rectangular admittance (Y_{RECT}) for solving series circuits in parallel. It should be noted that if the assumed voltage is one, I_{RECT} and Y_{RECT} are identical in meaning and method except for the names of the quantities. When $E = 1\underline{/0°}$, $I_{POLAR} = Y_{POLAR}$, $I_{RECT} = Y_{RECT}$, $I_R = G$, $I_C = B_C$, $I_L = B_L$, $-I_X = -B$, $+I_X = +B$ and $\pm I_X = \pm B$.

Note: Use formulas on following page to obtain I_{POLAR} then convert to I_{RECT} using above formulas.

All Series Circuit I Formulas

$$\mathbf{I}_{POLAR} = I\underline{/\pm\theta_I}$$

$$\mathbf{I}_{POLAR} = I\underline{/-(\pm\theta_Z)}$$

$$\mathbf{I}_{RECT} = [I\cos\theta_Z] - [I\sin(\pm\theta_Z)]\,j$$

$$I_R = I\cos\theta_Z$$

$$I_C = |I\sin(-\theta_Z)| \quad (I_C = -I_X)$$

$$I_L = I\sin(+\theta_Z) \quad (I_L = +I_X)$$

$$\pm I_X = I\sin(\pm\theta_Z)$$

$$\mathbf{I}_{RECT} = I_R - (\pm I_X)\,j$$

Parallel Equivalent Current or Equiv. parallel circuit current

Formula	Applicable Notes	Terms
$I = P/(E\cos\theta_Z)$	① ② ⑤	E P θ_Z
$I = (E\cos\theta_Z)/R$	① ② ⑤	E R θ_Z
$I = E/Z$	①	E Z θ_Z
$I = E\Big/\sqrt{R^2 + [(\omega L) - (\omega C)^{-1}]^2}$ $\pm\theta_Z = \tan^{-1}[(\omega L) - (\omega C)^{-1}]/R$	① ② ③ ④ ⑤ ⑥	E C L R
$I = E/\sqrt{R^2 + (X_L - X_C)^2}$ $\pm\theta_Z = \tan^{-1}[(X_L - X_C)/R]$	① ② ⑤ ⑥	E X_C X_L R
$I = \lvert(E\sin\theta_Z)/(X_L - X_C)\rvert$	① ② ④ ⑤ ⊗	E X_C X_L θ

Current & Phase,
Parallel Circuits

I

Resistive & Reactive
Currents
In Parallel

I = The magnitude and phase angle of the current developed by the application of voltage to a parallel circuit. ($\theta_E = 0°$)

$\mathbf{I}_{POLAR} = I\underline{/\theta_I} = I\underline{/\theta_Y} = I\underline{/-(\pm\theta_Z)}$

$\mathbf{I}_{RECT}$ = 1. The 0° and ±90° currents which have a resultant equal to $\mathbf{I}_{POLAR}$.
2. The resistive current and the reactive current in parallel.

$\mathbf{I}_{RECT} = I_R - (\pm I_X)j = [I\cos\theta_Z] - [I\sin(\pm\theta_Z)]\ j$

$\mathbf{I}_{RECT} = [I\cos\theta_I] + [I\sin(\pm\theta_I)]\ j$

$\mathbf{I}_{RECT} = [I\cos\theta_Y] + [I\sin(\pm\theta_Y)]\ j$

	Applicable Notes	Terms
$I = (EG)/(\cos\theta_Y)$	① ② ⑤	$E\ G\ \theta_Y$
$I = P/(E\cos\theta_Z)$	① ② ⑤	$E\ P\ \theta_Z$
$I = E/(R\cos\theta_Z)$	① ② ⑤	$E\ R\ \theta_Z$
$I = EY$	①	$E\ Y\ \theta_Y$
$I = E/Z$	①	$E\ Z\ \theta_Z$
$I = \sqrt{I_R^2 + (I_L - I_C)^2}$ $\pm\theta_I = \tan^{-1}[-(I_L - I_C)/I_R]$	① ② ⑤ ⑥	$I_R\ I_C\ I_L$
$I = \sqrt{P/(Z\cos\theta_Z)}$	① ② ⑤	$P\ Z\ \theta_Z$

Current & Phase, Parallel Circuits	Applicable Notes	Terms
$I = E\sqrt{G^2 + (B_L - B_C)^2}$ $\pm\theta_I = \tan^{-1}\left[-(B_L - B_C)/G\right]$	① ② ⑤ ⑥	$E\ B_C\ B_L\ G$
$I = \left\| \left[E(B_L - B_C)\right]/(\sin\theta_Y)\right\|$ $\pm\theta_I = \pm\theta_Y$	① ② ⑤ ⑥ ⊗	$E\ B_C\ B_L\ \theta_Y$
$I = E\sqrt{R^{-2} + \left[(\omega L)^{-1} - (\omega C)\right]^2}$ $\pm\theta_I = \tan^{-1}\left(-R\left[(\omega L)^{-1} - (\omega C)\right]\right)$	① ② ③ ④ ⑤ ⑥	$E\ C\ L\ R$
$I = \left(E\left[(\omega L)^{-1} - (\omega C)\right]\right)/(\sin\theta_Z)$ $\pm\theta_I = -(\pm\theta_Z)$	① ② ③ ④ ⑤ ⑥ ⊗	$E\ C\ L\ \theta_Z$
$I = E\sqrt{R^{-2} + (X_L^{-1} - X_C^{-1})^2}$ $\pm\theta_I = \tan^{-1}\left[-R(X_L^{-1} - X_C^{-1})\right]$	① ② ④ ⑤ ⑥	$E\ X_C\ X_L\ R$
$I = \left[E(X_L^{-1} - X_C^{-1})\right]/\left[\sin\theta_Z\right]$ $\pm\theta_I = -(\pm\theta_Z)$	① ② ④ ⑤ ⑥ ⊗	$E\ X_C\ X_L\ \theta_Z$

I — Parallel Complex Currents, Procedure Method

Terms – $I_1\angle\pm\theta_1$, $I_2\angle\pm\theta_2$, --- $I_n\angle\pm\theta_n$

Procedure for those who are uncomfortable when working with rectangular form currents. Maintains positive identity of reactive currents.

Procedure:

1. Convert each $I\angle\pm\theta_I$ to its equivalent parallel resistive current from:

   $I_{R_p} = I\cos\theta_I$
2. $(I_{R_p})_t = (I_{R_p})_1 + (I_{R_p})_2 \cdots + (I_{R_p})_n$
3. Convert each $I\angle\pm\theta_I$ with a negative angle to its equivalent parallel inductive current from:

   $I_{L_p} = I\sin|-\theta_I|$
4. $(I_{L_p})_t = (I_{L_p})_1 + (I_{L_p})_2 \cdots + (I_{L_p})_n$
5. Convert each $I\angle\pm\theta_I$ with a positive angle to its equivalent parallel capacitive current from:

   $I_{C_p} = I\sin(+\theta_I)$
6. $(I_{C_p})_t = (I_{C_p})_1 + (I_{C_p})_2 \cdots + (I_{C_p})_n$
7. Convert totals back to a single polar form current from:

   $$I_t = \sqrt{(I_{R_p})_t^2 + \left[(I_{L_p})_t - (I_{C_p})_t\right]^2}$$

   $$\pm\theta_{I_t} = \tan^{-1}\left(-\left[(I_{L_p})_t - (I_{C_p})_t\right]/\left[I_{R_p}\right]_t\right)$$

Formula Method | **Complex Currents, Sum & Differential**

$$I_t = \Big\{\big[(I_1 \cos\theta_1) + (I_2 \cos\theta_2) \cdots + (I_n \cos\theta_n)\big]^2 + \Big(\big[I_1 \sin(\pm\theta_1)\big] + \big[I_2 \sin(\pm\theta_2)\big] \cdots + \big[I_n \sin(\pm\theta_n)\big]\Big)^2\Big\}^{1/2}$$

$$\theta_{I_t} = \tan^{-1}\big[I \sin(\pm\theta)\big]_t \big/ \big[I \cos\theta\big]_t$$

Sum of $I_1\underline{/\theta_1}$, $I_2\underline{/\theta_2} \cdots$ and $I_n\underline{/\theta_n}$

$$I_t = \sqrt{\big[(I_1 \cos\theta_1) - (I_2 \cos\theta_2)\big]^2 + \Big(\big[I_1 \sin(\pm\theta_1)\big] - \big[I_2 \sin(\pm\theta_2)\big]\Big)^2}$$

$$\theta_{I_t} = \tan^{-1}\Big(\big[I_1 \sin(\pm\theta_1)\big] - \big[I_2 \sin(\pm\theta_2)\big]\Big) \Big/ \big[(I_1 \cos\theta_1) - (I_2 \cos\theta_2)\big]$$

Differential of $I_1\underline{/\theta_1}$ and $I_2\underline{/\theta_2}$

Current and Phase, Complex Circuits — I — Vector Algebra and/or Rectangular Form Method	Terms
$\mathbf{I}_{POLAR} = (\mathbf{E}_{POLAR})/(\mathbf{Z}_{POLAR})$ $I\angle\theta_I = (E\angle\theta_E)/(Z\angle\theta_Z)$ $I = E_g/Z, \quad \theta_I = 0^\circ - \theta_Z$	$\mathbf{E}_{POLAR}$ $\mathbf{Z}_{POLAR}$
$\mathbf{I}_{POLAR} = (\mathbf{E}_{POLAR}) \cdot (\mathbf{Y}_{POLAR})$ $I\angle\theta_I = (E\angle\theta_E) \cdot (Y\angle\theta_Y)$ $I = E_g Y, \quad \theta_I = 0^\circ + \theta_Y$	$\mathbf{E}_{POLAR}$ $\mathbf{Y}_{POLAR}$
$(\mathbf{I}_{RECT})_t = (\mathbf{I}_{RECT})_1 + (\mathbf{I}_{RECT})_2$ $(\mathbf{I}_{RECT})_t = \left[I_R + (\pm I_{90^\circ})j\right]_1 + \left[I_R + (\pm I_{90^\circ})j\right]_2$ $(\mathbf{I}_{RECT})_t = (I_{R1} + I_{R2}) + \left[(\pm I_{90^\circ})_1 + (\pm I_{90^\circ})_2\right]j$	$(\mathbf{I}_{RECT})_1$ $(\mathbf{I}_{RECT})_2$
$(\mathbf{I}_{RECT})_t = \left[(\mathbf{I}_{POLAR})_1\right]_{RECT} + \left[(\mathbf{I}_{POLAR})_2\right]_{RECT}$ $(\mathbf{I}_{RECT})_t = \left[I_1\angle\theta_1\right]_{RECT} + \left[I_2\angle\theta_2\right]_{RECT}$ $(\mathbf{I}_{RECT})_t = \left[(I_1 \cos\theta_1) + (I_2 \cos\theta_2)\right]$ $+ \left[(I_1 \sin \pm\theta_1) + (I_2 \sin \pm\theta_2)\right] j$ $(\mathbf{I}_{RECT})_t = (I_R)_t + \left[(\pm I_{90^\circ})_t\right] j$ $I_t = \sqrt{(I_R)_t^2 + (\pm I_{90^\circ})_t^2}$ $(\pm\theta_I)_t = \tan^{-1}\left[(\pm I_{90^\circ})_t/(I_R)_t\right]$	$(\mathbf{I}_{POLAR})_1$ $(\mathbf{I}_{POLAR})_2$

I_o Output Current & Phase

$$I_o = I_g / \left[(R_2/R_1) + 1\right]$$
$$\theta_{I_o} = \theta_Z = 0^\circ$$

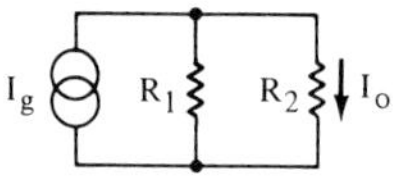

$$I_o = I_g / \left[R\sqrt{R^{-2} + X_C^{-2}}\right]$$
$$\theta_{I_o} = \theta_Z = \tan^{-1}(R/{-X_C})$$

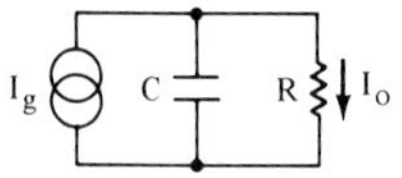

$$I_o = I_g / \left[R\sqrt{R^{-2} + X_L^{-2}}\right]$$
$$\theta_{I_o} = \theta_Z = \tan^{-1}(R/{+X_L})$$

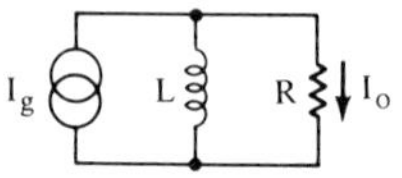

$$I_o = I_g / \left[X_C\sqrt{R^{-2} + X_C^{-2}}\right]$$
$$\theta_{I_o} = \theta_Z + 90^\circ$$
$$\theta_{I_o} = \left[\tan^{-1}(R/{-X_C})\right] + 90^\circ$$

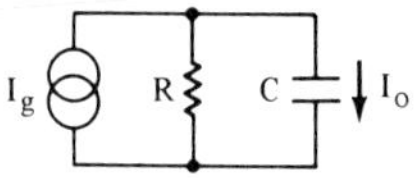

$$I_o = I_g / \left[X_L\sqrt{R^{-2} + X_L^{-2}}\right]$$
$$\theta_{I_o} = \theta_Z - 90^\circ$$
$$\theta_{I_o} = \left[\tan^{-1}(R/{+X_L})\right] - 90^\circ$$

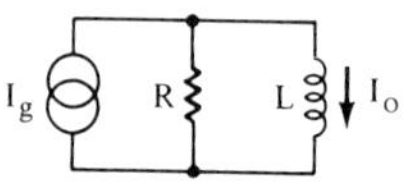

Note: —ↀ— = Infinite impedance current source

I_o Output Current & Phase

$$I_o = (I_g Z)/R$$
$$\theta_{I_o} = \theta_Z$$
$$I_o = I_g / \left[R\sqrt{R^{-2} + (X_L^{-1} - X_C^{-1})^2}\right]$$
$$\theta_{I_o} = \tan^{-1}\left[R(X_L^{-1} - X_C^{-1})\right]$$

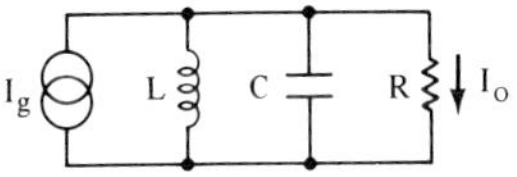

$$I_o = (I_g Z)/X_C$$
$$\theta_{I_o} = \theta_Z - (-90^\circ) \quad = \theta_Z + 90^\circ$$
$$I_o = I_g / \left[X_C\sqrt{R^{-2} + (X_L^{-1} - X_C^{-1})^2}\right]$$
$$\theta_{I_o} = 90^\circ + \tan^{-1}\left[R(X_L^{-1} - X_C^{-1})\right]$$

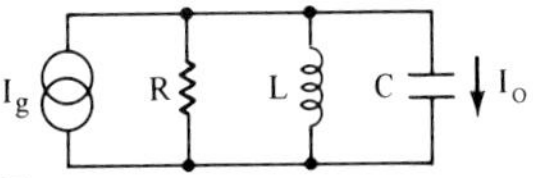

$$I_o = (I_g Z)/X_L$$
$$\theta_{I_o} = \theta_Z - (+90^\circ) \quad = \theta_Z - 90^\circ$$
$$I_o = I_g / \left[X_L\sqrt{R^{-2} + (X_L^{-1} - X_C^{-1})^2}\right]$$
$$\theta_{I_o} = -90^\circ + \tan^{-1}\left[R(X_L^{-1} - X_C^{-1})\right]$$

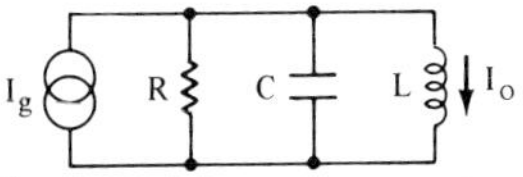

$$I_o = (I_g Z)/R$$
$$\theta_{I_o} = \theta_Z$$
$$I_o = I_g / \left[R\sqrt{R^{-2} + (X_L - X_C)^{-2}}\right]$$
$$\theta_{I_o} = \tan^{-1}\left[R/(X_L - X_C)\right]$$

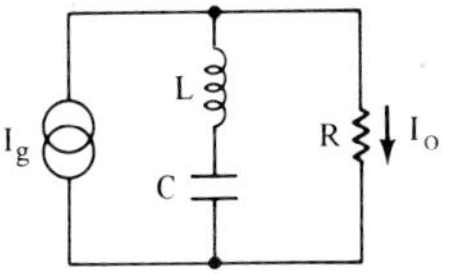

I_o Output Current & Phase

$$I_o = (I_g Z)/(X_L - X_C)$$

$$\theta_{I_o} = \theta_Z - (\pm 90^\circ)$$

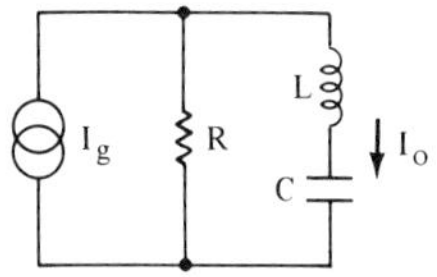

$$I_o = I_g/\left[(X_L - X_C)\sqrt{R^{-2} + (X_L - X_C)^{-2}}\right]$$

$$\theta_{I_o} = \left(\tan^{-1}\left[R/(X_L - X_C)\right]\right) - (\pm 90^\circ)$$

$$I_o = E_g/R$$

$$\theta_{I_o} = 0^\circ$$

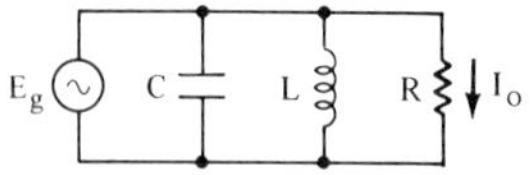

$$I_o = E_g/X_C$$

$$\theta_{I_o} = -(-90^\circ) = +90^\circ$$

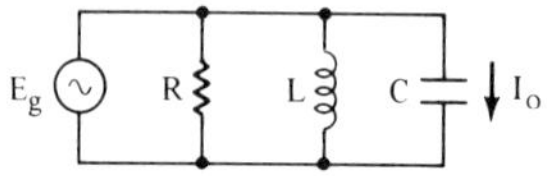

$$I_o = E_g/X_L$$

$$\theta_{I_o} = -(+90^\circ) = -90^\circ$$

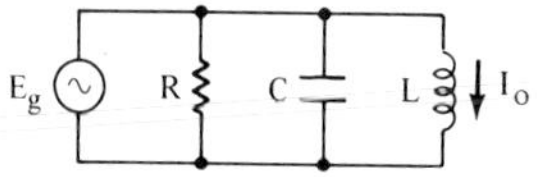

$$I_o = E_g/Z$$

$$\theta_{I_o} = -(\pm\theta_Z)$$

$$I_o = E_g/\sqrt{R^2 + (X_L - X_C)^2}$$

$$\theta_{I_o} = \tan^{-1}\left[-(X_L - X_C)/R\right]$$

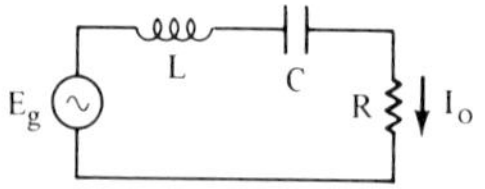

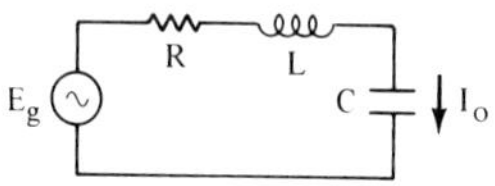

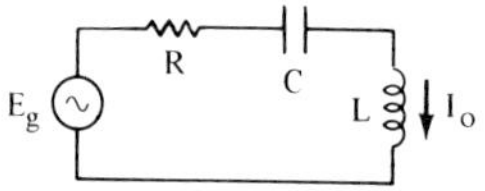

Vector Algebra AC Ohms Law

$\mathbf{E}_g = E_g\angle 0^\circ$ or $\mathbf{I}_g = I_g\angle 0^\circ$

$\mathbf{I} = \mathbf{E}_g/\mathbf{Z} = E_g/Z\angle 0^\circ - \theta_Z = -(\pm\theta_Z)$

$\mathbf{E} = \mathbf{I}_g\mathbf{Z} = I_gZ\angle 0^\circ + \theta_Z = \pm\theta_Z$

$\mathbf{Z} = \mathbf{E}_g/\mathbf{I} = E_g/I\angle 0^\circ - \theta_I = -(\pm\theta_I)$

$\mathbf{Z} = \mathbf{E}/\mathbf{I}_g = E/I_g\angle \theta_E - 0^\circ = \pm\theta_E$

Addition and Subtraction of Rect. Quantities
(See also – $\mathbf{Z}_{RECT}$, Addition and Subtraction)

$$\mathbf{I}_1 + \mathbf{I}_2 = \mathbf{I}_{1(RECT)} + \mathbf{I}_{2(RECT)}$$
$$= \left[I_R - (\pm I_X)j\right]_1 + \left[I_R - (\pm I_X)j\right]_2$$
$$= \left[(I_R)_1 + (I_R)_2\right] - \left[(\pm I_X)_1 + (\pm I_X)_2\right] j$$
$$\mathbf{I}_1 - \mathbf{I}_2 = \left[(I_R)_1 - (I_R)_2\right] - \left[(\pm I_X)_1 - (\pm I_X)_2\right] j$$
$$|+I_X| = I_L \quad |-I_X| = I_C$$

$$\mathbf{E}_1 + \mathbf{E}_2 = \mathbf{E}_{1(RECT)} + \mathbf{E}_{2(RECT)}$$
$$= \left[E_R + (\pm E_X)j\right]_1 + \left[E_R + (\pm E_X)j\right]_2$$
$$= \left[(E_R)_1 + (E_R)_2\right] + \left[(\pm E_X)_1 + (\pm E_X)_2\right] j$$
$$\mathbf{E}_1 - \mathbf{E}_2 = \left[(E_R)_1 - (E_R)_2\right] + \left[(\pm E_X)_1 - (\pm E_X)_2\right] j$$
$$|+E_X| = E_L \quad |-E_X| = E_C$$

Note: The rectangular current of a series circuit represents current through an equivalent parallel circuit. The rectangular voltage of a parallel circuit represents voltage across elements of an equivalent series circuit.

Output Current Vector Algebra

I_o

Applicable Notes

$\mathbf{E}_g = E_g\angle 0^\circ$

$\mathbf{Z}_i = \mathbf{Z}$

$\mathbf{I}_o = \mathbf{E}_g/\mathbf{Z}$

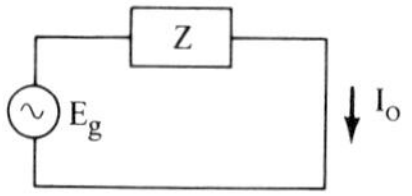

I_{VA} Notes ① ② ③

$\mathbf{I}_g = I_g\angle 0^\circ$

$\mathbf{Z}_i = [\mathbf{Z}_1^{-1} + \mathbf{Z}_2^{-1}]^{-1}$

$\mathbf{E}_g = \mathbf{I}_g\mathbf{Z}_i$

$\mathbf{I}_o = [\mathbf{I}_g\mathbf{Z}_i]/\mathbf{Z}_1$

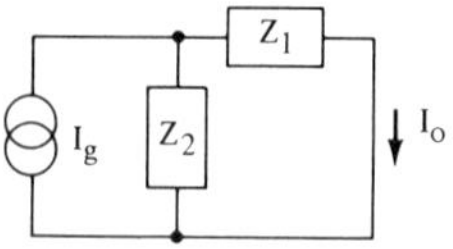

I_{VA} Notes ① ② ③

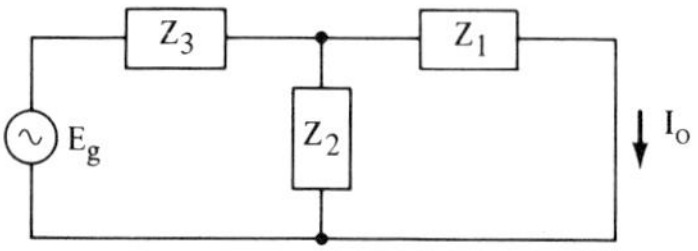

$\mathbf{E}_g = E\angle 0^\circ$

$\mathbf{Z}_i = \mathbf{Z}_3 + [\mathbf{Z}_2^{-1} + \mathbf{Z}_1^{-1}]^{-1}$

$\mathbf{I}_g = \mathbf{E}_g/\mathbf{Z}_i$

$\mathbf{I}_o = \mathbf{E}_g\Big/\left(\mathbf{Z}_i\left[(\mathbf{Z}_1/\mathbf{Z}_2) + 1\right]\right)$

I_{VA} Notes ① ② ③

I_{VA} *Notes:*

① E_g = Generator voltage
I_g = Generator current
Z_i = Input impedance

E_o = Output voltage
I_o = Output current
Z_o = Output impedance

Output Current
Vector Algebra

I_o

Applicable Notes

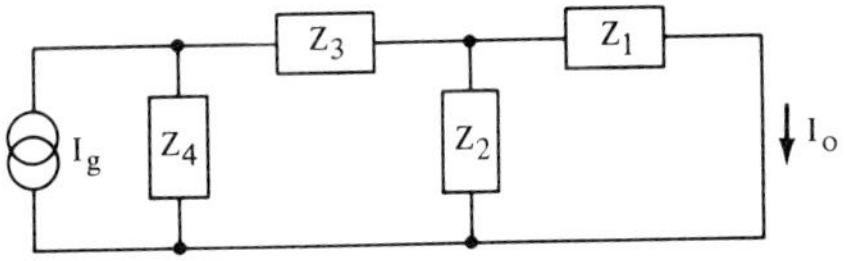

$\mathbf{I}_g = I_g \angle 0^\circ \quad \mathbf{E}_g = \mathbf{I}_g \mathbf{Z}_i$

$\mathbf{Z}_i = \left(\mathbf{Z}_4^{-1} + \left[\mathbf{Z}_3 + (\mathbf{Z}_2^{-1} + \mathbf{Z}_1^{-1})^{-1}\right]^{-1}\right)^{-1}$

$\mathbf{I}_o = \mathbf{I}_g \left[1 - (\mathbf{Z}_i/\mathbf{Z}_4)\right] / \left[(\mathbf{Z}_1/\mathbf{Z}_2) + 1\right]$

I_{VA} Notes ① ② ③

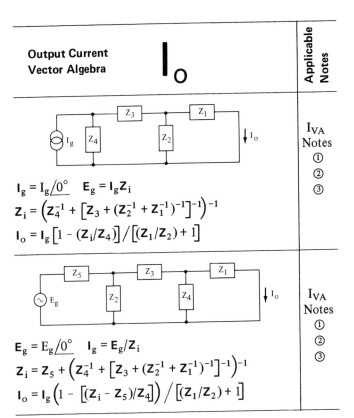

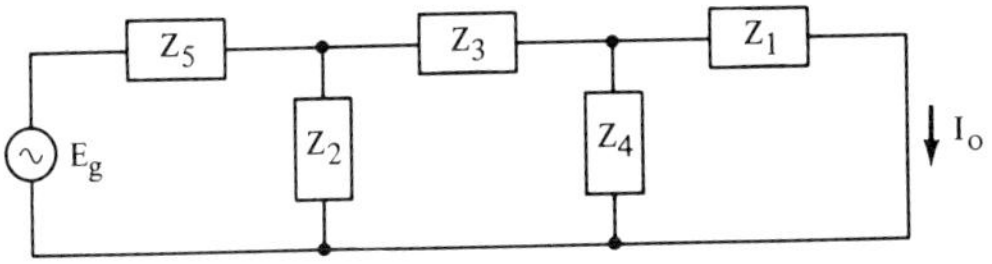

$\mathbf{E}_g = E_g \angle 0^\circ \quad \mathbf{I}_g = \mathbf{E}_g/\mathbf{Z}_i$

$\mathbf{Z}_i = \mathbf{Z}_5 + \left(\mathbf{Z}_4^{-1} + \left[\mathbf{Z}_3 + (\mathbf{Z}_2^{-1} + \mathbf{Z}_1^{-1})^{-1}\right]^{-1}\right)^{-1}$

$\mathbf{I}_o = \mathbf{I}_g \left(1 - \left[(\mathbf{Z}_i - \mathbf{Z}_5)/\mathbf{Z}_4\right]\right) / \left[(\mathbf{Z}_1/\mathbf{Z}_2) + 1\right]$

I_{VA} Notes ① ② ③

I_{VA} *Notes:*

② Z, Z_1, Z_2, Z_3, Z_4 and Z_5 may represent resistances, capacitances, inductances, series circuits, parallel circuits, unknown circuits or any combination.

③ All mathematical operations involving addition or subtraction must be performed in rectangular form. It is recommended that all mathematical operations involving multiplication or division be performed in polar form.

See also – **Z**, Vector Algebra   See – Z to Z^{-1} Conversion

$i_{N(th)}$ Thermal Noise Current

$i_{N(th)}$ = Symbol for thermal noise current. (other symbols for thermal noise current are I_N, I_{TH}, $I_{N(TH)}$, i_N, i_{th} etc)

$i_{N(th)}$ = Thermal noise (white noise) current of resistance. (thermal noise current is always rms current regardless of symbol)

$i_{N(th)} = \sqrt{(4kT_K\overline{BW})/R}$

k = Boltzmann constant $(1.38 \cdot 10^{-23}$ J/°K)

T_K = Temperature in Kelvin (°C + 273.15)

R = Resistance generating thermal noise.

BW = Noise bandwidth in hertz. (Bandwidth with zero noise contribution from frequencies outside of bandwidth. See—Active Circuits, Opamp, BW_{NOISE} for correction factors for noise measurement with standard filters.)

$i_{N(\sqrt{Hz})}$ = Symbol for noise current per root hertz. [formerly called root cycle $(\sqrt{\sim})$]

$i_{N(th)(\sqrt{Hz})} \simeq 1.287 \cdot 10^{-10} \sqrt{1/R}$ @ room temperature (BW_{NOISE} = 1 Hz)

Note: Above formulas do not include the excess noise current of resistors (that noise developed by dc voltage applied to resistors). Excess noise is 1/f noise and may be of significance at frequencies below 1 KHz.

j = Symbol for $\sqrt{-1}$

j = The imaginary unit of electrical complex numbers. The basic imaginary component of electrical rectangular form quantities. A unit identical to the mathematical imaginary unit i. A 90° indicator. A 90° multiplier. A mathematical quantity which rotates a number from the x axis (real numbers) to the y axis (imaginary numbers).

j = The imaginary unit used in all electronic calculations (instead of the mathematical unit i) to avoid confusion with the symbol for electrical current I or i.

$j = \sqrt{-1} = 1 \underline{/+90^\circ}$

$-j = -\sqrt{-1} = 1 \underline{/-90^\circ}$

$j^2 = -1 = 1 \underline{/\pm180^\circ}$ or $1 \underline{/+180^\circ}$

$-j^2 = +1 = 1 \underline{/0^\circ}$

$j^3 = -j = 1 \underline{/-90^\circ}$ or $1 \underline{/+270^\circ}$

$j^4 = +1 = 1 \underline{/0^\circ}$ or $1 \underline{/+360^\circ}$

J = Symbol for joules

J = A unit of work or work equivalent energy.

J = A unit of work equivalent to 1 watt • second.

J = A unit of work equal to .7376 foot • pounds.

J = A unit of work equal to .102 kg • meters or 10^7 ergs (dyne • centimeters).

J = A unit of work equivalent to $9.478 \cdot 10^{-4}$ Btu.

k — Coupling Coefficient

k = Symbol for coupling coefficient
(Capital K is sometimes used)

k = The ratio of mutual inductance to the square root of the product of the primary and the secondary inductances. The equivalent coupling coefficient provided by a discrete coupling element between two otherwise independent circuits.

$$k = M/\sqrt{L_1 L_2}$$

$$k = \omega_o^2 M \sqrt{C_1 C_2}$$

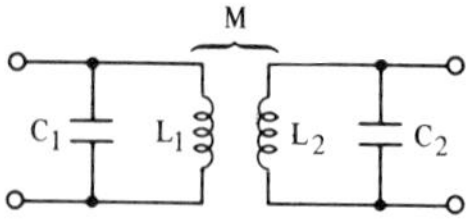

$$k = L_M \left[(L_M + L_1)(L_M + L_2)\right]^{-\frac{1}{2}}$$

$$k = \omega_o^2 L_M \sqrt{C_1 C_2}$$

$$k \approx L_M/\sqrt{L_1 L_2}$$

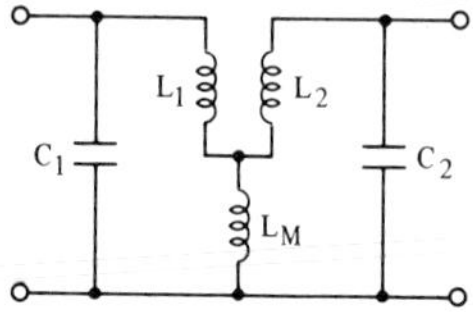

$$k = -\left((C_1 C_2)/\left[(C_M + C_1)(C_M + C_2)\right]\right)^{\frac{1}{2}}$$

$$k = -\left[(X_{C1} X_{CM}^{-1}) + 1\right]^{-1} \text{ when } C_1 = C_2$$

$$k \approx -\sqrt{C_1 C_2}/C_M$$

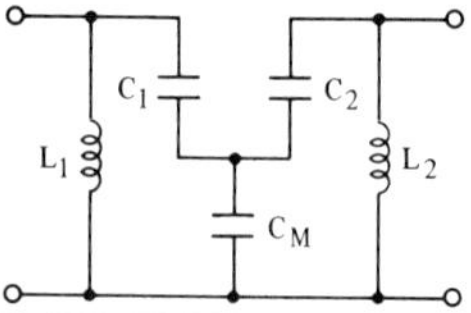

$$k = -\left[(C_1/C_M) + 1\right]^{-1} \text{ when } C_1 = C_2$$

$$k = -C_M/\sqrt{(C_M + C_1)(C_M + C_2)}$$

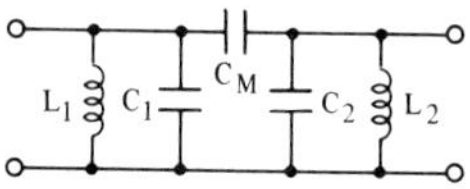

Note: Circuits exhibit double peaks above critical coupling.

k K

Kilo, Dielectric Constant, Kelvin

k = Symbol for kilo
(capital K is also used as a symbol for kilo)

k = A prefix symbol meaning 1000. A multiplier prefix used to indicate 10^3 units. Typical electronic uses include kilovolt (kV), kilowatt (kW), kilohertz (kHz) and kilohm (kΩ). The symbol for kilohms (kΩ) is often further abbreviated to k. In this form, it is most often capitalized. (K)

---

k = Symbol for dielectric constant (also K)

k = The capacitance multiplying effect of a specific material used as insulation between capacitor "plates" as compared to air. The ratio of capacitance of a given capacitor using a specific dielectric, to the capacitance of the capacitor using air as the dielectric.

k ≠ A true constant since k varies somewhat with frequency, temperature, etc.

Typical dielectric constants

Air = 1	Mica = 2.5 - 9.3
Paper = 2 - 3.5	PVC = 2.9
Ceramics = 10 - 10 k+	Polystyrene = 2.6
Fiber = 5.0	Waxes = 2.3 - 3.7

---

k = Boltzmann constant = $1.3805 \cdot 10^{-23}$ J/K°
(capital K is also used as a symbol for this constant)

K = Symbol for Kelvin (Kelvin temperature scale)

K = °C above absolute zero

K = °C + 273.15

L

L = Symbol for inductance. (self inductance)

L = In an inductor, in a coil, in a transformer, in a conductor or in any circuit where a varying electric current is flowing; that property which induces voltage in the same circuit from the varying magnetic field at a polarity which opposes the change of electric current.

Note that RC circuits and active circuit "inductances" which produce a lagging current do not meet the above definition and therefore cannot always perform in the same manner.

L = Inductance in henry (H) units.

Note that the basic unit is of convenient size for audio frequencies, but that millihenries (mH) and microhenries (μH) are more convenient at higher frequencies.

L = The symbol for an inductor on parts lists and schematics.

L = The symbol for inductive when used as a subscript.

L = One henry when a current change of one ampere per second develops one volt.

L_M = Mutual Inductance
(The symbol for mutual inductance is also M)

L_s = Series Circuit Inductance

L_p = Parallel Circuit Inductance

ℓ = Symbol for length

lf = Abbreviation for low frequency

Inductance, Series Circuits **L**	Applicable Notes	Terms
$L_t = L_1 + L_2 \cdots + L_n$ $L_2 = L_t - L_1$	①	L
$L_t = [(X_L)_1 + (X_L)_2 \cdots + (X_L)_n]/\omega$ $L_t = [(+X_1) + (+X_2) \cdots + (+X_n)]/\omega$	①	X_L +X
$L = R/(\omega D)$ Series reactive element must be inductive	①	D R
$L = (QR)/\omega$ Series reactive element must be inductive	①	Q R
$L = (R \tan \theta_Z)/\omega$ θ_Z must be a positive angle	①	R θ_Z
$L = (Z \sin \theta_Z)/\omega$ θ_Z must be a positive angle	①	Z θ_Z
Series to Parallel Conversion		
$L_p = [R_s^2/(\omega^2 L_s)] + (R_s/\omega)$	①	L_s R_s
$L_p = [\omega(Z^{-1} \sin \theta_Z)]^{-1}$ (θ_Z must be positive)	②	Z θ_Z

L *Notes:*

① B_L = Inductive susceptance, +B = Inductive susceptance, C = Capacitance, D = Dissipation Factor, E = rms Voltage, e = Instantaneous voltage, I = Current, L_M = Mutual inductance, L_p = Parallel circuit inductance, L_s = Series circuit inductance, ℓ = Length, M = Mutual Inductance, N = Number of turns, n (subscript) = Any number, Q = Quality, Merit or Storage Factor, R = Resistance, r = Radius, T = Time constant, W = Work, X_L = Inductive reactance, +X = Inductive reactance, Y = Admittance, Z = Impedance, θ = Phase angle, ω = Angular velocity = $2\pi f$, di/dt = Current rate of change.

② $x^{-1} = 1/x$, $|x|$ = Absolute value or magnitude of x

Inductance, Parallel Circuits **L**	Applicable Notes	Terms
$L_t = \left[\omega(B_{L1} + B_{L2} \cdots + B_{Ln})\right]^{-1}$	①	B_L
$L_t = \omega^{-1}\left[(+B_1) + (+B_2) \cdots + (+B_n)\right]^{-1}$	②	$+B$
$L_t = \left[L_1^{-1} + L_2^{-1} \cdots + L_n^{-1}\right]^{-1}$	① ②	L
$L_t = \omega^{-1}\left[(X_L^{-1})_t + (X_L^{-1})_2 \cdots + (X_L^{-1})_n\right]^{-1}$	①	X_L
$L_t = \omega^{-1}\left[(+X_1^{-1}) + (+X_2^{-1}) \cdots + (+X_n^{-1})\right]^{-1}$	②	$+X$
$L = (D\dot{R})/\omega$ Parallel reactive element must be inductive	①	$D\ R_p$
$L = \left\|\left[\omega G(\tan\theta_Y)\right]^{-1}\right\|$ θ_Y must be a negative angle	① ②	$G\ \theta_Y$
$L = R/(\omega Q)$ Parallel reactive element must be inductive	①	Q R
$L = R/(\omega \tan\theta_Z)$ θ_Z must be a positive angle	①	$R\ \theta_Z$
$L = [\omega Y \sin\|\theta_Y\|]^{-1}$ θ_Y must be a negative angle	① ②	$Y\ \theta_Y$
$L = Z/(\omega \sin\theta_Z)$ θ_Z must be a positive angle	①	$Z\ \theta_Z$
$L = E/(\omega I \sin\|\theta_I\|)$ θ_I must be a negative angle	① ②	$E\ I\ \theta_I$
Parallel to Series Conversion		
$L_s = \left[(\omega^2 L_p/R^2) + L_p^{-1}\right]^{-1}$	①	$L_p\ R_p$
$L_s = (Z\sin\theta_Z)/\omega$ (θ_Z must be positive)	②	$Z\ \theta_Z$

Inductance, Misc. Formulas **L**	Applicable Notes	Terms
$L = 1/(\omega B_L)$	①	B_L
$L_r = 1/(\omega^2 C)$ $L_r = X_C/\omega$ L required for resonance	①	C X_C
$L = X_L/\omega$	①	X_L
$L = (2W)/I^2$ (W = Work equivalent stored energy)	①	I W
$L = R/T$ (T = time constant)	①	R T
$L = -e/(di/dt)$ e = instantaneous voltage di/dt = rate of change in ampere/seconds	①	e $\frac{di}{dt}$
$L \approx (rN)^2/(9r + 10\ell)$ when magnetic path is air r = radius to center of winding N = number of turns ℓ = length of winding	①	ℓ N r
Coupled Series Inductances $L_t = L_1 + L_2 + 2M$ (fields aiding) $L_t = L_1 + L_2 - 2M$ (fields opposing)	①	L_1 L_2 M
Coupled Parallel Inductances $L_t = \left[(L_1 - M)^{-1} + (L_2 - M)^{-1}\right]^{-1}$ (fields opposing)	①	L_1 L_2 M

M

Mega, Mutual Inductance

M = Symbol for mega (also meg).

M = A prefix meaning one million. A multiplier prefix used to indicate 10^6 units.

Typical uses in electronics include megahertz (MHz), megawatt (MW), megavolt (MV) and megohm (MΩ).

Note: Megohm is often contracted to Meg and MΩ is often contracted to M.

M = Symbol for mutual inductance (The symbol L_M is also used)

M = The equivalent inductance common to both the primary and secondary windings of a transformer. In a circuit with two discrete inductors coupled by magnetic field interaction, the equivalent inductance common to both inductors.

$M = k\sqrt{L_p L_s}$

$M = kN_p N_s$

$M = (L_{ta} - L_{to})/4$

M *Notes:*

k = Coefficient of coupling.
L_p = Primary inductance.
L_s = Secondary inductance.
L_{ta} = Total inductance with primary and secondary windings connected series aiding.
L_{to} = Total inductance with primary and secondary windings connected series opposing.
N_p = Number of primary turns.
N_s = Number of secondary turns.

m — Flare Constant, Exponential Horns

m = Symbol for flare constant (flaring constant)

m = In acoustical horns, a constant used in formulas to determine the area, diameter or radius at any distance from the throat, e.g., $A = A_o \epsilon^{mx}$ or $S = S_o \epsilon^{mx}$. In an exponential horn of infinite length, a constant used in formulas to determine the frequency (flare cutoff frequency f_{FC}) below which no energy is coupled through the horn.

m = Flare constant expressed in units of inverse inches, inverse feet, inverse meters, etc.

$m = .6931/\ell_{2A}$ $m = .6931/\sqrt{\ell_{2d}}$ $m = \sqrt{.4804/\ell_{2d}}$ $m = \sqrt{.4804/\ell_{2r}}$ $m = (2.3025/\ell_{T-M})\left[\log(A_M/A_o)\right]$ $m = (4.605/\ell_{T-M})\left[\log(d_M/d_o)\right]$ $m = (4\pi)/\lambda_{FC}$ $m = (4\pi f_{FC})/v$	A, A_M, A_o, A_x are in square units. d, ℓ, r, v and x are in same units. m is in inverse of same units (units^{-1})

m *Notes:*

ℓ_{2A}, ℓ_{2d}, ℓ_{2r} = Length between centerline points of double area, diameter and radius respectively. ℓ_{T-M} = Throat to mouth length. A, A_M, A_o, A_x = Area, Area of mouth, throat and at distance x from throat respectively. d_o and d_M = Throat and mouth diameter. f_{FC} and λ_{FC} = Flare cutoff frequency and wavelength. ϵ = Base of natural logarithms = 2.718. v = Velocity of sound $\simeq$ 13630 in/sec, 1136 ft/sec or 346.3 meters/sec @ 25°C

n N

Nano, Number, Newton, Neper

n = Symbol for nano

n = Prefix symbol meaning 10^{-9} unit. One thousandth of a millionth unit.

Typical usage includes nanoamp (nA), nanovolt (nV), nanowatt (nW) and nanosecond (ns).

---

n = Symbol for an indefinite number

N = Symbol for number, number of turns, etc.

N = A pure number. Symbol for seldom used quantities where the natural symbol is in recognized use for another quantity. N_p = Number of turns of primary winding of a transformer. N_s = Number of turns of secondary winding of a transformer. N_{pp} = Number of pairs of poles in a motor or generator.

$N_p = (E_p N_s)/E_s$ $\quad N_s = (E_s N_p)/E_p$

$N_p = (I_s N_s)/I_p$ $\quad N_s = (I_p N_p)/I_s$

$N_{L1} = N_{Lt}\sqrt{L_1/L_t}$ (Tapped inductor turns)

$N_p = N_s\sqrt{Z_p/Z_s}$ $\quad N_s = N_p\sqrt{Z_s/Z_p}$

$N_{Z1} = N_{Zt}\sqrt{Z_1/Z_t}$ (Tapped secondary turns)

$N_{pp} = f/RPS$ (Pairs of poles in a generator or sync. motor)

N = Symbol for newton (SI unit of force)

N_p = Symbol for neper (logarithmic ratio unit)

$N_p = \ln\sqrt{P_2/P_1} = 8.686$ dB

$N_p = \ln(E_2/E_1) = \ln(I_2/I_1)$ when impedances are equal

NF NI

Noise Figure, Noise Index

NF = Symbol for noise figure.
(noise figure is also known as noise factor)

NF = The ratio in decibels of device output noise to ideal device output noise with all conditions of operation specified.

See—Active Circuits

NI = Symbol for noise index.

NI = The ratio, in decibels, of rms microvolts of excess noise in a decade of frequency, to the dc voltage applied to a resistor.

NI = Noise index expressed in decibels (dB).

$$NI = 20\left(\log\left[(10^6 E_{N(EX)})/(V_{dc})\right]\right)$$

$$NI = 20\left(\log\left[\frac{\text{(excess noise in microvolts rms)}}{\text{(applied dc voltage)}}\right]\right)$$

NI = −20 to +10 dB carbon composition

NI = −25 to −10 dB carbon film

NI = −40 to −15 dB metal film

NI = −40 to −15 dB wire wound

Notes:

① Excess noise is noise in excess of thermal noise.

② Excess noise is 1/f noise while thermal noise has equal output at all frequencies. (white noise)

③ $(E_N)_{EX} = \sqrt{(E_N)_t^2 - (E_N)_{th}^2}$ (all voltages rms)

o O Ø Subscript Only Zero and Letter O

O, o = Subscript symbol for output, open circuit, zero time, zero current, characteristic, etc.

o = Output in E_o, I_o, P_o, h_{ob}, h_{oe}, h_{oc}

o = Output in C_{ob}, g_{os}, P_{ob}, P_{oe}, Y_{oc}, Y_{os}

o = Output in C_{obo} and C_{oeo} (first o)

o = Open circuit in C_{obo} and C_{oeo} (last o)

o = Open circuit in C_{ibo}, C_{ieo}

O = Open circuit in BV_{CBO}, BV_{EBO}, LV_{CEO}

O = Open circuit in I_{CO}, I_{CBO}, I_{CEO}, I_{EBO}

O = Open circuit in V_{CBO}, V_{CEO}, V_{EBO}

O = Characteristic (impedance) in Z_O

O = Oscillation (frequency) in f_O

o = Resonant (frequency) in f_o (f_r is preferred)

o = Center (frequency of passband) in f_o and ω_o

o = Initial (at zero time) in E_o, etc.

O = Letter O in most printed material

0 = Zero in most printed material

Ø = The character used for many years to distinguish between zero and the letter O. Unfortunately, it has been used for both zero and the letter O. It also has been mistaken for the greek letter ϕ. The use of this character in formulas is not recommended.

P = Symbol for power

P = The rate at which energy is utilized to produce work. The rate at which work is done. The rate at which electrical energy is transformed to another form of energy such as heat, light, radiation, sound, mechanical work, potential energy in any form or any combination of any of the forms of energy.

P = Electrical power expressed or measured in watts (W)

Power is also expressed in dBm, microwatts (μW), milliwatts (mW), kilowatts (kW), megawatts (MW), etc.

P_{peak} = Instantaneous peak power

P = Effective or average power

$P = E_{dc} \cdot I_{dc} = E_{rms} \cdot I_{rms}$ (pure resistances only)

$P \neq E_{average} \cdot I_{average}$

$P_{sinewave}$ = Power produced by sinewave *voltage* and *current*, not the waveshape of the power. (Power waveshapes are rarely used except for rectangular waves where the waveshapes of power, voltage and current are identical.)

$P_{ac} = P_{dc}$ in heating effect and all other transformations of electrical energy

P = Zero in all purely reactive circuits

P = Zero when the phase angle of the current with respect to the voltage equals $\pm 90^\circ$

P

Power, DC Circuits

$P_t = P_1 + P_2 \cdots + P_n$ $P = E^2 G$ $P = EI$ $P = E^2/R$ $P = I^2/G$ $P = I^2 R$	General
$P_t = P_1 + P_2 \cdots + P_n$ $P_t = \left[(E_R)_1 + (E_R)_2 \cdots + (E_R)_n\right] I$ $P_t = \left[(E_R)_1 + (E_R)_2 \cdots + (E_R)_n\right]^2/(R_t)$ $P_t = E^2/(R_1 + R_2 \cdots + R_n)$ $P_t = I^2(R_1 + R_2 \cdots + R_n)$	Series Circuits
$P_t = P_1 + P_2 \cdots + P_n$ $P_t = E^2(G_1 + G_2 \cdots + G_n)$ $P_t = E(I_1 + I_2 \cdots + I_n)$ $P_t = E^2(R_1^{-1} + R_2^{-1} \cdots + R_n^{-1})$ $P_t = I_t^2/(G_1 + G_2 \cdots + G_n)$ $P_t = I_t^2/(R_1^{-1} + R_2^{-1} \cdots + R_n^{-1})$	Parallel Circuits

Note: $G = 1/R$ in all dc circuits

$P_{peak} = (E_R)_{peak} \cdot (I_{peak})$

$P_{peak} = (E_R)^2_{peak}/R$

$P_{peak} = (I_{peak})^2 R$ (all series circuits)

$P_{peak} = 2\ P_{average}$ (sinewave)

$P_{peak} = P_{average}$ (squarewave)

$P_{square} = 2\ P_{sine}$ (with same E_{peak} or I_{peak})

$P = (CE^2)/(2t)$ Power from a capacitor charge for time t)

$P = W/t$ (W = Work equivalent energy in joules or watt-seconds)

$P = (LI^2)/(2t)$ (Power for time t from energy stored in the field of an inductance)

P_{TH} = Thermal noise power (any value resistance)

$P_{TH} = K_B T_K BW$ (Available $P_{TH} = P_{TH}/4$)

K_B = Boltzmans constant = $1.38 \cdot 10^{-23}$ J/°K

T_K = Kelvin temperature, BW = Bandwidth

PWL = Power level in (acoustic) watts.

$PWL = \overline{SPL} + [20(\log r)] + .5$ dB = dB above 10^{-12} watt (Freefield conditions) $\overline{SPL}$ = Sound pressure level in dB above 20 $\mu N/m^2$, r = distance in feet

Power from Dissipation or Q Factor **P**	Terms	
$P = EI\cos(\tan^{-1} D^{-1})$	E I D	All Circuits
$P = EI\cos(\tan^{-1} Q)$	E I Q	
$P = \left[E^2 \cos(\tan^{-1} D^{-1})\right]/Z$	E Z D	
$P = \left[E^2 \cos(\tan^{-1} Q)\right]/Z$	E Z Q	
$P = I^2 Z\cos(\tan^{-1} D^{-1})$	I Z D	
$P = I^2 Z\cos(\tan^{-1} Q)$	I Z Q	
$P = \left[E\cos(\tan^{-1} D^{-1})\right]^2/\left[D(\omega C)^{-1}\right]$	E C D	Pure Series Circuit Only
$P = Q(\omega L)^{-1}\left[E\cos(\tan^{-1} Q)\right]^2$	E L Q	
$P = I^2 D(\omega C)^{-1}$	I C D	
$P = (I^2 \omega L)/Q$	I L Q	
$P = E^2 D\omega C$	E C D	Pure Parallel Circuit Only
$P = E^2/(Q\omega L)$	E L Q	
$P = \left[I\cos(\tan^{-1} D^{-1})\right]^2/(\omega CD)$	I C D	
$P = \omega LQ\left[I\cos(\tan^{-1} Q)\right]^2$	I L Q	

Power, Series Circuits **P**	Applicable Notes	Terms
$P_t = P_1 + P_2 \cdots + P_n$	① ②	P
$P_t = I\left[(E_R)_1 + (E_R)_2 \cdots + (E_R)_n\right]$	① ②	E_R I
$P_t = \left[(E_R)_1 + (E_R)_2 \cdots + (E_R)_n\right]^2/R_t$	① ②	E_R R
$P_t = I^2(R_1 + R_2 \cdots + R_n)$	① ②	I R
$P = EI\ pf$	①	E I pf
$P = EI \cos\theta_I$	① ②	E I θ_I
$P = (E\ pf)^2/R$	①	E R pf
$P = (E \cos\theta_Z)^2/R$	① ②	E R θ_Z
$P = (E^2\ pf)/Z$	①	E Z pf
$P = (E^2 \cos\theta_Z)/Z$	① ②	E Z θ_Z

P *Notes:*

① B_C = Capacitive susceptance, B_L = Inductive susceptance, C = Capacitance, D = Dissipation Factor, E = rms or dc Voltage, E_{peak} = Instantaneous peak voltage, G = Conductance, I = rms or direct current, I_{peak} = Instantaneous peak current, L = Inductance, pf = Power Factor, Q = Quality, Merit or Storage Factor, R = Resistance, t = Time, W = Work, X = Reactance, Y = Admittance, Z = Impedance, θ = Phase angle, ω = Angular velocity = $2\pi f$, tan = tangent, sin = sine, cos = cosine

Power, Series Circuits

P

	Applicable Notes	Terms
$P = I^2 Z \, pf$	①	I Z pf
$P = I^2 Z \cos \theta_Z$	① ②	I Z θ_Z
$P = (EZ^{-1})^2 \sqrt{Z^2 - [(\omega L) - (\omega C)^{-1}]^2}$	① ③	E C L Z
$P = (EZ^{-1})^2 \sqrt{Z^2 - (X_L - X_C)^2}$	① ② ③	E X_C X_L Z
$P = \left\| (E \, pf)^2 [\tan(\cos^{-1} pf)] [(\omega L) - (\omega C)^{-1}]^{-1} \right\|$	① ③ ④ ⊗	E C L pf
$P = [(E \cos \theta_Z)^2 (\tan \theta_Z)] / [(\omega L) - (\omega C)^{-1}]$	① ② ③ ⊗	E C L θ_Z
$P = \left\| (E \, pf)^2 [\tan(\cos^{-1} pf)] (X_L - X_C)^{-1} \right\|$	① ② ③ ④ ⊗	E X_C X_L pf
$P = [(E \cos \theta_Z)^2 (\tan \theta_Z)] / (X_L - X_C)$	① ② ⊗	E X_C X_L θ_Z
$P = I^2 \sqrt{Z^2 - [(\omega L) - (\omega C)^{-1}]^2}$	① ③	I C L Z
$P = I^2 \sqrt{Z^2 - (X_L - X_C)^2}$	① ②	I X_C X_L Z
$P = \left\| [I^2 (X_L - X_C)] / (\tan \theta_Z) \right\|$	① ② ⊗	I X_C X_L θ_Z

Power, Parallel Circuits **P**	Applicable Notes	Terms
$P_t = P_1 + P_2 \cdots + P_n$	① ②	P
$P = E^2(G_1 + G_2 \cdots + G_n)$	① ②	E G
$P = E^2(R_1^{-1} + R_2^{-1} \cdots + R_n^{-1})$	① ② ③	E R
$P = (I_G)_t^2/(G_1 + G_2 \cdots + G_n)$	① ②	I_G G
$P = (I_R)_t^2 (R_1^{-1} + R_2^{-1} \cdots + R_n^{-1})^{-1}$	① ② ③	I_R R
$P = EI_t\ pf$	① ②	E I_t pf
$P = EI_t \cos\theta_I$	① ②	E I_t θ_I
$P = E^2 Y\ pf$	①	E Y pf
$P = E^2 Y \cos\theta_Y$	① ②	E Y θ_Y
$P = (E^2\ pf)/Z$	①	E Z pf
$P = (E^2 \cos\theta_Z)/Z$	① ②	E Z θ_Z

P *Notes:*

② Subscripts C = capacitive, E = voltage, G = conductance, I = current, L = inductive, n = any number, R = resistive, t = total or equivalent, X = reactive, Y = admittance, Z = impedance

③ $x^{-1} = 1/x$, $x^{-2} = 1/x^2$, $|x|$ = Absolute value or magnitude of x

④ $\tan^{-1}$ = arc tangent
$\cos^{-1}$ = arc cosine

Power, Parallel Circuits **P**	Applicable Notes	Terms
$P = (I_t^2\ pf)/Y$	①	I_t Y pf
$P = (I_t^2 \cos\theta_Y)/Y$	① ②	I_t Y θ_Y
$P = I_t^2 Z\ pf$	① ②	I_t Z pf
$P = I_t^2 Z \cos\theta_Z$	① ②	I_t Z θ_Z
$P = E^2\sqrt{Y^2 - (B_L - B_C)^2}$	① ②	E B_C B_L Y
$P = E^2\sqrt{Z^{-2} - \left[(\omega L)^{-1} - (\omega C)\right]^2}$	① ③	E CL Z
$P = E^2\sqrt{Z^{-2} - (X_L^{-1} - X_C^{-1})^2}$	① ② ③	E X_C X_L Z
$P = \left\lvert\left[E^2(B_L - B_C)\right]/\left[\tan(\cos^{-1} pf)\right]\right\rvert$	① ② ③ ④ ⊗	E B_C B_L pf
$P = \left\lvert\left[E^2(B_L - B_C)\right]/(\tan\theta_Y)\right\rvert$	① ② ③ ⊗	E B_C B_L θ_Y
$P = \left\lvert E^2\left[(\omega L)^{-1} - (\omega C)\right]\left[\tan(\cos^{-1} pf)\right]^{-1}\right\rvert$	① ② ③ ④ ⊗	E CL pf
$P = \left\lvert E^2\left[(\omega L)^{-1} - (\omega C)\right]\left[\tan\theta_Z\right]^{-1}\right\rvert$	① ② ③ ⊗	E CL θ_Z

P *Notes:* ⊗ Division by zero, tangent of ±90° and purely reactive circuits are prohibited.

Power, Parallel Circuits **P**	Applicable Notes	Terms
$P = \left\| \left[E^2(X_L^{-1} - X_C^{-1})\right] / \left[\tan(\cos^{-1} pf)\right] \right\|$	① ② ③ ④ ⊗	$E\ X_C\ X_L\ pf$
$P = \left\| E^2(X_L^{-1} - X_C^{-1})(\tan\theta_Z)^{-1} \right\|$	① ② ③ ⊗	$E\ X_C\ X_L\ \theta_Z$
$P = (I_t Y^{-1})^2 \sqrt{Y^2 - (B_L - B_C)^2}$	① ② ③	$I\ B_C\ B_L\ Y$
$P = (I_t Z)^2 \sqrt{Z^{-2} - \left[(\omega L)^{-1} - (\omega C)\right]^2}$	① ② ③	$I\ CL\ Z$
$P = (I_t Z)^2 \sqrt{Z^{-2} - (X_L^{-1} - X_C^{-1})^2}$	① ② ③	$I\ X_C\ X_L\ Z$
$P = \left\| \left[(I\ pf)^2 \tan(\cos^{-1} pf)\right] / (B_L - B_C) \right\|$	① ② ③ ④ ⊗	$I\ B_C\ B_L\ pf$
$P = \left\| \left[(I \cos\theta_Y)^2 \tan\theta_Y\right] / (B_L - B_C) \right\|$	① ② ③ ⊗	$I\ B_C\ B_L\ \theta_Y$
$P = \left\| \left[(I\ pf)^2 \tan(\cos^{-1} pf)\right] / \left[(\omega L)^{-1} - (\omega C)\right] \right\|$	① ③ ④ ⊗	$I\ CL\ pf$
$P = \left\| \left[(I \cos\theta_Z)^2 \tan\theta_Z\right] / \left[(\omega L)^{-1} - (\omega C)\right] \right\|$	① ② ③ ⊗	$I\ CL\ \theta_Z$
$P = \left\| \left[(I\ pf)^2 \tan(\cos^{-1} pf)\right] / (X_L^{-1} - X_C^{-1}) \right\|$	① ② ③ ④ ⊗	$I\ X_C\ X_L\ pf$
$P = \left\| \left[(I \cos\theta_Z)^2 \tan\theta_Z\right] / (X_L^{-1} - X_C^{-1}) \right\|$	① ② ③ ⊗	$I\ X_C\ X_L\ \theta_Z$

p PF pf

Pico, Power Factor

p = Symbol for pico (pronounced peeko).

p = Prefix symbol meaning 10^{-12} unit. Replaces old $\mu\mu$ prefix.

Typical usage includes picofarad (pF), picosecond (ps), picoampere (pA), and picowatt (pW).

PF = Symbol for power factor.

pf = Symbol for power factor. (other symbols for power factor include: F_p, $\cos\theta$, PF, P.F. and p.f.)

pf = The ratio of actual power of an alternating current to apparent power. The ratio of power in watts to volt-amperes. The cosine of the phase angle of alternating current with respect to the voltage.

pf = Power factor expressed as a decimal or as a percentage.

pf $\simeq$ The inverse of Q factor when $Q > 7$

pf = A measurement more often than a calculation.

pf = The cosine of the phase angle when the angle is positive or negative, when the phase angle is current with respect to voltage or voltage with respect to current and when the angle represents the phase of impedance or admittance.

pf = A decimal number between zero and one, or a percentage between 0 and 100.

pf = One in purely resistive circuits and zero in purely reactive circuits

pf = The ratio of resistance to impedance

Power Factor, Series Circuits **pf**	Applicable Notes	Terms
$pf = \cos\theta \quad (\theta = \theta_E, \theta_I \text{ or } \theta_Z)$	①	θ
$pf = R/Z$	①	R Z
$pf = \left(R^{-2}\left[(\omega L) - (\omega C)^{-1}\right]^2 + 1\right)^{-\frac{1}{2}}$	① ②	C L R
$pf = \sqrt{1 - \left(\left[(\omega L) - (\omega C)^{-1}\right]/Z\right)^2}$	① ②	C L Z
$pf = P/(EI)$	①	E I P
$pf = (RI)/E$	①	E I R
$pf = (PZ)/E^2$	①	E P Z
$pf = P/(I^2Z)$	①	I P Z
$pf = \left(\left[(X_L - X_C)/R\right]^2 + 1\right)^{-\frac{1}{2}}$	① ②	R X_C X_L
$pf = \sqrt{1 - \left[(X_L - X_C)/Z\right]^2}$	①	X_C X_L Z

pf *Notes:*

① B_C = Capacitive Susceptance, B_L = Inductive Susceptance, C = Capacitance, E = rms Voltage, G = Conductance, I = rms Current, L = Inductance, P = Power, R = Resistance, R_p = Parallel Resistance, X_C = Capacitive Reactance, X_L = Inductive Reactance, Y = Admittance, Z = Impedance, θ = Phase Angle, ω = Angular Velocity, $\omega = 2\pi f$

Power Factor, Parallel Circuits — pf	Applicable Notes	Terms
$pf = \cos\theta \quad (\theta = \theta_E, \theta_I, \theta_Y \text{ or } \theta_Z)$	①	θ
$pf = G/Y$	①	G Y
$pf = Z/R_p$	①	R_p Z
$pf = \left(\left[(B_L - B_C)/G\right]^2 + 1\right)^{-\frac{1}{2}}$	① ②	B_C B_L G
$pf = \sqrt{1 - \left[(B_L - B_C)/Y\right]^2}$	①	B_C B_L Y
$pf = (EG)/I$	①	E I G
$pf = P/(EI)$	①	E I P
$pf = E/(IR_p)$	①	E I R_p
$pf = (PZ)/E^2$	①	E P Z
$pf = \left(\left[R_p(X_L^{-1} - X_C^{-1})\right]^2 + 1\right)^{-\frac{1}{2}}$	① ②	R_p X_C X_L
$pf = \sqrt{1 - \left[Z(X_L^{-1} - X_C^{-1})\right]^2}$	①	X_C X_L Z

pf *Notes:*

② $x^{-1} = 1/x$, $x^{-2} = 1/x^2$, $x^{-\frac{1}{2}} = 1/\sqrt{x}$, cos = cosine

Q — Q Factor, Quality Factor

Q = Symbol for Q Factor, Merit Factor, Storage Factor, Energy Factor, Magnification Factor and Quality Factor. (All names refer to the same factor. "Q" Factor is preferred)

Q = 1. The ratio of energy stored to the energy dissipated in inductors, coils, tuned circuits, and transformers. (Dissipation Factor which is the inverse of Q is commonly used for capacitors and dielectrics).
2. The tangent of the phase angle of alternating current with respect to the voltage in inductors.
3. In inductors at a given frequency, the ratio of reactance to the equivalent series resistance.

Q = A number from zero to infinity. (usually between 10 and 100)

Q = A factor used to calculate equivalent series or parallel resistance and a factor used to predict the voltage or current magnification of LC resonant circuits.

Real or Equivalent Resistance in Series with Reactance	Real or Equivalent Resistance in Parallel with Reactance	
$Q = (\omega L_s)/R_s$ $Q = (X_L)_s/R_s$	$Q = R_p/(\omega L_p)$ $Q = R_p/(X_L)_p$	**Inductors**
$Q = 1/D_s$ $Q = 1/(\omega C_s R_s)$ $Q = (X_C)_s/R_s$	$Q = 1/D_p$ $Q = \omega C_p R_p$ $Q = R_p/(X_C)_p$	**Capacitors**

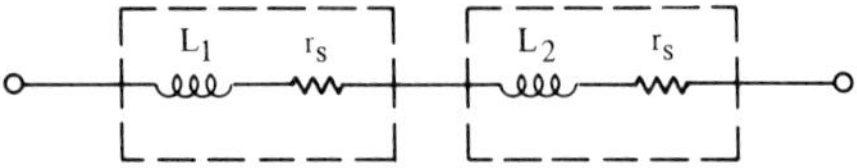

$Q_1 = \omega L_s/r_s$ $\quad$ $Q_2 = \omega L_s/r_s$

$$Q_t = (L_1 + L_2)/\left[(L_1/Q_1) + (L_2/Q_2)\right]$$

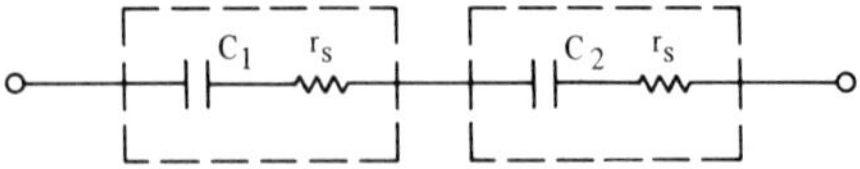

$D_1 = r_s\omega C_s$ $\quad$ $D_2 = r_s\omega C_s$

$$Q_t = (C_1 + C_2)/\left[(C_1D_2) + (C_2D_1)\right]$$

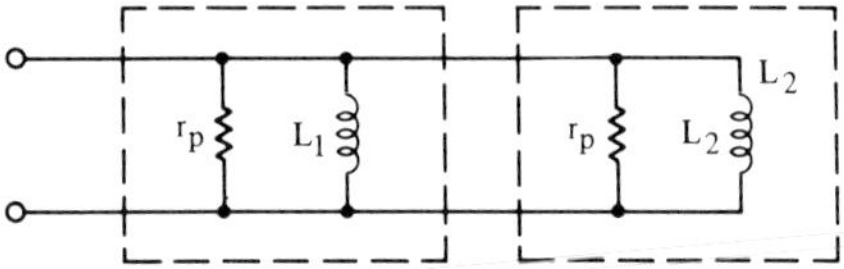

$Q_1 = r_p/\omega L_p$ $\quad$ $Q_2 = r_p/\omega L_p$

$$Q_t = (L_1^{-1} + L_2^{-1})/\left[(L_1Q_1)^{-1} + (L_2Q_2)^{-1}\right]$$

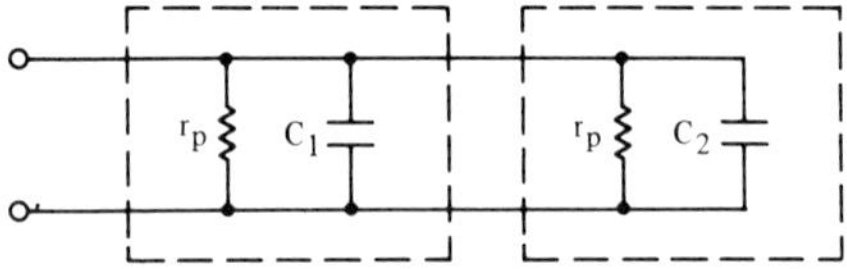

$D_1 = (\omega C_p r_p)^{-1}$ $\quad$ $D_2 = (\omega C_p r_p)^{-1}$

$$Q_t = (C_1 + C_2)/\left[(C_1D_1) + (C_2D_2)\right]$$

Note: For series circuits C, D, L & Q must be C_s, D_s, L_s & Q_s. For parallel circuits C, D, L & Q must be C_p, D_p, L_p & Q_p. See Q Notes ③ & ④

Series Resonant Circuits	# Q	**Resonant Circuit Q Factor**

$Q = \infty$

$Z = 0, \quad BW = 0$

$f_r = \left(2\pi\sqrt{LC}\right)^{-1}$

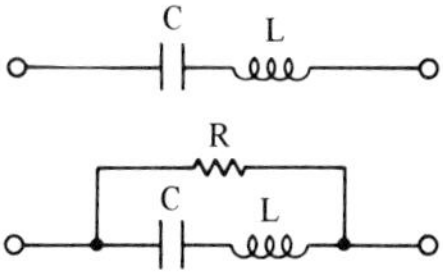

Ideal C &L

$Q = (Q_s^{-1} + D_s)^{-1}$

$Q = f_r/(f_2 - f_1)_{-3\,dB}$

$f_r = \left(2\pi\sqrt{LC}\right)^{-1}$

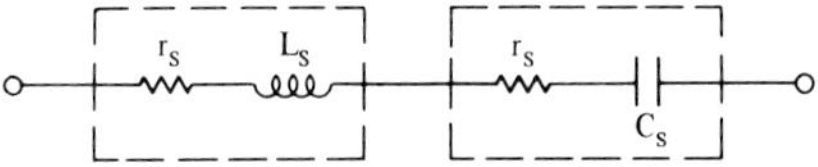

$Q_s = \omega_r L_s/r_s$ $\qquad$ $D_s = r_s\omega_r C_s$

$Q = L/\left(R\sqrt{LC}\right)$

$Q = (2\pi f_r L)/R$

$Q = f_r/(f_2 - f_1)_{-3\,dB}$

$f_r = \left(2\pi\sqrt{LC}\right)^{-1}$

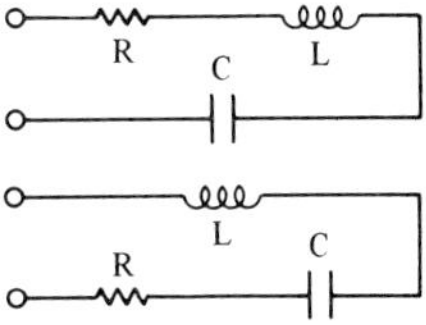

$Q = \sqrt{LC}\,\left[(RC)^{-1} + (R/L)\right]$

$Q \approx \sqrt{LC}/(CR)$

$Q = (f_r)_{DEF.1}/(f_2 - f_1)_{-3\,dB}$

$(f_r)_{DEF.1} = \left[(LC) - (L/R)^2\right]^{-\frac{1}{2}}/(2\pi)$

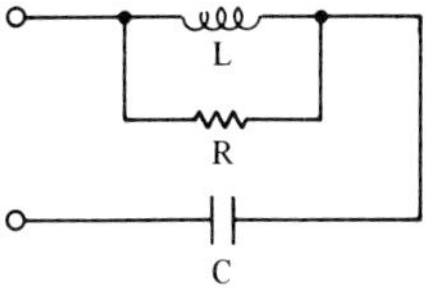

Q *Notes:* ① BW = Bandwidth, C = Capacitance, D = Dissipation Factor, f_r = Frequency of Resonance, L = Inductance, R = Resistance, X = Reactance

Series Resonant Circuits

Q

Resonant Circuit Q Factor

$$Q = \left[(RC) + (L/R)\right]/\sqrt{LC}$$

$$Q \approx L/\left(R\sqrt{LC}\right)$$

$$Q = (f_r)_{DEF.1}/(f_2 - f_1)_{-3\,dB}$$

$$(f_r)_{DEF.1} = \left[(LC)^{-1} - (CR)^{-2}\right]^{\frac{1}{2}}/(2\pi)$$

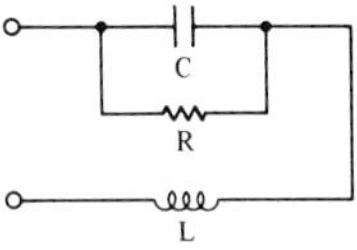

$$Q = \left[Q_S^{-1} + D_s + \left(R\sqrt{L_sC_s}/L\right)\right]^{-1}$$

$$Q = L/\left[\sqrt{L_sC_s}(R + r_{Ls} + r_{Cs})\right]$$

$$Q = f_r/(f_2 - f_1)_{-3\,dB}$$

$$f_r = \left(2\pi\sqrt{L_sC_s}\right)^{-1}$$

$$r_L = (\omega_r L_s)/Q_s, \quad r_C = D_s/\omega_r C_s$$

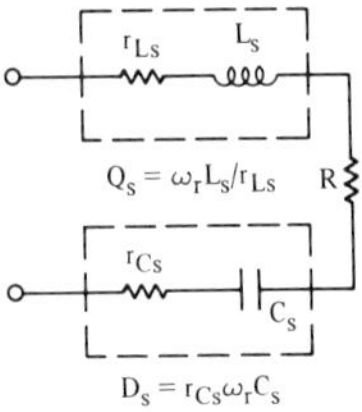

Note ③

$$Q = \left[\omega_r L_s/(R_{Ls} + R_{Cs})\right] \text{ Note ⑧}$$

$$Q \approx \left[2\pi f_r L(R_L^{-1} + R_C^{-1})\right]^{-1}$$

$$Q = (f_r)_{DEF.1}/(f_2 - f_1)_{-3\,dB}$$

$$(f_r)_{DEF.1} = \sqrt{\left[(R_C^2C)^{-1} - L^{-1}\right]/\left[(L/R_L^2) - C\right]}/(2\pi)$$

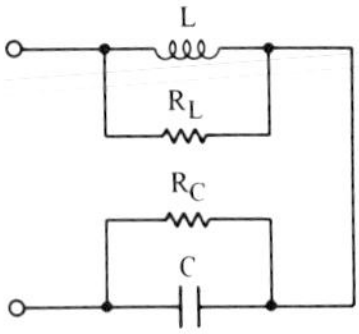

Q *Notes:*

① Cont. $\pi = 3.1416$, ω = Angular Velocity ($2\pi f$) ω_r = Resonant Angular Velocity ($2\pi f_r$)

② $x^{-1} = 1/x$, $x^{\frac{1}{2}} = \sqrt{x}$, $x^{-\frac{1}{2}} = 1/\sqrt{x}$

③ D, Q, L and C do not have exactly the same value when capacitors and inductors are measured in the parallel mode. L_sC_s, D_sQ_s = Series mode.

$Q = \infty$

$Z = \infty, \quad BW = 0$

$f_r = \left(2\pi\sqrt{LC}\right)^{-1}$

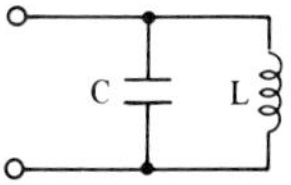

Ideal C & L

Note ④

$Q = (Q_p^{-1} + D_p)^{-1}$ Note ⑤

$Q = f_r/(f_2 - f_1)_{-3\,dB}$

$f_r = \left(2\pi\sqrt{L_p C_p}\right)^{-1}$

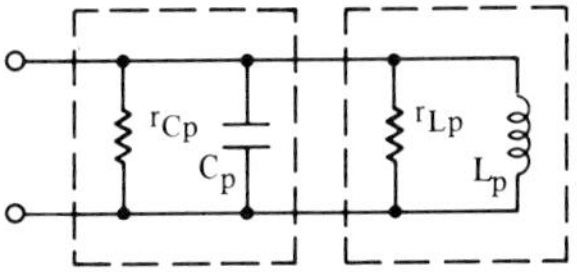

$D_p = (r_{Cp}\omega_r C_p)^{-1}$ $Q_p = r_{Lp}/\omega_r L_p$

Note ⑥

$Q = \left(R\sqrt{LC}\right)/L$

$Q = R/(2\pi f_r L)$

$Q = f_r/(f_2 - f_1)_{-3\,dB} = f_r/BW$

$f_r = \left(2\pi\sqrt{LC}\right)^{-1}$

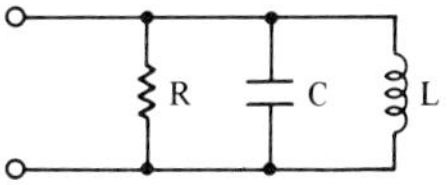

$Q = \sqrt{(L/CR^2) - 1}$ exception = $\sqrt{-x}$

$Q = f_r/(f_2 - f_1)_{-3\,dB} = f_r/BW$

$(f_r)_{DEF.1} = \sqrt{(LC)^{-1} - (R/L)^2}/(2\pi)$

exception = $\sqrt{-x}$

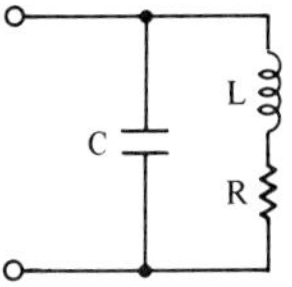

Q *Notes:*

④ D, Q, L and C do not have exactly the same value when measured in the series mode. D_p, Q_p, L_p and C_p = parallel mode.

$$Q = \sqrt{(L/CR^2) - 1}$$

$$Q = (f_r)_{DEF.1}/(f_2 - f_1)_{-3\,dB}$$

$$(f_r)_{DEF.1} = \left[(LC) - (CR)^2\right]^{-\frac{1}{2}}/(2\pi)$$

exception = $x^{-\frac{1}{2}}$

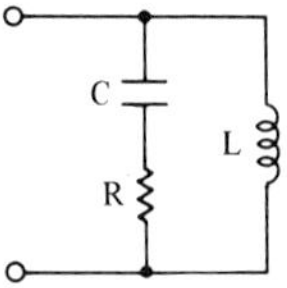

$$Q = \left[(L_p/C_pQ_p) + D_p + \left(L_p/R\sqrt{L_pC_p}\right)\right]^{-1}$$ Note ⑤ Note ⑥

$$Q = f_r/(f_2 - f_1)_{-3\,dB}$$

$$f_r = \left(2\pi\sqrt{L_pC_p}\right)^{-1}$$

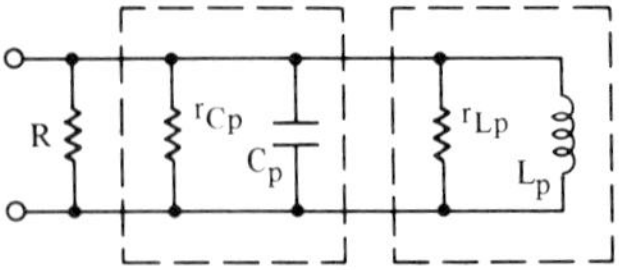

$D_p = (\omega C_p r_p)^{-1}$ $Q_p = r_p/\omega L_p$

$$Q = \left[\omega_r L_p(R_{Lp}^{-1} + R_{Cp}^{-1})\right]^{-1}$$ Note ⑨

$$Q \approx \sqrt{L/\left[C(R_L + R_C)^2\right]}$$

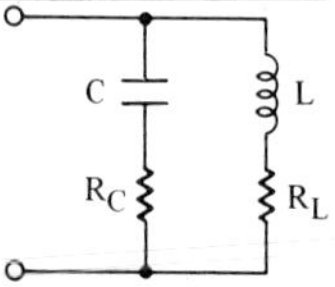

$$(f_r)_{DEF.1} = \sqrt{\left[C^{-1} - (R_L^2/L)\right]/\left[L - (R_C^2 C)\right]}\Big/(2\pi)$$

exception = $\sqrt{-x}$

Q *Notes:*

⑤ C_p, D_p, L_p & Q_p = Parallel or equivalent parallel values. C_s, D_s, L_s & Q_s = Series or equivalent series values.

⑥ r_s = Equivalent series resistance derived from Q_s or D_s. r_p = Equivalent parallel resistance derived from Q_p or D_p.

⑦ Def. 1 = Resonant frequency definition 1. – See f_r. $(f_2 - f_1)_{-3dB}$ = 3 dB down bandwidth (half power)

⑧ L_s = Equivalent series inductance, R_{Ls} = Equivalent series resistance of inductor resistor. R_{Cs} = Equivalent series resistance of capacitor resistor.

⑨ L_p, R_{Cp} & R_{Lp} = Parallel equivalent of series quantities.

Q q

Electric Charge

Q = Symbol for quantity of electric charge

Q = Quantity of electric charge. The amount of excess electrons or the amount of holes (deficiency of electrons).

Q = Electric charge expressed in coulomb (C) units. (Many in electronics feel uncomfortable in using the symbol C for coulombs since the unit symbol C (coulombs) is seldom used and the capacitance symbol (C) is often used)

Q = Electric charge in units equal to $6.242 \cdot 10^{18}$ electrons

Q = The product of current and time in ampere · seconds

$Q = CE$

$Q = It$

$Q = (2W)/E$ (W = work equivalent energy in joules or watt-seconds)

$Q = \sqrt{2CW}$

Charge of capacitor C, t seconds after application of voltage E to series RC circuit.

$Q = EC\left[1 - \epsilon^{\frac{-t}{RC}}\right]$ (ϵ = ln base = 2.71828)

Q = Schematic Symbol for transistor. See – Active circuits

---

q = The electric charge of one electron or $1.6 \cdot 10^{-19}$ coulombs. (symbol e is also used for q)

Notes:

$x^{(-y/z)} = (x^{-1})^{(y/z)} = \sqrt[z]{(x^{-1})^y}$

(your scientific calculator will perform correctly with a negative exponent)

R = Symbol for resistance

R = That property which opposes the flow of electric current by the transformation of electrical energy into heat or other forms of energy. The total opposition to the flow of direct current at a given voltage. The non-reactive part of the total opposition to alternating current of a given voltage. The real part of impedance. The reciprocal of conductance in purely resistive or in dc circuits.

R = Resistance in units of ohms (Ω). (Ω = Greek letter capital omega) kΩ = 1000 ohms, MΩ = 1,000,000 ohms. kΩ is often contracted to K and MΩ is often contracted to M.

R = Parts list symbol for resistor.

R = $R\angle 0^\circ$ in terms of polar impedance

R = R + j0 in terms of rectangular impedance

R *Notes:*

① B = Susceptance, C = Capacitance, D = Dissipation Factor, E = dc or rms voltage, f = Frequency, G = Conductance, I = rms or direct current, L = Inductance, P = Power, Q = Quality Factor, X = Reactance, Y = Admittance, Z = Impedance, Δ = Delta, θ = Phase Angle, π = Pi, ω = Angular Velocity, Ω = Ohm

② Subscripts:
C = capacitive, E = voltage, I = current, L = inductive, n = any number, p = parallel circuit, r = resonant, R = resistive, s = series circuit, t = total or equivalent, x = unknown, Y = admittance, Z = impedance 1, 2, 3 = first, second, third, A, B, C = first, second, third counterparts

Resistance, DC Circuits — R	Terms	
$R_t = R_1 + R_2 \cdots + R_n$ $R_x = R_t - R_1$	R	Series Circuits
$R_t = \left[(E_R)_1 + (E_R)_2 \cdots + (E_R)_n\right]/I$	E I	
$R_t = (E_R)_t^2/(P_1 + P_2 \cdots + P_n)$ $R_t = \left[(E_R)_1 + (E_R)_2 \cdots + (E_R)_n\right]^2/P_t$	E P	
$R_t = (P_1 + P_2 \cdots + P_n)/I^2$	I P	
$R_t = (G_1 + G_2 \cdots + G_n)^{-1}$	G	Parallel Circuits
$R_t = (R_1 R_2)/(R_1 + R_2)$ $R_t = (R_1^{-1} + R_2^{-1} \cdots + R_n^{-1})^{-1}$ $R_x = (R_1 R_t)/(R_1 - R_t)$ $R_x = (R_t^{-1} - R_1^{-1})^{-1}$	R	
$R_t = E/\left[(I_R)_1 + (I_R)_2 \cdots + (I_R)_n\right]$	E I	
$R_t = E^2/(P_1 + P_2 \cdots + P_n)$	E P	
$R_t = P_t/\left[(I_R)_1 + (I_R)_2 \cdots + (I_R)_n\right]^2$	I P	

R *Notes:*

③ sin = sine, cos = cosine, tan = tangent
④ $x^{-1} = 1/x$, $x^{-2} = 1/x^2$, $x^{-\frac{1}{2}} = 1/\sqrt{x}$

Equivalent Resistance from D and Q

$R_{EQUIV.}$

Equivalent Resistance

R_s = Equiv. Series Resistance	R_p = Equiv. Parallel Resistance	Terms
$R_s = D/(\omega C)$	$R_p = (\omega CD)^{-1}$	D C
$R_s = \omega LD$	$R_p = (\omega L)/D$	D L
$R_s = X_C D$	$R_p = X_C/D$	D X_C
$R_s = X_L D$	$R_p = X_L/D$	D X_L
$R_s = (\omega CQ)^{-1}$	$R_p = \omega CQ$	Q C
$R_s = (\omega L)/Q$	$R_p = Q/(\omega L)$	Q L
$R_s = X_C/Q$	$R_p = Q/X_C$	Q X_C
$R_s = X_L/Q$	$R_p = Q/X_L$	Q X_L
Series Resonant Circuits	Parallel Resonant Circuits	
$R_s = [D_C/(2\pi f_r C)] + [(2\pi f_r L)/Q_L]$	$R_p = \left([2\pi f_r C D_C] + [(2\pi f_r L)/Q_L]\right)^{-1}$	D Q C L
$R_s = [D_C X_{C(@f_r)}] + [(X_{L(@f_r)})/Q_L]$	$R_p = \left([D/X_{C(@f_r)}] + [(X_{L(@f_r)})/Q_L]\right)^{-1}$	D Q X_C X_L

Special Note: $f_r = (2\pi\sqrt{LC})^{-1}$

Resistance, Series AC Circuits **R**	Applicable Notes	Terms
$R_t = R_1 + R_2 \cdots + R_n$ $R_x = R_t - R_1$	②	R
$R = \sqrt{Z^2 - (\omega C)^{-2}}$	① ④ ⑦	C Z
$R = \left\| \left[\tan \theta_Z(\omega C)\right]^{-1} \right\|$	① ② ③ ④ ⑤ ⑦	C θ_Z
$R = E_R/I$	① ②	E_R I
$R = E_R^2/P$	① ②	E_R P
$R = P/I^2$	①	I P
$R = \sqrt{Z^2 - (\omega L)^2}$	① ⑦	L Z
$R = (\omega L)/(\tan \theta_Z)$	① ② ③ ⑦	L θ_Z
$R = \sqrt{Z^2 - X_C^2}$	① ②	X_C Z
$R = \left\|X_C/(\tan \theta_Z)\right\|$	① ② ③ ⑤	X_C θ_Z
$R = \sqrt{Z^2 - X_L^2}$	① ②	X_L Z
$R = X_L/(\tan \theta_Z)$	① ② ③	X_L θ_Z
$R = Z \cos \theta_Z$	① ② ③	Z θ_Z

R *Notes:*

⑤ $|x|$ = absolute value or magnitude of x

Resistance, Series AC Circuits **R**	Applicable Notes	Terms
$R = \sqrt{Z^2 - [(\omega L) - (\omega C)^{-1}]^2}$	① ④ ⑦	C L Z
$R = \left\| [(\omega L) - (\omega C)^{-1}]/(\tan \theta_Z) \right\|$	① ② ③ ④ ⑥ ⑦ ⊗	C L θ_Z
$R = (E \cos \theta_I)/I$	① ② ③	E I θ_I
$R = (E_t \cos \theta_I)^2/P$	① ② ③	E_t P θ_I
$R = \sqrt{Z^2 - (X_L - X_C)^2}$	① ②	X_C X_L Z
$R = \left\| (X_L - X_C)/(\tan \theta_Z) \right\|$	① ② ③ ⑤ ⑥ ⊗	X_C X_L θ_Z
Series to Parallel Conversion		
$R_p = Z/(\cos \theta_Z)$	① ②	Z θ_Z
$R_p = [(X_L - X_C)_s^2/R_s] + R_s$	③	X_C X_L R

R *Notes:*

⑥ Phase angle may be θ_Z or θ_Y also $\theta_E - \theta_I$ or $\theta_I - \theta_E$
⊗ Division by zero at resonance prohibited ($\tan 0° = 0$)
⑦ $\omega = 2\pi f$

Resistance, Parallel AC Circuits **R**	Applicable Notes	Terms
$R = 1/G$ $R_t = (G_1 + G_2 \cdots G_n)^{-1}$ $R_x = (G_t - G_1)^{-1}$	① ② ④	G
$R_t = (R_1 R_2)/(R_1 + R_2)$ $R_t = (R_1^{-1} + R_2^{-1})^{-1}$ $R_t = (R_1^{-1} + R_2^{-1} \cdots + R_n^{-1})^{-1}$ $R_x = (R_1 R_t)/(R_1 - R_t)$ $R_x = (R_t^{-1} - R_1^{-1})^{-1}$	① ② ④	R
$R = [Y^2 - B^2]^{-\frac{1}{2}}$	① ④	B Y
$R = \left|(\tan\theta)/B\right|$	① ③ ⑤	B θ
$R = \left[Y^2 - (\omega C)^2\right]^{-\frac{1}{2}}$	① ④ ⑦	C Y
$R = \left[Z^{-2} - (\omega C)^2\right]^{-\frac{1}{2}}$	① ④ ⑦	C Z
$R = \left|(\tan\theta)/(\omega C)\right|$	① ③ ⑤ ⑦	C θ
$R = E/I_R$	① ②	E I_R
$R = E^2/P$	①	E P
$R = P/I_R^2$	① ②	I_R P

Resistance, Parallel AC Circuits **R**	Applicable Notes	Terms
$R = \left[Y^2 - (\omega L)^{-2}\right]^{-\frac{1}{2}}$	① ④ ⑦	L Y
$R = \left[Z^{-2} - (\omega L)^{-2}\right]^{-\frac{1}{2}}$	① ④ ⑦	L Z
$R = \left\|\omega L(\tan\theta)\right\|$	① ③ ⑤ ⑥ ⑦	L θ
$R = [Z^{-2} - X^{-2}]^{-\frac{1}{2}}$	① ④	X Z
$R = \left\|X(\tan\theta)\right\|$	① ③ ⑤ ⑥	X θ
$R = [Y\cos\theta_Y]^{-1}$	① ② ③ ④	Y θ_Y
$R = Z/(\cos\theta_Z)$	① ② ③ ⊗	Z θ_Z
$R = \left[Y^2 - (B_L - B_C)^2\right]^{-\frac{1}{2}}$	① ② ④	B_C B_L Y
$R = \left\|(\tan\theta)/(B_L - B_C)\right\|$	① ② ③ ⑤ ⊗	B_C B_L θ
$R = \left(Y^2 - \left[(\omega L)^{-1} - (\omega C)\right]^2\right)^{-\frac{1}{2}}$	① ④ ⑦	C L Y
$R = \left(Z^{-2} - \left[(\omega L)^{-1} - (\omega C)\right]^2\right)^{-\frac{1}{2}}$	① ④ ⑦	C L Z
$R = \left\|(\tan\theta)/\left[(\omega L)^{-1} - (\omega C)\right]\right\|$	① ③ ④ ⑤ ⑦ ⊗	C L θ

Resistance, Parallel AC Circuits — R	Applicable Notes	Terms
$R = EI_t(\cos\theta)$	① ② ③ ⑥	E I_t θ
$R = P/\left[I_t(\cos\theta)\right]$	① ② ③ ⑥	I_t P θ
$R = \left[Z^{-2} - (X_L^{-1} - X_C^{-1})^2\right]^{-\frac{1}{2}}$	① ② ④	X_C X_L Z
$R = \left\|(\tan\theta)/(X_L^{-1} - X_C^{-1})\right\|$	① ② ③ ④ ⑤ ⑥ ⊗	X_C X_L θ

Series Equivalent Resistance of a Parallel Circuit. [Parallel to Series Conversion (Transformation)]	Applicable Notes	Terms
$R_s = (\cos\theta_Y)/Y$	① ② ③	Y θ_Y
$R_s = Z\cos\theta_Z$	① ② ③	Z θ_Z
$R_s = G/\left[G^2 + (B_L - B_C)^2\right]$	① ②	B_C B_L G
$R_s = \left[R_p(X_L^{-1} - X_C^{-1})^2 + R_p^{-1}\right]^{-1}$	① ② ④	X_C X_L R

Series Circuits in Parallel

$$R_t = \left[(R_1 + R_2 \cdots + R_n)_1^{-1} + (R_1 + R_2 \cdots + R_n)_2^{-1} \cdots + (R_1 + R_2 \cdots + R_n)_n^{-1}\right]^{-1}$$

Parallel Circuits in Series

$$R_t = (R_1^{-1} + R_2^{-1} \cdots + R_n^{-1})_1^{-1} + (R_1^{-1} + R_2^{-1} \cdots + R_n^{-1})_2^{-1} \cdots + (R_1^{-1} + R_2^{-1} \cdots + R_n^{-1})_n^{-1}$$

Delta to Wye (Δ to Y) Transformation

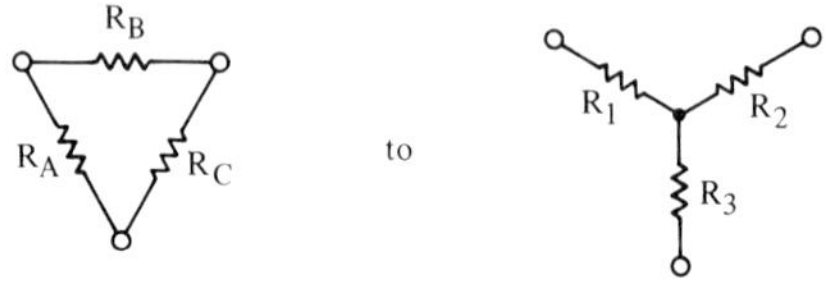

π section to T section Transformation

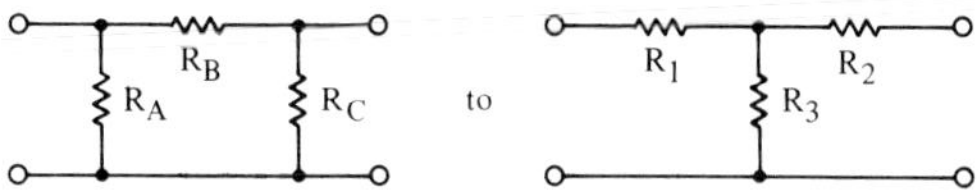

$$R_1 = (R_A R_B)/(R_A + R_B + R_C)$$

$$R_2 = (R_B R_C)/(R_A + R_B + R_C)$$

$$R_3 = (R_A R_C)/(R_A + R_B + R_C)$$

Notes Applicable to this page ① ② ④

$R_A = \left[(R_1 R_2) + (R_2 R_3) + (R_1 R_3)\right]/R_2$ $R_B = \left[(R_1 R_2) + (R_2 R_3) + (R_1 R_3)\right]/R_3$ $R_C = \left[(R_1 R_2) + (R_2 R_3) + (R_1 R_3)\right]/R_1$	Y to Δ or T to π Transformation

Second **s S** Siemen

s = Symbol for second

s = Basic unit of time. 9 192 631 770 transitions between the two hyperfine levels of the ground state of the cesium-133 atom.

s = 10^{12} ps, 10^{9} ns, 10^{6} μs and 10^{3} ms

---

s = 1/3600 of an angular degree (decimals preferred)

---

s = Symbol for spacing

S = Symbol for siemens

S = Basic SI unit of conductance (G), susceptance (B) and admittance (Y) [The mho (Ω^{-1} or ℧) predominates for this unit in the USA]

S = The reciprocal of resistance

S = $1/\Omega$ = mho

S = Abbreviation of signal (Sig is preferred).

S = Symbol for standing wave ratio.
(not recommended–use SWR or VSWR)

S = Symbol for cross-sectional area.
(the preferred symbol is A)

---

s = Subscript symbol for series and secondary

---

s = Subscript symbol for source and short-circuited

t = Symbol for time.

t = The duration of an event.

t = Time measured in seconds. (s or sec.) [time is expressed in picoseconds (ps), nanoseconds (ns), microseconds (μs), milliseconds (ms), seconds (s), minutes (min.), hours (hr) etc]

$t = 1/f$	Duration of one complete cycle of a periodic wave or of a periodic event.
$t = (CE)/I$	Time required to charge capacitance C to voltage E with current I.
$t = Q/I$	Time required to accumulate charge Q in a capacitance with current I.

Time required to charge capacitance C to voltage e through resistance R from source voltage E.

$$t = -RC\left(\ln\left[1 - (e/E)\right]\right)$$

Time required to discharge capacitance C through resistance R from voltage E to voltage e.

$$t = RC\left(\ln\left[1 - (e/E)\right]^{-1}\right)$$

Time after application of voltage E to series inductance L and resistance R for current to rise from zero to current i.

$$t = -LR^{-1}\left(\ln\left[1 - (iRE^{-1})\right]\right)$$

t *Notes:*

$x^{-1} = 1/x$, $\ln(x) = \log_\epsilon(x)$

t T Temperature, Telsa, Tera

t = Symbol for "customary" temperature (int'l).

T = Symbol for Kelvin temperature (int'l).

T = Symbol for temperature (USA common usage).

T_C = Symbol for temperature in degrees Celsius (°C).

T_F = Symbol for temperature in degrees Fahrenheit (°F).

T_K = Symbol for temperature in Kelvin (K).

$T_C = (T_F - 32)/1.8$

$T_C = T_K - 273.15$

$T_F = 1.8T_C + 32$

$T_F = 1.8T_K - 459.67$

$T_K = T_C + 273.15$

T_K = °C above absolute zero

Temperature determination of copper wire and copper wire windings by resistance measurement.

$$T_2 = \left[(R_2/R_1)(T_1 + 234.5)\right] - 234.5$$

R_1 = Resistance at known temperature T_1

R_2 = Resistance at unknown temperature T_2

T_1, T_2 = Temperature in °C

T = Symbol for telsa [SI unit of magnetic flux density (magnetic induction)]

T = Symbol for tera (unit prefix meaning 10^{12} units)

T TC Time Constant, Temperature Coefficient

T = Symbol for time constant. [other symbols include: t_C, Tc, TC, RC, script greek letter tau (τ) etc.]

T = 1. The time required for a capacitance to discharge through a resistance to 36.8% of the initial voltage or for the current to fall to 36.8% of the initial current.

2. The time required for a capacitance to charge through a resistance to 63.2% of the final voltage or for the current to fall to 36.8% of the initial current.

3. The time required for the voltage developed by cutoff of current through an inductor to fall to 36.8% of the maximum value.

4. The time required after application of voltage for the current through a series connected inductance and resistance to rise to 63.2% of the final value. [The exact value of 36.8% and 63.2% is $100\ \epsilon^{-1}$ and $100(1 - \epsilon^{-1})$]

T = RC also $(M\Omega) \cdot (\mu F)$

T = L/R

TC = Symbol for temperature coefficient (other symbols include α)

TC = In circuit elements or materials, a factor used to determine the changes in characteristics with changes in its temperature.

TC = A factor in decimal, percentage or parts per million form per degree temperature change. (temperature coefficient is almost always in °C)

Notes:

ϵ = Base of natural logarithms (2.71828 ---), $\epsilon^{-1} = 1/\epsilon$. One part per million = .0001%.

U

Mu Substitute, Unit

u = Typewritten substitute for greek letter mu (μ)
See—μ

u, U = Abbreviation of unit, ultra, etc.

V

Velocity

v = Symbol for velocity

v = Rate of motion in a given direction. A vector quantity having both magnitude (speed) and direction with respect to a reference.

v = Velocity measured in various linear units per second.

$v = f\lambda$ (f = frequency, λ = wavelength)

Velocity of sound in air

$v \simeq (1051 + 1.1T_F)$ ft/sec. (1136 @ 77°F)

$v \simeq (331.4 + .6T_C)$ meters/sec. (346.3 @ 25°C)

Velocity of sound in fresh water

$v = 1557 - [.245(74 - T_C)]$ meters/sec.

Velocity of electromagnetic waves in vacuum. (including light)

$v = 2.997\,925 \cdot 10^8$ m/s (use symbol c for light)

V

Volt, Voltage, Volume

V = Symbol for volt (unit of electromotive force)

V = Symbol for electromotive force
(See—E for passive circuits. See also—V in active circuit sections)

V = The basic unit of electromotive force, potential or voltage. The electric force required to develop a current of one ampere in a circuit with an impedance of one ohm.

V = Unit often used with multiplier prefixes

$\mu V = 10^{-6}$ V   mV $= 10^{-3}$ V

kV $= 10^{3}$ V   MV $= 10^{6}$ V

V = $\pm V_{dc}$ or $V_{rms(magnitude)}$ (exceptions noted)

V_{BE}, V_{CC}, V_{CE}, etc.—See Active Circuits

V = Symbol for volume (cubic content)

V = The amount of space in three dimensions.

V = Volume measured in various units such as cubic inches (in^3), cubic feet (ft^3), cubic centimeters (cm^3), cubic meters (m^3), etc.

Volume required for Helmholtz resonator. (ported hollow sphere or box)

$V = d[1948.7/f_r]^2$ (d in x units, V in x^3 units)

d = diameter of port.

f_r = frequency of resonance in hertz

W

Watt, Work, Energy

W = Symbol for watt.

W = Basic unit of electric power. A unit of power equal to a current of one ampere through a resistance of one ohm. ($P = I^2R$)

W = Unit often used with multiplier prefixes

$\mu W = 10^{-6}$ Watts $mW = 10^{-3}$ Watts

$kW = 10^{3}$ Watts $MW = 10^{6}$ Watts

W = Symbol for work.

W = Symbol for energy. (Energy is potential work.) (The energy symbol E is rarely used in electronics thus avoiding confusion with emf symbol E.)

W = The product of power and time.

W = Work or energy in joule (J) units in electronics. (joules = watts · seconds) Other units include kilowatt hour (kWh), foot-pound (ft · lbf), erg (erg) etc.

Energy stored in a capacitor charge

$W = .5CE^2$

$W = Q^2/(2C)$

$W = .5QE$

$W = .5LI^2$ Energy stored in the field of an inductance.

X = Symbol for reactance

X = That property of inductances and capacitances which opposes the flow of alternating current by storage of electrical energy. The imaginary part of impedance. The reciprocal of susceptance in purely parallel circuits. The non-resistive part of the total opposition to the flow of alternating current.

X = Reactance expressed in ohm (Ω) units.

$X = X_{magnitude}$

X_C = Magnitude of capacitive reactance

X_L = Magnitude of inductive reactance

$-X$ = Reactance identified as capacitive, not a real negative quantity.

$+X$ = Reactance identified as inductive, not a real positive quantity.

$-X = |-X| = X_C$

$+X = |+X| = X_L$

$\mathbf{X}$ = Complete description of reactance

$\mathbf{X}_C = X_C \angle -90^\circ$ in terms of polar impedance

$\mathbf{X}_L = X_L \angle +90^\circ$ in terms of polar impedance

$\mathbf{X}_C = 0 - jX_C = 0 + (-X_C)\,j$ (rectangular impedance)

$\mathbf{X}_L = 0 + jX_L = 0 + (+X_L)\,j$ (rectangular impedance)

Note that in $(-X_C)\,j$ and $(+X_L)\,j$, X_C and X_L have become real negative and positive quantities with the same signs that are assigned to the magnitude quantities.

Reactance, General and Misc. **X**	Applicable Notes	
$X_C = (\omega C)^{-1} = 1/(2\pi f C)$ $X_L = \omega L = 2\pi f L$ $X = R_s/D$ $X = R_s Q$ $X = Z$ when $R_s = 0$ $X_L - X_C = \pm X$ $\lvert +X \rvert = X_L$ $\lvert -X \rvert = X_C$ $X_L - X_C = 0$ @ resonance	① ② ③ ⑤	**Series Circuits**
$X_C = B_C^{-1} = 1/B_C$ $X_L = B_L^{-1} = 1/B_L$ $X_C = (\omega C)^{-1} = 1/(2\pi f C)$ $X_L = \omega L = 2\pi f L$ $X = R_p D$ $X = Q/R_p$ $X_L^{-1} - X_C^{-1} = \pm X_p^{-1} = \pm B$ $\lvert +X_p^{-1} \rvert = X_L^{-1} = B_L$ $\lvert -X_p^{-1} \rvert = X_C^{-1} = B_C$ $[X_L^{-1} - X_C^{-1}]^{-1} = \infty$ @ resonance	① ② ③ ⑤ ⓓ	**Parallel Circuits**

Reactance, Series Circuits — X	Applicable Notes	Terms
$(X_C)_t = \omega^{-1}(C_1^{-1} + C_2^{-1} \cdots + C_n^{-1})$	① ②	C
$-X_t = \omega^{-1}(C_1^{-1} + C_2^{-1} \cdots + C_n^{-1})$	③ ⑤	
$(X_L)_t = \omega(L_1 + L_2 \cdots + L_n)$	① ②	L
$+X_t = \omega(L_1 + L_2 \cdots + L_n)$	③ ⑤	
$(X_C)_t = (X_C)_1 + (X_C)_2 \cdots + (X_C)_n$	① ②	X_C
$-X_t = (-X_1) + (-X_2) \cdots + (-X_n)$	③ ⑤	$-X$
$(X_L)_t = (X_L)_1 + (X_L)_2 \cdots + (X_L)_n$	① ②	X_L
$+X_t = (+X_1) + (+X_2) \cdots + (+X_n)$	③ ⑤	$+X$
$\pm X = (\omega L) - (\omega C)^{-1}$	① ③ ⑤	C L
$\lvert X \rvert = \sqrt{Z^2 - R^2}$	① ③	R Z
$\pm X = R\left[\tan(\pm\theta_Z)\right]$	① ② ④ ⑤	R θ_Z
$\pm X = X_L - X_C$	① ② ⑤	X_C X_L
$\pm X = Z\left[\sin(\pm\theta_Z)\right]$	① ② ④ ⑤	Z θ_Z

Reactance, Series Circuits **X**	Applicable Notes	Terms
$\lvert X\rvert = \sqrt{(E/I)^2 - (P/I^2)^2}$	① ③	E I P
$\lvert X\rvert = \sqrt{(E/I)^2 - R^2}$	① ③	E I R
$\pm X = (E/I)\left[-\sin(\pm\theta_I)\right]$	① ② ④ ⑤	E I θ_I
$\pm X = P^{-1}(E\cos\theta)^2\left[\tan(\pm\theta_Z)\right]$	① ② ③ ④ ⑤ ⊗	E P θ_Z
$\lvert X\rvert = \sqrt{Z^2 - (P/I^2)^2}$	① ③	I P Z
$\pm X = (P/I^2)\left[-\tan(\pm\theta_I)\right]$	① ② ④ ⑤	I P θ_I
Series to Parallel Conversion		
$\pm X_p = Z\left[\sin(\pm\theta_Z)\right]^{-1}$	① ② ④ ⑤ ⓓ	Z θ_Z
$\pm X_p = \pm X_s^{-1}(\pm X_s^2 + R_s^2)$	① ② ④ ⑤ ⓓ	R_s $\pm X_s$

Reactance, Parallel Reactive Elements X	Applicable Notes	Terms
$(X_C)_t = \left[(B_C)_1 + (B_C)_2 \cdots + (B_C)_n\right]^{-1}$	① ②	B_C
$(X_C)_t = \left[(-B_1) + (-B_2) \cdots + (-B_n)\right]^{-1}$	③	$-B$
$(X_L)_t = \left[(B_L)_1 + (B_L)_2 \cdots + (B_L)_n\right]^{-1}$	① ②	B_L
$(X_L)_t = \left[(+B_1) + (+B_2) \cdots + (+B_n)\right]^{-1}$	③	$+B$
$(X_C)_t = \left[\omega(C_1 + C_2 \cdots + C_n)\right]^{-1}$	① ② ③	C
$(X_L)_t = \left[\omega^{-1}(L_1^{-1} + L_2^{-1} \cdots + L_n^{-1})\right]^{-1}$	① ② ③	L
$(X_C)_t = \left[(X_C)_1^{-1} + (X_C)_2^{-1} \cdots + (X_C)_n^{-1}\right]^{-1}$	① ②	X_C
$(X_C)_t = \left[(-X_1)^{-1} + (-X_2)^{-1} \cdots + (-X_n)^{-1}\right]^{-1}$	③	$-X$
$(X_L)_t = \left[(X_L)_1^{-1} + (X_L)_2^{-1} \cdots + (X_L)_n^{-1}\right]^{-1}$	① ②	X_L
$(X_L)_t = \left[(+X_1)^{-1} + (+X_2)^{-1} \cdots + (+X_n)^{-1}\right]^{-1}$	③	$+X$
$\pm X = [B_L - B_C]^{-1}$	① ② ③ ⑤ ⓓ	B_C B_L
$\pm X = \left[(\omega L)^{-1} - (\omega C)\right]^{-1}$	① ② ③ ⑤ ⓓ	C L
$\pm X = [X_L^{-1} - X_C^{-1}]^{-1}$	① ② ③ ⑤ ⓓ	X_C X_L

Reactance, Parallel Circuits **X**	Applicable Notes	Terms
$\lvert X \rvert = [Y^2 - G^2]^{-\frac{1}{2}}$	① ③	G Y
$\pm X = \left[-G \tan(\pm\theta_Y)\right]^{-1}$	① ② ③ ④ ⑤ ⓓ	G θ_Y
$\lvert X \rvert = \left[Z^{-2} - R^{-2}\right]^{-\frac{1}{2}}$	① ③	R Z
$\pm X = R\left[\tan(\pm\theta_Z)\right]^{-1}$	① ② ③ ④ ⑤ ⓓ	R θ_Z
$\pm X = \left[-Y \sin(\pm\theta_Y)\right]^{-1}$	① ② ③ ④ ⑤ ⓓ	Y θ_Y
$\pm X = \left[Z^{-1} \sin(\pm\theta_Z)\right]^{-1}$	① ② ③ ④ ⑤ ⓓ	Z θ_Z
$\lvert X \rvert = \left[(I/E)^2 - G^2\right]^{-\frac{1}{2}}$	① ③	E I G
$\pm X = E[I_L - I_C]^{-1}$	① ② ③ ⑤ ⓓ	E I_C I_L
$\lvert X \rvert = \left[(I/E)^2 - R^{-2}\right]^{-\frac{1}{2}}$	① ③	E I R
$\pm X = -E\left[I \sin(\pm\theta_I)\right]^{-1}$	① ② ③ ⑤ ⓓ	E I θ_I

Reactance, Parallel Circuits X	Applicable Notes	Terms
$\lvert X\rvert = \left[Z^{-2} - (P/E^2)^2\right]^{-\frac{1}{2}}$	① ③	E P Z
$\pm X = (E^2/P)\left[\tan(\pm\theta_Z)\right]^{-1}$	① ② ③ ④ ⑤ ⓓ	E P θ_Z
Parallel to Series Conversion		
$\pm X_s = Z\left[\sin(\pm\theta_Z)\right]$	① ② ③	Z θ_Z
$\pm X_s = \left[(\pm X_p/R_p^2) + (\pm X_p)^{-1}\right]^{-1}$	④ ⑤ ⓓ	R_p $\pm X_p$

X *Notes:*

① General: B = Susceptance, C = Capacitance, D = Dissipation factor, E = Voltage, f = Frequency, G = Conductance, I = Current, L = Inductance, P = Power, Q = Q factor, R = Resistance, X = Reactance, Y = Admittance, Z = Impedance, θ = Phase angle, ω = Angular velocity, $\omega = 2\pi f$, $\omega = 6.283 \text{---} f$

② Subscripts:
c = Capacitive, E = Voltage, I = Current, L = Inductive, n = Any number, p = Parallel circuit, R = Resistive, s = Series circuit, t = Total or equiv., X = Reactive, Y = Admittance, Z = Impedance

③ Mathematics:
$x^{-1} = 1/x$, $x^{-2} = 1/x^2$, $x^{\frac{1}{2}} = \sqrt{x}$, $x^{-\frac{1}{2}} = 1/\sqrt{x}$, $|x|$ = Magnitude of x, ∞ = Infinite

④ tan = tangent, sin = sine, cos = cosine, $\tan^{-1}$ = arc tangent, $\sin^{-1}$ = arc sine

⑤ Reminders:
$\pm B$, $\pm X$, $\pm\theta$—use the sign of the quantity.
$|+B| = B_L$, $|-B| = B_C$, $|+X| = X_L$, $|-X| = X_C$.
$+\theta_Z$ = Inductive circuit, $-\theta_Z$ = Capacitive circuit.

ⓓ The reciprocal of zero may be manually converted to infinity.
$\infty \cdot x = \infty$ when $x \neq 0$, $\infty/x = \infty$ when $x \neq \infty$
The reciprocal of infinity may be manually converted to zero.
$0 \cdot x = 0$ when $x \neq \infty$, $0/x = 0$ when $x \neq 0$

⊗ Division by zero is prohibited.

Y

Admittance Definitions

Y = Symbol for admittance

Y = The total ease of alternating current flow at a given frequency and voltage. The reciprocal of impeddance. A quantity which in rectangular form is as useful for parallel circuits as impedance is for series circuits. The resultant of conductance and susceptance in parallel. The resultant of reciprocal resistance and reciprocal reactance in parallel.

Y = Admittance expressed in siemens (S) or mho (Ω^{-1}) units.

$Y = |Y| = Y_{MAGNITUDE}$

θ_Y = Phase angle of admittance

$\mathbf{Y}_{POLAR} = Y\underline{/\pm\theta_Y} = Z^{-1}\underline{/-(\pm\theta_Z)}$

$\mathbf{Y}_{RECT} = G - (\pm B)\,j = R_p^{-1} - (\pm X_p^{-1})\,j$

$\mathbf{Y}_{RECT}$ =
1. The rectangular form of admittance
2. The complex number form of admittance
3. The mathematical equivalent of conductance (G) and susceptance (B) in parallel
4. The mathematical equivalent of reciprocal resistance (R^{-1}) and reciprocal reactance (X^{-1}) in parallel.

$\mathbf{Y}_{RECT}$ = An easy method of transforming a series circuit to a parallel equivalent circuit.

$\mathbf{Y}_{RECT}$ = Complex quantity used to solve problems involving complex parallel circuits.

$\mathbf{Y}_{RECT}$ = A quantity that is identical to rectangular assumed current when the assumed voltage is one.

Y *Notes:*

① General:

B = Susceptance	C = Capacitance	D = Dissipation factor
E = rms Voltage	f = Frequency	G = Conductance
I = rms Current	j = Imaginary number	L = Inductance
P = Power	Q = Q factor	R = Resistance
X = Reactance	Y = Admittance	Z = Impedance

ϵ = Base of natural logarithms π = Circum. to diam. ratio

θ = Phase angle ω = Angular Velocity

② Subscripts:

C = capacitive	E = voltage	I = current
L = inductive	n = any number	p = parallel circuit
R = resistive	s = series circuit	t = total or equiv.
X = reactive	Y = admittance	Z = impedance

③ Constants:

$j = \sqrt{-1}$ = mathematical i = 90° multiplier

$\epsilon = 2.718$ --- $\epsilon^{-1} = .36788$ ---

$\pi = 3.1416$ $2\pi = 6.283$ ---

$\omega = 2\pi f$ $\omega = 6.283$ --- f

④ Algebra:

$x^{-1} = 1/x$ $x^{-2} = 1/x^2$ $x^{\frac{1}{2}} = \sqrt{x}$

$x^{-\frac{1}{2}} = 1/\sqrt{x}$ $|x|$ = absolute value or magnitude of x

⑤ Trigonometry:

sin = sine	cos = cosine	tan = tangent
$\sin^{-1}$ = arc sine	$\cos^{-1}$ = arc cosine	$\tan^{-1}$ = arc tangent

⑥ Reminders:

$\pm\theta$ --- Use the sign of the angle

$\pm X, \pm I_X, \pm E_X, \pm B$ --- + identifies the quantity as inductive
– identifies the quantity as capacitive

(As terms in formulas, these quantities must be used as real positive or negative quantities)

⑦ Cosine θ:

The cosine of either a positive or a negative angle is positive, therefore, $\cos\theta_Z = \cos\theta_Y = \cos\theta_E = \cos\theta_I$

Admittance, Series Circuits **Y**	Applicable Notes	Terms
$Y = Z^{-1} = 1/Z$	① ④	Z
$Y = \left(R^2 + \left[(\omega L) - (\omega C)^{-1}\right]^2\right)^{-\frac{1}{2}}$	① ③ ④	CL R
$Y = \left\|(\sin\theta)/\left[(\omega L) - (\omega C)^{-1}\right]\right\|$	① ③ ④ ⑤ ⑦ ⊗	CL θ
$Y = I/E$	①	E I
$Y = \left[R^2 + (X_L - X_C)^2\right]^{-\frac{1}{2}}$	① ② ④	R $X_C X_L$
$Y = (\cos\theta)/R$	① ⑤ ⑦	R θ
$Y = \left\|(\sin\theta)/(X_L - X_C)\right\|$	① ② ⑤ ⑦ ⊗	$X_C X_L$ θ
$Y = P/(E^2 \cos\theta)$	① ⑤ ⑦	E P θ
$Y = (I^2 \cos\theta)/P$	① ⑤ ⑦	I P θ

Y *Notes:*

ⓓ The reciprocal of zero may be manually converted to infinity. $\infty \cdot x = \infty$ when $x \neq 0$, $\infty/x = \infty$ when $x \neq \infty$
The reciprocal of infinity may be manually converted to zero. $0 \cdot x = 0$ when $x \neq \infty$, $0/x = 0$ when $x \neq 0$

⊗ Division by zero is prohibited. A zero divisor will occur at resonance and/or in purely reactive circuits.

Admittance, Parallel Circuits $\quad Y$	Applicable Notes	Terms
$Y = Z^{-1} = 1/Z$	① ④	Z
$Y = \sqrt{G^2 + (B_L - B_C)^2}$	① ②	$B_C B_L$ G
$Y = \|(B_L - B_C)/(\sin \theta_Y)\|$	① ② ④ ⑤ ⊗	$B_C B_L$ θ_Y
$Y = \sqrt{R^{-2} + \left[(\omega L)^{-1} - (\omega C)\right]^2}$	① ③ ④	CL R
$Y = \left\|\left[(\omega L)^{-1} - (\omega C)\right]/(\sin \theta_Z)\right\|$	① ② ③ ④ ⑤ ⊗	CL θ_Z
$Y = I/E$	①	E I
$Y = \|G/(\cos \theta_Y)\|$	① ② ④	G θ_Y
$Y = \sqrt{R^{-2} + (X_L^{-1} - X_C^{-1})^2}$	① ② ④	R $X_C X_L$
$Y = [R \cos \theta_Z]^{-1}$	① ② ④ ⑤	R θ_Z
$Y = P/(E^2 \cos \theta_Z)$	① ② ⑤	P E θ_Z
$Y = (I_t^2 \cos \theta_Z)/P$	① ② ⑤	I_t P θ_Z

Series Circuit Polar Admittance Formulas

$$\mathbf{Y}_{POLAR} = Y\underline{/\pm\theta_Y} = Z^{-1}\underline{/-(\pm\theta_Z)}$$

Polar Impedance is Preferred

Series Circuit Rectangular Admittance Formulas

Special Note: Rectangular admittance is intrinsically a parallel circuit quantity. The rectangular admittance of a series circuit is the mathematical equivalent of reciprocal resistance and reciprocal reactance in *parallel.*

$$\mathbf{Y}_{RECT} = G - (\pm B)\,j \quad \text{where} \quad |+B| = B_L, \quad |-B| = B_C$$

$$\mathbf{Y}_{RECT} = R_p^{-1} - (\pm X^{-1})\,j \quad \text{where} \quad |+X| = X_L, \quad |-X| = X_C$$

$$\mathbf{Y}_{RECT} = \left[Y \cos\theta_Y\right] - \left(Y \sin\left[-(\pm\theta_Y)\right]\right) j$$

$$\mathbf{Y}_{RECT} = \left[Z^{-1} \cos\theta_Z\right] - \left[Z^{-1} \sin(\pm\theta_Z)\right] j$$

$$G = R_p^{-1} = Y\cos\theta_Y = Z^{-1}\cos\theta_Z \qquad \textit{Note } ⓓ$$

$$\pm B = B_L - B_C = X_{L_p}^{-1} - X_{C_p}^{-1}$$

$$\pm B = Y \sin\left[-(\pm\theta_Y)\right] = Z^{-1}\sin(\pm\theta_Z)$$

Note: Rectangular admittance is identical to rectangular current produced by a voltage of one except for the names of quantities. When $E = 1$, $\mathbf{I}_{POLAR} = \mathbf{Y}_{POLAR}$, $\mathbf{I}_{RECT} = \mathbf{Y}_{RECT}$, $I_R = G$, $I_{X_L} = B_L$, $I_{X_C} = B_C$, $\pm I_X = \pm B$.

Note: The use of $\mathbf{Y}_{POLAR}$ is not recommended unless used as a means of identification of a parallel quantity. Convert directly from $\mathbf{Z}_{POLAR}$ to $\mathbf{Y}_{RECT}$

See also—Z, θ, G and B

Admittance and Phase, Series Circuits — **Y**	Applicable Notes	Terms
$Y = \left(R^2 + \left[(\omega L) - (\omega C)^{-1}\right]^2\right)^{-\frac{1}{2}}$ $\angle\pm\theta_Y = \tan^{-1}\left(-\left[(\omega L) - (\omega C)^{-1}\right]/R\right)$ $\mathbf{Y}_{RECT} = G - (\pm B)\,j$ $G = \left[(\pm X_s^2/R_s) + R_s\right]^{-1}$ $\pm B = \left[(R_s^2/\pm X_s) + (\pm X_s)\right]^{-1}$ $\pm X_s = (\omega L_s) - (\omega C_s)^{-1}$	① ② ③ ④ ⑤ ⑥ ⓓ	C L R
$Y = \left[R^2 + (X_L - X_C)^2\right]^{-\frac{1}{2}}$ $\angle\pm\theta_Y = \tan^{-1}\left[-(X_L - X_C)/R\right]$ $\mathbf{Y}_{RECT} = R_p^{-1} - (\pm X_p^{-1})\,j$ $R_p^{-1} = R_s/\left[R_s^2 + (\pm X_s)^2\right]$ $\pm X_p^{-1} = \pm X_s/\left[(\pm X_s)^2 + R_s^2\right]$ $\pm X_s = X_L - X_C$	① ② ③ ④ ⑤ ⑥ ⓓ	X_C X_L R
$Y = Z^{-1}, \quad \angle\pm\theta_Y = -(\pm\theta_Z)$ $\mathbf{Y}_{RECT} = G - (\pm B)\,j$ $G = Z^{-1}\cos\theta_Z$ $\pm B = Z^{-1}\sin(\pm\theta_Z)$	① ② ③ ④ ⑤ ⑥	Z $\pm\theta_Z$

Admittance and Phase, Parallel Circuits

Y

$Y_{POLAR} = Y\underline{/\pm\theta_Y}$

$Y_{RECT} = G - (\pm B)\,j$

Note: $G = R_p^{-1}$, $\pm B = (X_L)_p^{-1} - (X_C)_p^{-1}$ ⓓ	Applicable Notes	Terms
$Y = \sqrt{G^2 + (B_L - B_C)^2}$ $\pm\theta_Y = \tan^{-1}\left[-(B_L - B_C)/G\right]$ $Y_{RECT} = G - (B_L - B_C)\,j$	① ② ⑤ ⑥	$B_C\ B_L\ G$
$\pm\theta_Y = \sin^{-1}\left[-(B_L - B_C)/Y\right]$ $Y_{RECT} = \sqrt{Y^2 - (B_L - B_C)^2} - (B_L - B_C)\,j$	① ② ⑤ ⑥	$B_C\ B_L\ Y$
$Y = \left|(B_L - B_C)/(\sin\theta_Y)\right|$ $Y_{RECT} = \left(-(B_L - B_C)/\left[\tan(\pm\theta_Y)\right]\right) - (B_L - B_C)\,j$	① ② ⑤ ⑥ ⊗	$B_C\ B_L\ \theta_Y$
$Y = \sqrt{R^{-2} + \left[(\omega L)^{-1} - (\omega C)\right]^2}$ $\pm\theta_Y = \tan^{-1}\left(R\left[(\omega L)^{-1} - (\omega C)\right]\right)$ $Y_{RECT} = R^{-1} - \left[(\omega L)^{-1} - (\omega C)\right]j$	① ② ③ ④ ⑤ ⑥	C L R
$Y = Z^{-1}$ $\pm\theta_Y = \sin^{-1}\left(-Z\left[(\omega L)^{-1} - (\omega C)\right]\right)$ $Y_{RECT} = \sqrt{Z^{-2} - \left[(\omega L)^{-1} - (\omega C)\right]^2} - \left[(\omega L)^{-1} - (\omega C)\right]j$	① ② ③ ④ ⑤ ⑥	C L Z

Admittance and Phase, Parallel Circuits

Y

$$\mathbf{Y}_{POLAR} = Y\underline{/\pm\theta_Y}$$

$$\mathbf{Y}_{RECT} = G - (\pm B)\,j$$

Note: $G = R_p^{-1}$, $\pm B = (X_L)_p^{-1} - (X_C)_p^{-1}$ ⓓ

	Applicable Notes	Terms
$Y = \left\lvert \left[(\omega L)^{-1} - (\omega C)\right] / (\sin\theta_Y) \right\rvert$ $\mathbf{Y}_{RECT} = \left(-\left[(\omega L)^{-1} - (\omega C)\right] / \left[\tan(\pm\theta_Y)\right]\right) - \left[(\omega L)^{-1} - (\omega C)\right] j$	① ② ③ ④ ⑤ ⑥ ⊗	$C\ L\ \theta_Y$
$Y = G/(\cos\theta_Y)$ $\mathbf{Y}_{RECT} = G - \left[-G\tan(\pm\theta_Y)\right] j$	① ② ⑤ ⑥	$G\ \theta_Y$
$Y = \sqrt{R^{-2} + (X_L^{-1} - X_C^{-1})^2}$ $\pm\theta_Y = \tan^{-1}\left[-R(X_L^{-1} - X_C^{-1})\right]$ $\mathbf{Y}_{RECT} = R^{-1} - (X_L^{-1} - X_C^{-1})\,j$	① ② ④ ⑤ ⑥	$R\ X_C\ X_L$
$Y = [R\cos\theta_Z]^{-1}$ $\pm\theta_Y = -(\pm\theta_Z)$ $\mathbf{Y}_{RECT} = R^{-1} - \left[R^{-1}\tan(\pm\theta_Z)\right] j$	① ② ④ ⑤ ⑥	$R\ \theta_Z$
$Y = Z^{-1}$ $\pm\theta_Y = \sin^{-1}\left[-Z(X_L^{-1} - X_C^{-1})\right]$ $\mathbf{Y}_{RECT} = \sqrt{Z^{-2} - (X_L^{-1} - X_C^{-1})^2} - (X_L^{-1} - X_C^{-1})\,j$	① ② ④ ⑤ ⑥	$X_C\ X_L\ Z$

Admittance and Phase, Parallel Circuits

Y

$$Y_{POLAR} = Y\angle{\pm\theta_Y}$$

$$Y_{RECT} = G - (\pm B)\,j$$

Note: $G = R_p^{-1}$, $\pm B = (X_L)_p^{-1} - (X_C)_p^{-1}$ ⓓ	Applicable Notes	Terms
$Y = \left\|(X_L^{-1} - X_C^{-1})/(\sin\theta_Z)\right\|$ $Y_{RECT} = \left(\left[X_L^{-1} - X_C^{-1}\right]/\left[\tan(\pm\theta_Z)\right]\right) - (X_L^{-1} - X_C^{-1})\,j$	① ② ④ ⑤ ⑥ ⊗	$X_C\ X_L\ \theta_Z$
$Y_{RECT} = \left[Y\cos\theta_Y\right] - \left[-Y\sin(\pm\theta_Y)\right]j$	① ② ⑤ ⑥	$Y\ \theta_Y$
$Y = Z^{-1}$, $\pm\theta_Y = -(\pm\theta_Z)$ $Y_{RECT} = \left[Z^{-1}\cos\theta_Z\right] - \left[Z^{-1}\sin(\pm\theta_Z)\right]j$	① ② ④ ⑤ ⑥	$Z\ \theta_Z$
$Y = I_t/E$, $\pm\theta_Y = -(\pm\theta_Z)$ $Y_{RECT} = \left[(I_t/E)\cos\theta_Z\right] - \left[(I_t/E)\sin(\pm\theta_Z)\right]j$	① ② ⑤ ⑥	$E\ I\ \theta_Z$
$Y = P/(E^2\cos\theta_Z)$, $\pm\theta_Y = -(\pm\theta_Z)$ $Y_{RECT} = (P/E^2) - \left[(P/E^2)\tan(\pm\theta_Z)\right]j$	① ② ⑤ ⑥	$E\ P\ \theta_Z$
$Y = (I_t^2\cos\theta_Z)/P$, $\pm\theta_Y = -(\pm\theta_Z)$ $Y_{RECT} = \left[(I_t\cos\theta_Z)^2/P\right] - \left[Y\sin(\pm\theta_Z)\right]j$	① ② ⑤ ⑥	$I_t\ P\ \theta_Z$

Complex Networks In Parallel | Y | Parallel Complex Network Admittance

Note: $\mathbf{Y}_{RECT} = G - (\pm B)\,j$ The sign of $(\pm B)\,j$ is real.	Terms
$(\mathbf{Y}_{RECT})_t = (\mathbf{Y}_{RECT})_1 + (\mathbf{Y}_{RECT})_2 \cdots + (\mathbf{Y}_{RECT})_n$ $G_t = G_1 + G_2 \cdots + G_n$ $\pm B_t = \pm B_1 \pm B_2 \cdots \pm B_n$ $(\mathbf{Y}_{RECT})_t = G_t - (\pm B_t)\,j$	$(\mathbf{Y}_{RECT})_1$ $(\mathbf{Y}_{RECT})_2$ $(\mathbf{Y}_{RECT})_n$

$\mathbf{Y}_{RECT}$ Procedure applies to any circuit in parallel with others.

1. Convert each series and each parallel circuit to polar impedance using applicable formulas.
2. Convert each polar impedance to rectangular admittance from:

$$\mathbf{Y}_{RECT} = \left[\cos\theta_Z/Z\right] - \left[\sin(\pm\theta_Z)/Z\right] j$$

3. The quantities inside the brackets represent G and ±B. Maintain the sign of B inside brackets. Do not simplify to ±jB.
4. Algebraically sum all ±B quantities. Sum all G quantities.
5. Convert to total polar impedance if desired from:

$$Z_t = \left[G_t^2 + (\pm B_t)^2\right]^{-\frac{1}{2}}$$

$$(\pm\theta_Z)_t = \tan^{-1}\left[(\pm B_t)/G_t\right]$$

Conversions To Rectangular Admittance

Y_{RECT}

Z_{POLAR} To Y_{RECT}

$$Z_{POLAR} = Z\angle \pm\theta_Z, \quad Y_{RECT} = G - (\pm B)\,j$$

$$Y_{RECT} = \left[Z^{-1}\cos\theta_Z\right] - \left[Z^{-1}\sin(\pm\theta_Z)\right] j$$

Z_{RECT} To Y_{RECT}

$$Z_{RECT} = R_s + (\pm X_s)\,j, \quad Y_{RECT} = G - (\pm B)\,j$$

$$Y_{RECT} = \left[R_s/(\pm X_s^2 + R_s^2)\right] - \left[\pm X_s/(\pm X_s^2 + R_s^2)\right] j$$

Series R and X To Y_{RECT}

$$R_s = \text{Series } R_t, \quad \pm X_s = \text{Series } (X_L - X_C)_t$$

$$Y_{RECT} = \left[R_s/(\pm X_s^2 + R_s^2)\right] - \left[\pm X_s/(\pm X_s^2 + R_s^2)\right] j$$

Parallel R and X To Y_{RECT}

$$R_p = \text{Parallel } R_t, \quad \pm X_p = \text{Parallel } (X_L^{-1} - X_C^{-1})_t^{-1}$$

$$Y_{RECT} = (R_p)_t^{-1} - (\pm X_p)_t^{-1}\,j$$

Note ⓓ

Y_{POLAR} To Y_{RECT}

$$Y_{POLAR} = Y\angle \pm\theta_Y, \quad Y_{RECT} = G - (\pm B)\,j$$

$$Y_{RECT} = \left[Y\cos\theta_Y\right] - \left[-Y\sin(\pm\theta_Y)\right] j$$

Conversions From Rectangular Admittance

Y_{RECT}

Y_{RECT} To Z_{POLAR}

$$Y_{RECT} = G - (\pm B)\,j, \quad Z_{POLAR} = Z\underline{/\pm\theta_Z}$$

$$Z_{POLAR} = \left[G^2 + (\pm B)^2\right]^{-\frac{1}{2}} \underline{/\tan^{-1}[\pm B/G]}$$

Y_{RECT} To Z_{RECT}

$$Y_{RECT} = G - (\pm B)\,j, \quad Z_{RECT} = R_s + (\pm X_s)\,j$$

$$Z_{RECT} = \left[G/(\pm B^2 + G^2)\right] + \left[\pm B/(\pm B^2 + G^2)\right] j$$

Y_{RECT} To Y_{POLAR}

$$Y_{RECT} = G - (\pm B)\,j, \quad Y_{POLAR} = Y\underline{/\pm\theta_Y}$$

$$Y_{POLAR} = \sqrt{G^2 + (\pm B)^2}\,\underline{/\tan^{-1}[-(\pm B/G)]}$$

Y_{RECT} To Equiv. Series R and X

$$Y_{RECT} = G - (\pm B)\,j$$

$$R_s = G/(\pm B^2 + G^2), \quad \pm X_s = \pm B/(\pm B^2 + G^2)$$

$$|-X_s| = X_C, \qquad |+X_s| = X_L$$

Y_{RECT} To Equiv. Parallel R and X

$$Y_{RECT} = G - (\pm B)\,j$$

$$R_p = G^{-1}, \quad \pm X_p = \pm B^{-1}$$

$$|-X_p| = X_C, \quad |+X_p| = X_C$$

Note ⓓ

Vector Algebra AC Ohms Law

$\mathbf{E}_g = E_g\underline{/0^\circ}$ or $\mathbf{I}_g = I_g\underline{/0^\circ}$ $(1 = 1\underline{/0^\circ})$

$\mathbf{E} = \mathbf{I}_g/\mathbf{Y} = I_g/Y\underline{/0^\circ - \theta_Y = -(\pm\theta_Y)}$

$\mathbf{I} = \mathbf{E}_g\mathbf{Y} = E_gY\underline{/0^\circ + \theta_Y = \pm\theta_Y}$

$\mathbf{Y} = 1/\mathbf{Z} = 1/Z\underline{/0^\circ - \theta_Z = -(\pm\theta_Z)}$

$\mathbf{Y} = \mathbf{I}/\mathbf{E}_g = I/E_g\underline{/\theta_I - 0^\circ = \pm\theta_I}$

$\mathbf{Y} = \mathbf{I}_g/\mathbf{E} = I_g/E\underline{/0^\circ - \theta_E = -(\pm\theta_E)}$

$\mathbf{Z} = 1/\mathbf{Y} = 1/Y\underline{/0^\circ - \theta_Y = -(\pm\theta_Y)}$

Addition and Subtraction of Rectangular Admittance

$$\mathbf{Y}_1 + \mathbf{Y}_2 = \mathbf{Y}_{1\,(RECT)} + \mathbf{Y}_{2\,(RECT)}$$

$$= \left[G - (\pm B)\,j\right]_1 + \left[G - (\pm B)\,j\right]_2$$

$$= \left[G_1 + G_2\right] - \left[(\pm B_1) + (\pm B_2)\right] j$$

$$\mathbf{Y}_1 - \mathbf{Y}_2 = \left[G_1 - G_2\right] - \left[(\pm B_1) - (\pm B_2)\right] j$$

$$G_t = (R_p^{-1})_t$$

$$\pm B_t = (\pm X_p^{-1})_t$$

$$|-B| = B_C \qquad |+B| = B_L$$

$$B_C = (X_C)_p^{-1} \qquad B_L = (X_L)_p^{-1}$$ *Note* ⓓ

See also—Z, Vector Algebra
See also—B, G, θ

ADMITTANCE
Vector Algebra

Y

Applicable Notes

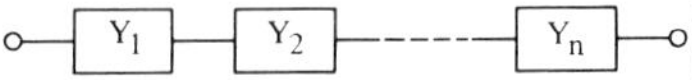

$$Y_t = [Y_1^{-1} + Y_2^{-1} \cdots + Y_n^{-1}]^{-1}$$

Y_{VA} - ① - ②

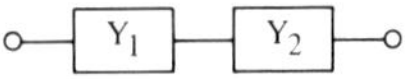

$$Y_2 = [Y_t^{-1} - Y_1^{-1}]^{-1}$$

Y_{VA} - ① - ②

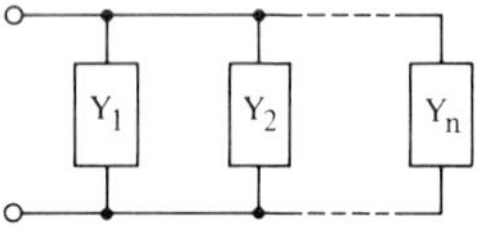

$$Y_t = Y_1 + Y_2 \cdots + Y_n$$

Y_{VA} - ① - ②

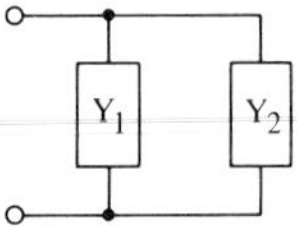

$$Y_2 = Y_t - Y_1$$

Y_{VA} - ① - ②

Y_{VA} *Notes:*

① Admittance is a complex quantity requiring the mathematical operations of addition and subtraction to be performed in rectangular form. Rectangular form quantities may be multiplied like other binomials except that $j^2 = -1$. Reciprocals or other division by rectangular form quantities requires the divisor to be rationalized by multiplication of both the divisor and the dividend by the conjugate of the divisor. (The conjugate of $G - Bj$ is $G + Bj$). When using a calculator, it is easier to convert rectangular quantities to polar form for multiplication and division then reconverting to rectangular form for addition and subtraction.

ADMITTANCE

Vector Algebra

Y

Applicable Notes

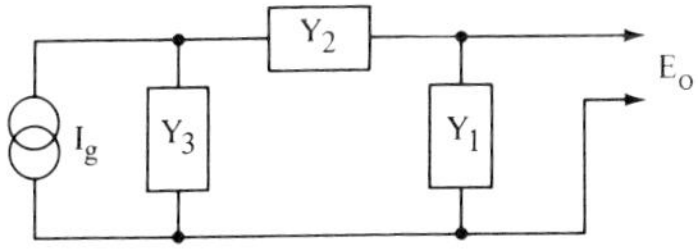

$\mathbf{Y}_i = \mathbf{Y}_3 + (\mathbf{Y}_2^{-1} + \mathbf{Y}_1^{-1})^{-1}$

$\mathbf{Y}_o = \mathbf{Y}_1 + (\mathbf{Y}_2^{-1} + \mathbf{Y}_3^{-1})^{-1}$

$\mathbf{E}_o = \mathbf{I}_g \left[1 - (\mathbf{Y}_3/\mathbf{Y}_i)\right] / \mathbf{Y}_1$

Y_{VA} - ① - ②

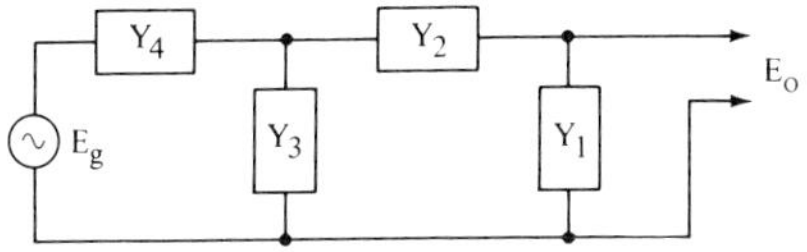

$\mathbf{Y}_i = \mathbf{Y}_4^{-1} + \left[\mathbf{Y}_3 + (\mathbf{Y}_2^{-1} + \mathbf{Y}_1^{-1})^{-1}\right]^{-1}$

$\mathbf{Y}_o = \mathbf{Y}_1 + \left[\mathbf{Y}_2^{-1} + (\mathbf{Y}_3 + \mathbf{Y}_4)^{-1}\right]^{-1}$

$\mathbf{E}_o = \mathbf{E}_g \left[1 - (\mathbf{Y}_i/\mathbf{Y}_4)\right] / \left[(\mathbf{Y}_1/\mathbf{Y}_2) + 1\right]$

Y_{VA} - ① - ②

Y_{VA} *Notes:*

② B_C or $-B$ = Capacitive Susceptance, B_L or $+B$ = Inductive Susceptance, $\mathbf{E}_g$ = Generator Voltage ∗, $\mathbf{E}_o$ = Output Voltage ∗, G = Conductance (Parallel Circuit Reciprocal Resistance), $\mathbf{I}_g$ = Generator Current ∗, $\mathbf{I}_o$ = Output Current ∗, R_p^{-1} = Parallel Circuit Reciprocal Resistance (Conductance), $\pm X_p^{-1}$ = Parallel Circuit Reciprocal Reactance (Susceptance), $\mathbf{Y}_i$ = Input Admittance ∗, $\mathbf{Y}_o$ = Output Admittance ∗, $\mathbf{Z}_i$ = Input Impedance ∗, $\mathbf{Z}_o$ = Output Impedance ∗, ∗ = Vector (Phasor) characteristic.

Z = Symbol for impedance

Z = The total opposition to the flow of alternating current of a given frequency. A complex quantity having components of resistance and reactance. The ratio of applied alternating voltage to the alternating current flow through a circuit.

Z = Impedance expressed in ohm (Ω) units.

$Z = Z_{MAGNITUDE} = |Z|$

θ_Z = Phase angle of impedance

$\mathbf{Z}$ = Complete description of impedance which includes both magnitude and phase angle information.

$\mathbf{Z}_{POLAR}$ = Polar form of impedance = $Z\angle\pm\theta_Z$

$\mathbf{Z}_{POLAR}$ = The vectorial resultant of resistance (0°) and reactance (±90°).

$\mathbf{Z}_{RECT}$ = Rectangular form of impedance or the complex number form of impedance.

$\mathbf{Z}_{RECT}$ = The 0° (resistance) and ±90° (reactance) vectors in complex number form which have a resultant equal to polar impedance.

$\mathbf{Z}_{RECT}$ = The mathematical equivalent of resistance and reactance in series. (The series equivalent of a parallel circuit)

$\mathbf{Z}_{RECT} = R \pm jX = R + (\pm X)j = R + (X_L - X_C)j$
where $|+X| = X_L$ and $|-X| = X_C$

$(\mathbf{Z}_{RECT})^{-1} = \mathbf{Y}_{RECT} = G - (\pm B)j = R_p^{-1} - (\pm X_p^{-1})j$

$(\mathbf{Z}_{RECT})^{-1}$ = A parallel equivalent circuit. See—$\mathbf{Y}_{RECT}$

Impedance, Series Circuits **Z**	Applicable Notes	Terms
$Z = \sqrt{R^2 + [(\omega L) - (\omega C)^{-1}]^2}$	① ④ ⑦	C L R
$Z = \left\| [(\omega L) - (\omega C)^{-1}]/(\sin \theta_Z) \right\|$	① ② ③ ④ ⑦ ⊗	C L θ_Z
$Z = E/I$	①	E I
$Z = \sqrt{R^2 + (X_L - X_C)^2}$	① ③	R $X_C X_L$
$Z = R/(\cos \theta_Z)$	① ② ③	R θ_Z
$Z = \left\| (X_L - X_C)/(\sin \theta_Z) \right\|$	① ② ③ ⊗	$X_C X_L$ θ_Z
$Z = \sqrt{E_R^2 + (E_L - E_C)^2}/I$	① ③	E_R $E_C E_L$ I
$Z = (E^2 \cos \theta_E)/P$	① ② ③	E P θ_E
$Z = P/(I^2 \cos \theta_I)$	① ② ③	I P θ_I

Z *Notes:*

① B = Susceptance, C = Capacitance, E = rms Voltage, G = Conductance, I = rms Current, L = Inductance, P = Power, R = Resistance, X = Reactance, Y = Admittance, Z = Impedance, θ = Phase angle, ω = Angular velocity

Impedance, Parallel Circuits **Z**	Applicable Notes	Terms
$Z = Y^{-1} = 1/Y$	① ④	Y
$Z = \left[G^2 + (B_L - B_C)^2\right]^{-\frac{1}{2}}$	① ③ ④	$B_C B_L$ G
$Z = \left\|(\sin\theta_Y)/(B_L - B_C)\right\|$	① ② ③ ⊗	$B_C B_L$ θ_Y
$Z = \left(R^{-2} + \left[(\omega L)^{-1} - (\omega C)\right]^2\right)^{-\frac{1}{2}}$	① ④ ⑦	CL R
$Z = \left\|(\sin\theta_Z)/\left[(\omega L)^{-1} - (\omega C)\right]\right\|$	① ② ③ ④ ⑦ ⊗	CL θ_Z
$Z = E/I$	①	E I
$Z = (\cos\theta_Y)/G$	① ② ③	G θ_Y
$Z = \left[R^{-2} + (X_L^{-1} - X_C^{-1})^2\right]^{-\frac{1}{2}}$	① ③ ④	R $X_C X_L$
$Z = R\cos\theta_Z$	① ② ③	R θ_Z
$Z = \left\|(\sin\theta_Z)/(X_L^{-1} - X_C^{-1})\right\|$	① ② ③ ④ ⊗	$X_C X_L$ θ_Z

Z *Notes:*

② cos = cosine, sin = sine, tan = tangent, $\cos^{-1}$ = arc cosine, $\sin^{-1}$ = arc sine, $\tan^{-1}$ = arc tangent

Impedance and Phase, Single Elements **Z**	**Applicable Notes**	**Terms**
$Z_{POLAR} = B_C^{-1} \angle -90° = \|-B^{-1}\| \angle -90°$ $Z_{RECT} = 0 + (-B^{-1})\,j$	① ③ ④ ⑤ ⑦	B_C or $-B$
$Z_{POLAR} = B_L^{-1} \angle +90° = \|+B^{-1}\| \angle +90°$ $Z_{RECT} = 0 + (+B^{-1})\,j$	① ③ ④ ⑤ ⑦	B_L or $+B$
$Z_{POLAR} = (\omega C)^{-1} \angle -90°$ $Z_{RECT} = 0 + (-\omega C)^{-1}\,j$	① ④ ⑦	C
$Z_{POLAR} = G^{-1} \angle 0°$ $Z_{RECT} = G^{-1} + 0j$	① ④ ⑥ ⑦	G
$Z_{POLAR} = (\omega L) \angle +90°$ $Z_{RECT} = 0 + (+\omega L)\,j$	① ⑦	L
$Z_{POLAR} = R \angle 0°$ $Z_{RECT} = R + 0j$	① ⑦	R
$Z_{POLAR} = X_C \angle -90° = \|-X\| \angle -90°$ $Z_{RECT} = 0 + (-X)\,j$	① ③ ⑦	X_C or $-X$
$Z_{POLAR} = X_L \angle +90° = \|+X\| \angle +90°$ $Z_{RECT} = 0 + (+X)\,j$	① ③ ⑦	X_L or $+X$

Z *Notes:*

③ Subscripts c = capacitive, E = voltage, I = current, L = inductive, n = any number, p = parallel circuit, s = series circuit, t = total or equivalent, Y = admittance, Z = impedance

$Z/\underline{\theta_Z}$ Impedance, Series Circuits **Z_{POLAR}**	Applicable Notes	Terms
$Z = \sqrt{R^2 + [(\omega L) - (\omega C)^{-1}]^2}$ $\pm\theta_Z = \tan^{-1}\left([(\omega L) - (\omega C)^{-1}]/R\right)$	① ② ③ ④ ⑦ ⑧	CL R
$\pm\theta_Z = \sin^{-1}\left([(\omega L) - (\omega C)^{-1}]/Z\right)$	① ② ③ ④ ⑦ ⑧	CL Z
$Z = \left\| [(\omega L) - (\omega C)^{-1}]/(\sin\theta_Z) \right\|$	① ② ③ ④ ⑦ ⊗	CL θ_Z
$Z = \sqrt{R^2 + (X_L - X_C)^2}$ $\pm\theta_Z = \tan^{-1}[(X_L - X_C)/R]$	① ② ③ ⑧	R $X_C X_L$
$\|\theta_Z\| = \tan^{-1}[\sqrt{Z^2 - R^2}/R]$	① ② ③ ④	R Z

Z *Notes:*

④ $x^{-1} = 1/x$, $x^{\frac{1}{2}} = \sqrt{x}$, $x^{-2} = 1/x^2$, $x^{-\frac{1}{2}} = 1/\sqrt{x}$, $|x| = x$ magnitude or the absolute value of x

⑤ Series resistance must equal zero.

⑥ Series reactance must equal zero.

⑦ $\omega = 2\pi f \approx 6.28f$ (f = frequency), $j = \sqrt{-1}$ = mathematical i = 90° multiplier = imaginary quantity = y axis quantity = reactive quantity

⑧ Reminders: $\pm\theta$–Use the sign of the phase angle $\pm X$, $\pm B$–treat the signs as real in all calculations except when converting to X_L, X_C, B_L or B_C.
The signs of $\pm X$ and $\pm B$ identify these reactive quantities as inductive or capacitive.
$|+X| = X_L$, $|-X| = X_C$, $|+B| = B_L$, $|-B| = B_C$
X_L, X_C, B_L and B_C are magnitudes, while $\pm X$ and $\pm B$ as used in formulas are "real" quantities.

$Z\underline{/\theta_Z}$ Impedance, Series Circuits **Z POLAR**	Applicable Notes	Terms
$Z = R/(\cos\theta_Z)$	① ② ③	$R \ \theta_Z$
$\pm\theta_Z = \sin^{-1}\left[(X_L - X_C)/Z\right]$	① ② ③ ⑧	$X_C X_L \ Z$
$Z = \left\|(X_L - X_C)/(\sin\theta_Z)\right\|$	① ② ③ ⊗	$X_C X_L \ \theta_Z$
$Z = E/I$ $\pm\theta_Z = \pm\theta_E$	① ③ ⑧ ⑨	$E \ I \ \theta_E$
$Z = \sqrt{E_R^2 + (E_L - E_C)^2}/I$ $\pm\theta_Z = \tan^{-1}\left[(E_L - E_C)/E_R\right]$	① ② ③ ⑧	$E_C E_L \ E_R \ I$
$Z = (E^2 \cos\theta_E)/P$ $\pm\theta_Z = \pm\theta_E$	① ② ③ ⑧ ⑨	$E \ P \ \theta_E$

Z *Notes:*

⑨ The phase angle of Z, Y, I and E ($\theta_Z, \theta_Y, \theta_I$ and θ_E) in a given circuit represent the same one and only one phase angle. $\pm\theta_Z = \pm\theta_E = -(\pm\theta_Y) = -(\pm\theta_I)$. The author does not recommend this use of θ_E and θ_I where each uses the other as the reference phase. The author uses the generator E_g or I_g as the reference. See also–θ

$Z\angle\theta_Z$ Impedance, Parallel Circiuts Z_{POLAR}	Applicable Notes	Terms
$Z = \left[G^2 + (B_L - B_C)^2\right]^{-\frac{1}{2}}$ $\pm\theta_Z = \tan^{-1}\left[(B_L - B_C)/G\right]$	① ② ③ ④ ⑧	$B_C B_L$ G
$Z = Y^{-1}$ $\pm\theta_Z = \sin^{-1}\left[(B_L - B_C)/Y\right]$	① ② ③ ④ ⑧	$B_C B_L$ Y
$Z = \left\|(\sin\theta_Y)/(B_L - B_C)\right\|$ $\pm\theta_Z = -(\pm\theta_Y)$	① ② ③ ④ ⑧ ⊗	$B_C B_L$ θ_Y
$Z = \left(R^{-2} + \left[(\omega L)^{-1} - (\omega C)\right]^2\right)^{-\frac{1}{2}}$ $\pm\theta_Z = \tan^{-1}\left(R\left[(\omega L)^{-1} - (\omega C)\right]\right)$	① ② ③ ④ ⑦ ⑧	CL R
$\pm X_p = \left[(\omega L)^{-1} - (\omega C)\right]^{-1}$ $\pm\theta_Z = \sin^{-1}\left[Z/\pm X_p\right]$	① ② ③ ④ ⑦ ⑧ ⓓ	CL Z
$Z = \left\|(\sin\theta_Z)/\left[(\omega L)^{-1} - (\omega C)\right]\right\|$	① ② ③ ④ ⑦ ⊗	CL θ_Z
$Z = (\cos\theta_Y)/G$ $\pm\theta_Z = -(\pm\theta_Y)$	① ② ③ ⑧	G θ_Y

$Z/\underline{\theta_Z}$ **Impedance, Parallel Circuits** **Z POLAR**	**Applicable Notes**	**Terms**
$Z = R \cos \theta_Z$	① ② ③	$R \ \theta_Z$
$Z = \left\|(\sin \theta_Z)/(X_L^{-1} - X_C^{-1})\right\|$	① ② ③ ④ ⊗	$X_C X_L \ \theta_Z$
$Z = Y^{-1}$ $\pm\theta_Z = -(\pm\theta_Y)$	① ③ ④ ⑧	$Y \ \theta_Y$
$Z = E/\sqrt{I_R^2 + (I_L - I_C)^2}$ $\pm\theta_Z = \tan^{-1}\left[(I_L - I_C)/I_R\right]$	① ② ③ ⑧ ⑨	$E \ I_R \ I_C I_L$
$Z = E/I$ $\pm\theta_Z = -(\pm\theta_I)$	① ③ ⑧ ⑨	$E \ I \ \theta_I$

Z *Notes:*

ⓓ Mathematics and calculators do not allow a division by zero or infinity. In formulas noted ⓓ however, the reciprocal of zero may be manually converted to infinity and the reciprocal of infinity may be manually converted to zero. The following additional manual operations may also be performed as required:

$x \cdot \infty = \infty$ when $x \neq 0$, $x/\infty = 0$ when $x \neq \infty$
$x \cdot 0 = 0$ when $x \neq \infty$, $x/0 = \infty$ when $x \neq 0$
$0^x = 0$ when $x \neq 0$, $\infty^x = \infty$ when $x \neq 0$

Calculators require the substitution of a very small number such as 10^{-30} for zero and of a very large number such as 10^{30} for infinity to perform these operations. All very small resultants must then be accepted as zero and all very large resultants must be accepted as infinity. Extreme care must be exercised to avoid accidental violation of the listed exceptions whenever more than one zero and/or infinity appear in the same formula. The arc tangent of infinity may be obtained from a calculator by also substituting a very large number for infinity.

⊗ Division by zero is prohibited. At circuit resonance, a zero divisor and a zero dividend will be presented. The division of zero by zero is always prohibited.

Polar Impedances In Series

$$Z_t = \Big\{\Big([Z_1 \cos(\theta_Z)_1] + [Z_2 \cos(\theta_Z)_2] \cdots + [Z_n \cos(\theta_Z)_n]\Big)^2 + \Big([Z_1 \sin(\pm\theta_Z)_1] + [Z_2 \sin(\pm\theta_Z)_2] \cdots + [Z_n \sin(\pm\theta_Z)_n]\Big)^2\Big\}^{\frac{1}{2}}$$

$$\pm\theta_{Z_t} = \tan^{-1}\left[\frac{\Big([Z_1 \sin(\pm\theta_Z)_1] + [Z_2 \sin(\pm\theta_Z)_2] \cdots + [Z_n \sin(\pm\theta_Z)_n]\Big)}{\Big([Z_1 \cos(\theta_Z)_1] + [Z_2 \cos(\theta_Z)_2] \cdots + [Z_n \cos(\theta_Z)_n]\Big)}\right]$$

Unknown Series Impedance

$$Z_x = \sqrt{\Big([Z_t \cos(\theta_Z)_t] - [Z_1 \cos(\theta_Z)_1]\Big)^2 + \Big([Z_t \sin(\pm\theta_Z)_t] - [Z_1 \sin(\pm\theta_Z)_1]\Big)^2}$$

$$\pm\theta_{Z_x} = \tan^{-1}\Big[\Big([Z_t \sin(\pm\theta_Z)_t] - [Z_1 \sin(\pm\theta_Z)_1]\Big) \Big/ \Big([Z_t \cos(\theta_Z)_t] - [Z_1 \cos(\theta_Z)_1]\Big)\Big]$$

Polar Impedances in Parallel

$$Z_t = \left[\left(\left[Z_1^{-1}\cos(\theta_Z)_1\right] + \left[Z_2^{-1}\cos(\theta_Z)_2\right] \cdots + \left[Z_n^{-1}\cos(\theta_Z)_n\right]\right)^2 + \left(\left[Z_1^{-1}\sin(\pm\theta_Z)_1\right] + \left[Z_2^{-1}\sin(\pm\theta_Z)_2\right] \cdots + \left[Z_n^{-1}\sin(\pm\theta_Z)_n\right]\right)^2\right]^{-\frac{1}{2}}$$

$$\pm\theta_{Z_t} = \tan^{-1}\left[\frac{\left(\left[Z_1^{-1}\sin(\pm\theta_Z)_1\right] + \left[Z_2^{-1}\sin(\pm\theta_Z)_2\right] \cdots + \left[Z_n^{-1}\sin(\pm\theta_Z)_n\right]\right)}{\left(\left[Z_1^{-1}\cos(\theta_Z)_1\right] + \left[Z_2^{-1}\cos(\theta_Z)_2\right] \cdots + \left[Z_n^{-1}\cos(\theta_Z)_n\right]\right)}\right]$$

Unknown Parallel Polar Impedance

$$Z_x = \left[\left(\left[Z_t^{-1}\cos(\theta_Z)_t\right] - \left[Z_1^{-1}\cos(\theta_Z)_1\right]\right)^2 + \left(\left[Z_t^{-1}\sin(\pm\theta_Z)_t\right] - \left[Z_1^{-1}\sin(\pm\theta_Z)_1\right]\right)^2\right]^{-\frac{1}{2}}$$

$$\pm\theta_{Z_x} = \tan^{-1}\left[\left(\left[Z_t^{-1}\sin(\pm\theta_Z)_t\right] - \left[Z_1^{-1}\sin(\pm\theta_Z)_1\right]\right) \Big/ \left(\left[Z_t^{-1}\cos(\theta_Z)_t\right] - \left[Z_1^{-1}\cos(\theta_Z)_1\right]\right)\right]$$

Z Parallel Circuits In Series, Procedure Method

Procedure:

1. Convert each parallel circuit to polar form impedances using $\mathbf{Z}_{POLAR}$, parallel circuit formulas.
2. Convert each polar impedance to equivalent series resistance and reactance from:

   $R_s = Z \cos \theta_Z \quad \pm X_s = Z \sin(\pm\theta_Z)$

   [If your calculator has the polar to rectangular conversion feature (P-R), enter Z_{POLAR} as the polar coordinates. The calculator x axis output is R_s and the calculator y axis output is $\pm X_s$]
3. Sum all R_s quantities and algebraically sum all $\pm X_s$ quantities.
   [If your calculator has multiple memories, sum all R_s quantities into one memory and all $\pm X_s$ quantities into a second memory.]
4. Convert the total equivalent series resistance and the total equivalent series reactance to total polar impedance from:

   $Z_t = \sqrt{(R_s)_t^2 + (\pm X_s)_t^2}$

   $(\pm\theta_Z)_t = \tan^{-1}\left[(\pm X_s)_t/(R_s)_t\right]$

   [If your calculator has the rectangular to polar conversion feature (R-P), enter $(R_s)_t$ as the x coordinate and $(\pm X_s)_t$ as the y coordinate. Calculator output will be polar impedance coordinates]

Z Series Circuits In Parallel, Procedure Method

Procedure:

1. Convert each series circuit to polar form impedance using $\mathbf{Z}_{POLAR}$, series circuit formulas.
2. Convert each polar impedance to equivalent parallel reciprocal resistance and equivalent parallel reciprocal reactance. [Note: parallel reciprocal resistance is also known as conductance (G) and parallel reciprocal reactance is also known as susceptance (B)]

   $R_p^{-1} = G = Z^{-1} \cos \theta_Z$

   $\pm X_p^{-1} = \pm B = Z^{-1} \sin(\pm\theta_Z)$

   [If your calculator has the polar to rectangular conversion feature, enter Z^{-1} and $\pm\theta_Z$ as the polar coordinates. The calculator x coordinate output is R_p^{-1} or G and the calculator y coordinate output is $\pm X_p^{-1}$ or $\pm B$.]
3. Sum all R_p^{-1}(G) quantities and algebraically sum all $\pm X_p^{-1}$ ($\pm B$) quantities.
   [If your calculator has multiple memories, sum all R_p^{-1} (G) quantities into one memory and sum all $\pm X_p^{-1}$ ($\pm B$) quantities into a second memory.]
4. Convert the total equivalent parallel reciprocal resistance [$(R_p^{-1})_t$ or G_t] and the total equivalent parallel reciprocal reactance [$(\pm X_p^{-1})_t$ or $\pm B_t$] to total polar impedance from:

   $Z_t = 1/\sqrt{G_t^2 + (\pm B_t)^2}$

   $(\pm\theta_Z)_t = \tan^{-1}[\pm B_t/G_t]$

   [If your calculator has the rectangular to polar conversion feature (R-P), enter G_t as the x coordinate and $\pm B_t$ as the y coordinate. Calculator output will be $Z^{-1}(\pm\theta_Z)$. Convert the magnitude to Z with the 1/x key]

See also – Note ⓓ

Polar Impedance Definitions and Vector Algebra

Z_{POLAR}

$Z_{POLAR} = Z\underline{/\pm\theta_Z}$

Z = Magnitude of impedance

$\pm\theta_Z$ = "Phase" angle of impedance

$\pm\theta_Z$ = The vectorial resultant angle when the magnitude of series resistance is placed at 0°, the magnitude of inductive reactance is placed at +90° and the magnitude of capacitive reactance is placed at −90°.

$\pm\theta_Z$ = That angle which has a tangent equal to the series reactance divided by the series resistance; the reactance having a positive sign if inductive and a negative sign if capacitive.

Z_{POLAR} = Impedance in a form where multiplication and division operations may be performed almost as easily as with ordinary numbers. (vector algebra)

$(Z_{POLAR})_1 \cdot (Z_{POLAR})_2 = Z_1 Z_2 \underline{/(\pm\theta_Z)_1 + (\pm\theta_Z)_2}$

$(Z_{POLAR})_1 / (Z_{POLAR})_2 = Z_1/Z_2 \underline{/(\pm\theta_Z)_1 - (\pm\theta_Z)_2}$

$E_{POLAR} = I\,(Z_{POLAR}) = IZ\underline{/0^\circ + (\pm\theta_Z)}$

$I_{POLAR} = E/(Z_{POLAR}) = E/Z\underline{/0^\circ - (\pm\theta_Z)}$

$Y_{POLAR} = 1/(Z_{POLAR}) = 1/Z\underline{/0^\circ - (\pm\theta_Z)}$

$\pm\theta_Z = \pm\theta_E = -(\pm\theta_I) = -(\pm\theta_Y)$

Z_{POLAR} = The form used most often as the final resultant when simplifying complex circuits. It is equally "correct" however for the final resultant to be $Z_{RECT.}$, Y_{POLAR} or $Y_{RECT.}$. Both forms of **Z** are series equivalent while both forms of **Y** are parallel equivalent.

$Z_{POLAR} = Z_{RECT} = [Y_{POLAR}]^{-1} = [Y_{RECT}]^{-1}$

Rectangular Impedance Definitions, Sum and Difference

$\mathbf{Z}_{RECT}$

$\mathbf{Z}_{RECT} = R_s + (\pm X_s)j$

R_s = Actual or equivalent total series resistance

$\pm X_s$ = Actual or equivalent net series reactance where:

$|+X_s| = X_L$ and $|-X_s| = X_C$

$\pm X_s = X_L - X_C$

$\mathbf{Z}_{RECT} = R_s + (\pm X_s)j$ regardless of actual circuit configuration. The rectangular impedance of a parallel circuit represents the equivalent series circuit. (Note: Equivalent circuit values will vary with frequency.)

$\mathbf{Z}_{RECT}$ = Impedance in a form where multiple impedances in series may be summed as easily as multiple resistances and multiple reactances in series. The series connected impedances may be any combination of individual series, parallel and unknown circuits.

$[\mathbf{Z}_{RECT}]_{TOTAL}$ = The sum of the resistive quantities and the algebraic sum of the reactive quantities.

$\mathbf{Z}_{RECT}$ = The form necessary to perform any mathematical operation involving the addition or subtraction of impedances.

$\mathbf{Z}_{RECT}$ = The form used by some for all mathematical operations and for the final resultant. (Not recommended. Use $\mathbf{Z}_{POLAR}$ for all multiplication and division and for the final resultant)

$\mathbf{Z}_{RECT}$ may be converted (transformed) at any time to $\mathbf{Z}_{POLAR}$ or $[\mathbf{Z}_{RECT}]^{-1}$ using the appropriate formula. ($[\mathbf{Z}_{RECT}]^{-1} = \mathbf{Y}_{RECT}$)

$\mathbf{Z}_{RECT} = \mathbf{Z}_{POLAR} = [\mathbf{Y}_{POLAR}]^{-1} = [\mathbf{Y}_{RECT}]^{-1}$

Z_{RECT} Rectangular Impedance Notes

1. This handbook contains no circuit elements to $\mathbf{Z}_{RECT}$ direct formulas. It is intended that all simple circuits should be converted to polar impedance and then converted as necessary to and from $\mathbf{Z}_{RECT}$ and $[\mathbf{Z}_{RECT}]^{-1}$.
2. The author apologizes to readers with good working knowledge of $\mathbf{Y}_{RECT}$ for the use of $[\mathbf{Z}_{RECT}]^{-1}$, however $\mathbf{Y}_{RECT}$ is a necessary part of this section, $[\mathbf{Z}_{RECT}]^{-1}$ fits the format better and many engineers as well as most technicians are very uncomfortable with $\mathbf{Y}_{RECT}$.
3. The use of $[\mathbf{Z}_{POLAR}]^{-1}$ or $\mathbf{Y}_{POLAR}$ is not recommended. Do not confuse yourself or others by continually changing the signs of the angles. Convert directly from $\mathbf{Z}_{POLAR}$ to $\mathbf{Y}_{RECT}$.
4. Use the rectangular forms $R_s + (\pm X_s)j$ and $G - (\pm B)j$ as shown and do not simplify. The plus sign will identify the complex quantity as impedance or voltage and as a series equivalent quantity while the minus sign identifies the complex quantity as reciprocal impedance, admittance or current and also as a parallel equivalent quantity. Note also that in this form the sign of the reactive quantity within the parentheses is real and does not change during inversions. This maintains identification of the reactive quantity as inductive (+) or as capacitive (−) at all times.
5. If any reader is uncomfortable with all rectangular quantities, direct conversion of series resistive and series reactive quantities to equivalent parallel reciprocal resistive and reciprocal reactive quantities is recommended. This conversion and its reverse allows the simplification of any series, parallel or series-parallel combination of impedances to a single impedance. This method is as fast as any other and there is less chance of error.

Reciprocal Rectangular Impedance Definitions, Sum and Difference $[\mathbf{Z}_{RECT}]^{-1}$

$[\mathbf{Z}_{RECT}]^{-1} = [R_s + (\pm X_s)j]^{-1} = R_p^{-1} - (\pm X_p^{-1})j$

The reciprocal of $\mathbf{Z}_{RECT.}$ (intrinsically a series or series equivalent quantity) is intrinsically a parallel or parallel equivalent quantity.

$[\mathbf{Z}_{RECT}]^{-1} = \mathbf{Y}_{RECT}$ = Rectangular admittance

$\mathbf{Y}_{RECT} = G - (\pm B)j$ See also—$\mathbf{Y}_{RECT}$

G = Total parallel conductance or the equivalent parallel conductance

$\pm B$ = Total parallel susceptance or the equivalent parallel susceptance where: $|+B| = B_L$, $|-B| = B_C$

$\pm B = B_L - B_C$

$[\mathbf{Z}_{RECT}]^{-1} = R_p^{-1} - (\pm X_p^{-1})j$ regardless of actual circuit configuration. The rectangular admittance of a series circuit represents the equivalent parallel circuit with the resistance and reactance in reciprocal form. (Note: Equiv. circuit values vary with freq.)

$[\mathbf{Z}_{RECT}]^{-1}$ = Reciprocal impedance in a form where complex quantities in parallel may be simplified as easily as multiple resistances and multiple reactances in series. The complex quantities in parallel may represent any combination of individual series, parallel or unknown circuit configurations.

$[\mathbf{Z}_{RECT}]^{-1}_{TOTAL}$ = The sum of the reciprocal resistances and the algebraic sum of the reciprocal reactances.

$[\mathbf{Z}_{RECT}]^{-1}$ may be inverted back to rectangular or polar impedance at any time using the appropriate formula.

$[\mathbf{Z}_{RECT}]^{-1} = [\mathbf{Z}_{POLAR}]^{-1} = \mathbf{Y}_{RECT} = \mathbf{Y}_{POLAR}$

See also – Note ⓓ

Z_{RECT} Rectangular Impedances In Series

$$[Z_{RECT}]_{TOTAL} = [Z_{RECT}]_1 + [Z_{RECT}]_2 \cdots + [Z_{RECT}]_n$$

Note: Rectangular impedance represents equivalent series resistance and reactance, however the actual circuit configuration may be series, parallel or unknown.

The first number in the rectangular impedance quantity is equivalent series resistance, the real part of impedance, the 0° component of impedance or the x axis coordinate of impedance.

The second number in the rectangular impedance quantity represents equivalent series reactance, the imaginary part of impedance, the ±90° component of impedance or the y axis coordinate of impedance.

When summing rectangular impedances, all resistive (R_s) components and all reactive ($\pm X_s$) components must be summed separately. The reactive components ($\pm X_s$) must also be summed algebraically. (Use the rectangular form $R_s + (\pm X_s)\,j$ not $R_s \pm jX_s$)

$$[Z_{RECT}]_1 = (R_s)_1 + (\pm X_s)_1\, j$$

$$[Z_{RECT}]_2 = (R_s)_2 + (\pm X_s)_2\, j$$

$$[Z_{RECT}]_n = (R_s)_n + (\pm X_s)_n\, j$$

$$[Z_{RECT}]_{TOTAL} = (R_s)_{TOTAL} + (\pm X_s)_{TOTAL}\, j$$

Note: $Z_{POLAR} = \sqrt{(R_s)_t^2 + (\pm X_s)_t^2}\ \angle \tan^{-1}\left[(\pm X_s)_t/(R_s)_t\right]$

$$Z_{RECT} = \left[Z \cos\theta_Z\right] + \left[Z \sin(\pm\theta_Z)\right] j$$

$[Z_{RECT}]^{-1}$

Reciprocal Rectangular Impedances In Parallel

$$[Z_{RECT}]_t^{-1} = [Z_{RECT}]_1^{-1} + [Z_{RECT}]_2^{-1} \cdots + [Z_{RECT}]_n^{-1}$$

$$[Z_{RECT}]^{-1} = Y_{RECT}$$

$$[Y_{RECT}]_t = [Y_{RECT}]_1 + [Y_{RECT}]_2 \cdots + [Y_{RECT}]_n$$

$$[Z_{RECT}]^{-1} = R_p^{-1} - (\pm X_p^{-1})\,j$$

$$Y_{RECT} = G - (\pm B)\,j$$

$$G = R_p^{-1}, \quad \pm B = \pm X_p^{-1}$$

$$[Z_{RECT}]_1^{-1} = (R_p^{-1})_1 \quad - (\pm X_p^{-1})_1 \quad j$$

$$[Z_{RECT}]_2^{-1} = (R_p^{-1})_2 \quad - (\pm X_p)_2 \quad j$$

$$[Z_{RECT}]_n^{-1} = (R_p^{-1})_n \quad - (\pm X_p^{-1})_n \quad j$$

$$[Z_{RECT}]_{TOTAL}^{-1} = (R_p^{-1})_{TOTAL} - (\pm X_p^{-1})_{TOTAL}\, j$$

$$[Y_{RECT}]_1 = G_1 \quad - (\pm B_1) \quad j$$

$$[Y_{RECT}]_2 = G_2 \quad - (\pm B_2) \quad j$$

$$[Y_{RECT}]_n = G_n \quad - (\pm B_n) \quad j$$

$$[Y_{RECT}]_{TOTAL} = G_{TOTAL} \quad - (\pm B)_{TOTAL} \quad j$$

Note: $Z_{POLAR} = \left[G_t^2 + (\pm B_t)^2\right]^{-\frac{1}{2}} \underline{/\tan^{-1}[\pm B_t/G_t]}$

$$[Z_{RECT}]^{-1} = \left[Z^{-1}\cos\theta_Z\right] - \left[Z^{-1}\sin(\pm\theta_Z)\right] j$$

See also – Note ⓓ

Z_{POLAR} To Z_{RECT}

$$Z_{RECT} = [Z \cos\theta_Z] + [Z \sin(\pm\theta_Z)]\, j$$

Z_{POLAR} To Y_{RECT} or $[Z_{RECT}]^{-1}$

$$Y_{RECT} = G - (\pm B)\, j = R_p^{-1} - (\pm X_p^{-1})\, j$$

$$Y_{RECT} = [Z^{-1} \cos\theta_Z] - [Z^{-1} \sin(\pm\theta_Z)]\, j$$

$$[Z_{RECT}]^{-1} = Y_{RECT}$$

Z_{POLAR} To Y_{POLAR}

$$Y_{POLAR} = Z^{-1} \angle -(\pm\theta_Z)$$

Z_{POLAR} To Series R and X

$$R_s = Z \cos\theta_Z \qquad \pm X_s = Z \sin(\pm\theta_Z)$$

$$|+X_s| = X_L \qquad |-X_s| = X_C$$

Z_{POLAR} To Parallel R and X

$$R_p = Z/(\cos\theta_Z) \qquad \pm X_p = Z/[\sin(\pm\theta_Z)]$$

$$|+X_p| = X_L \qquad |-X_p| = X_C$$

See also – Note ⓓ

Conversion To Polar Impedance

Z_{POLAR}

Conversions

Equiv. Series R and X To $\mathbf{Z}_{POLAR}$

$$Z = \sqrt{R_s^2 + (\pm X_s)^2}$$

$$\pm\theta_Z = \tan^{-1}[\pm X_s/R_s]$$

Equiv. Parallel R and X To $\mathbf{Z}_{POLAR}$

$$Z = \left[R_p^{-2} + (\pm X_p)^{-2}\right]^{-\frac{1}{2}}$$

$$\pm\theta_Z = \tan^{-1}[R_p/\pm X_p]$$

$\mathbf{Z}_{RECT}$ To $\mathbf{Z}_{POLAR}$

$$\mathbf{Z}_{RECT} = R_s + (\pm X_s)\,j$$

$$Z = \sqrt{R_s^2 + (\pm X_s)^2}$$

$$\pm\theta_Z = \tan^{-1}[\pm X_s/R_s]$$

$[\mathbf{Z}_{RECT}]^{-1}$ or $\mathbf{Y}_{RECT}$ To $\mathbf{Z}_{POLAR}$

$$[\mathbf{Z}_{RECT}]^{-1} = \mathbf{Y}_{RECT} = G - (\pm B)\,j$$

$$= R_p^{-1} - (\pm X_p^{-1})\,j$$

$$Z = \left[G^2 + (\pm B)^2\right]^{-\frac{1}{2}}$$

$$\pm\theta_Z = \tan^{-1}[\pm B/G]$$

See also – Note ⓓ

Z_{RECT} To Z_{POLAR}

$$Z_{RECT} = R_s + (\pm X_s)\,j$$

$$Z_{POLAR} = \sqrt{R_s^2 + (\pm X_s)^2}\ \underline{/\tan^{-1}[\pm X_s/R_s]}$$

Z_{RECT} To Y_{POLAR}

$$Z_{RECT} = R_s + (\pm X_s)\,j$$

$$Y_{POLAR} = \left[R_s^2 + (\pm X_s)^2\right]^{-\frac{1}{2}}\ \underline{/\tan^{-1}\left[-(\pm X_s/R_s)\right]}$$

Z_{RECT} To Y_{RECT} or $[Z_{RECT}]^{-1}$

$$Z_{RECT} = R_s + (\pm X_s)\,j$$

$$Y_{RECT} = G - (\pm B)\,j = [Z_{RECT}]^{-1} = R_p^{-1} - (\pm X_p^{-1})\,j$$

$$Y_{RECT} = \left[R_s/(\pm X_s^2 + R_s^2)\right] - \left[\pm X_s/(\pm X_s^2 + R_s^2)\right] j$$

Z_{RECT} To Equiv. Parallel R and X

$$Z_{RECT} = R_s + (\pm X_s)\,j$$

$$R_p = (\pm X_s^2 + R_s^2)/R_s$$

$$\pm X_p = (\pm X_s^2 + R_s^2)/\pm X_s$$

$$|+X_p| = X_L \quad |-X_p| = X_C$$

See also – Note ⓓ

Z_{POLAR} To Z_{RECT}

$$Z_{POLAR} = Z \angle \pm\theta_Z$$

$$Z_{RECT} = \left[Z \cos \theta_Z\right] + \left[Z \sin(\pm\theta_Z)\right] j$$

Y_{POLAR} To Z_{RECT}

$$Y_{POLAR} = Y \angle \pm\theta_Y$$

$$Z_{RECT} = \left[Y^{-1} \cos \theta_Y\right] + \left[-Y^{-1} \sin(\pm\theta_Y)\right] j$$

Y_{RECT} or $[Z_{RECT}]^{-1}$ To Z_{RECT}

$$Y_{RECT} = G - (\pm B) j$$

$$Z_{RECT} = \left[G/(\pm B^2 + G^2)\right] + \left[\pm B/(\pm B^2 + G^2)\right] j$$

Series R and X To Z_{RECT}

$$Z_{RECT} = R_s + (\pm X_s) j$$

Parallel R and X To Z_{RECT}

$$Z_{RECT} = \left[(R_p/\pm X_p^2) + R_p^{-1}\right]^{-1} + \left[(\pm X_p/R_p^2) + (\pm X_p^{-1})\right]^{-1} j$$

G and B to Z_{RECT}

$$Z_{RECT} = \left[G/(\pm B^2 + G^2)\right] + \left[\pm B/(\pm B^2 + G^2)\right] j$$

See also – Note ⓓ

Conversion From Reciprocal Rectangular Impedance

$[\mathbf{Z}_{RECT}]^{-1}$

$[\mathbf{Z}_{RECT}]^{-1}$ or $\mathbf{Y}_{RECT}$ To $\mathbf{Z}_{POLAR}$

$$[\mathbf{Z}_{RECT}]^{-1} = \mathbf{Y}_{RECT} = G - (\pm B)\,j$$

$$\mathbf{Z}_{POLAR} = \left[G^2 + (\pm B)^2\right]^{-\frac{1}{2}} \angle \tan^{-1}\left[\pm B/G\right]$$

$[\mathbf{Z}_{RECT}]^{-1}$ or $\mathbf{Y}_{RECT}$ To $\mathbf{Z}_{RECT}$

$$[\mathbf{Z}_{RECT}]^{-1} = \mathbf{Y}_{RECT} = G - (\pm B)\,j$$

$$\mathbf{Z}_{RECT} = \left[G/(\pm B^2 + G^2)\right] + \left[\pm B/(\pm B^2 + G^2)\right] j$$

$[\mathbf{Z}_{RECT}]^{-1}$ or $\mathbf{Y}_{RECT}$ To $\mathbf{Y}_{POLAR}$

$$[\mathbf{Z}_{RECT}]^{-1} = \mathbf{Y}_{RECT} = G - (\pm B)\,j$$

$$\mathbf{Y}_{POLAR} = \sqrt{G^2 + (\pm B)^2} \angle \tan^{-1}\left[-(\pm B/G)\right]$$

$[\mathbf{Z}_{RECT}]^{-1}$ or $\mathbf{Y}_{RECT}$ To Equiv. Series R and X

$$[\mathbf{Z}_{RECT}]^{-1} = \mathbf{Y}_{RECT} = G - (\pm B)\,j$$

$$R_s = G/(\pm B^2 + G^2) \qquad \pm X_s = \pm B/(\pm B^2 + G^2)$$

$[\mathbf{Z}_{RECT}]^{-1}$ or $\mathbf{Y}_{RECT}$ To Equiv. Parallel R and X

$$[\mathbf{Z}_{RECT}]^{-1} = \mathbf{Y}_{RECT} = G - (\pm B)\,j$$

$$R_p = G^{-1} \qquad \pm X_p = \pm B^{-1}$$

See also – Note ⓓ

Conversions To Reciprocal Rectangular Impedance

$[Z_{RECT}]^{-1}$

Z_{POLAR} To Y_{RECT} or $[Z_{RECT}]^{-1}$

$$Z_{POLAR} = Z\angle\pm\theta_Z$$

$$[Z_{RECT}]^{-1} = Y_{RECT} = G - (\pm B)j$$

$$[Z_{RECT}]^{-1} = [Z^{-1}\cos\theta_Z] - [Z^{-1}\sin(\pm\theta_Z)]j$$

Z_{RECT} To Y_{RECT} or $[Z_{RECT}]^{-1}$

$$Z_{RECT} = R_s + (\pm X_s)j$$

$$[Z_{RECT}]^{-1} = Y_{RECT} = G - (\pm B)j$$

$$[Z_{RECT}]^{-1} = [R_s/(\pm X_s^2 + R_s^2)] - [\pm X_s/(\pm X_s^2 + R_s^2)]j$$

Series R and X To Y_{RECT} or $[Z_{RECT}]^{-1}$

$$[Z_{RECT}]^{-1} = Y_{RECT} = G - (\pm B)j$$

$$[Z_{RECT}]^{-1} = [R_s/(\pm X_s^2 + R_s^2)] - [\pm X_s/(\pm X_s^2 + R_s^2)]j$$

Parallel R and X To Y_{RECT} or $[Z_{RECT}]^{-1}$

$$[Z_{RECT}]^{-1} = R_p^{-1} - (\pm X_p^{-1})j$$

Y_{POLAR} To Y_{RECT} or $[Z_{RECT}]^{-1}$

$$[Z_{RECT}]^{-1} = [Y\cos\theta_Y] - [-Y\sin(\pm\theta_Y)]j$$

See also – Note ⓓ

Rules of Vector Algebra

$$\mathbf{Z}_1 \cdot \mathbf{Z}_2 = Z_1 Z_2 \underline{/(\pm\theta_Z)_1 + (\pm\theta_Z)_2}$$

$$\mathbf{Z}_1/\mathbf{Z}_2 = Z_1/Z_2 \underline{/(\pm\theta_Z)_1 - (\pm\theta_Z)_2}$$

$$(+1) \cdot \mathbf{Z} = Z \underline{/0^\circ + (\pm\theta_Z) = \pm\theta_Z}$$

$$(+1)/\mathbf{Z} = 1/Z \underline{/0^\circ - (\pm\theta_Z) = -(\pm\theta_Z)}$$

$$\mathbf{Z}_1 + \mathbf{Z}_2 = [\mathbf{Z}_{RECT}]_1 + [\mathbf{Z}_{RECT}]_2$$

$$\mathbf{Z}_1 + \mathbf{Z}_2 = [R_s + (\pm X_s)j]_1 + [R_s + (\pm X_s)j]_2$$

$$\mathbf{Z}_1 + \mathbf{Z}_2 = [(R_s)_1 + (R_s)_2] + [(\pm X_s)_1 + (\pm X_s)_2]j$$

$$\mathbf{Z}_1 - \mathbf{Z}_2 = [(R_s)_1 - (R_s)_2] + [(\pm X_s)_1 - (\pm X_s)_2]j$$

$$\mathbf{Z} + (+1) = [R_s + 1] + [\pm X_s]j$$

$$\mathbf{Z} - (+1) = [R_s - 1] + [\pm X_s]j$$

$$\mathbf{Z}_1^{-1} + \mathbf{Z}_2^{-1} = [\mathbf{Z}_{RECT}]_1^{-1} + [\mathbf{Z}_{RECT}]_2^{-1}$$

$$\mathbf{Z}_1^{-1} + \mathbf{Z}_2^{-1} = [\mathbf{Y}_{RECT}]_1 + [\mathbf{Y}_{RECT}]_2$$

$$\mathbf{Z}_1^{-1} + \mathbf{Z}_2^{-1} = [G - (\pm B)j]_1 + [G - (\pm B)j]_2$$

$$\mathbf{Z}_1^{-1} + \mathbf{Z}_2^{-1} = [G_1 + G_2] - [(\pm B)_1 + (\pm B)_2]j$$

$$\mathbf{Z}_1^{-1} - \mathbf{Z}_2^{-1} = [G_1 - G_2] - [(\pm B)_1 - (\pm B)_2]j$$

$$\mathbf{Z}^{-1} + (+1) = [G + 1] - [\pm B]j$$

$$\mathbf{Z}^{-1} - (+1) = [G - 1] - [\pm B]j$$

See–**Z** Conversion Formulas

IMPEDANCE Vector Algebra	**Z**	Applicable Notes
$Z_t = Z_1 + Z_2 \cdots + Z_n$	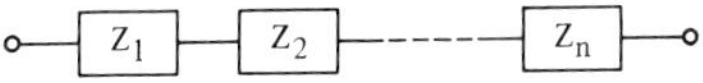	VA-1 VA-2 VA-3 VA-5
$Z_2 = Z_t - Z_1$	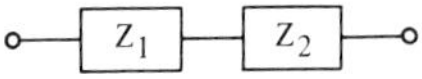	VA-1 VA-2 VA-3 VA-5
$Z_t = [Z_1^{-1} + Z_2^{-1} \cdots + Z_n^{-1}]^{-1}$ $Y_t = Y_1 + Y_2 \cdots + Y_n$	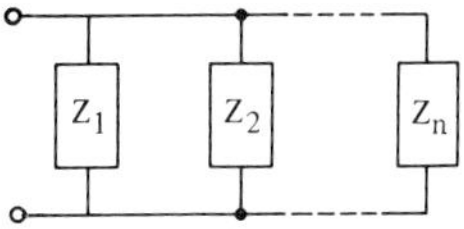	VA-1 VA-2 VA-3 VA-4 VA-5
$Z_2 = [Z_t^{-1} - Z_1^{-1}]^{-1}$ $Y_2 = Y_t - Y_1$	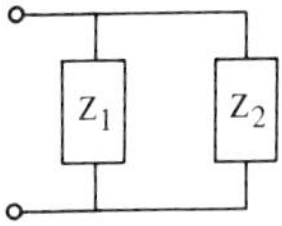	VA-1 VA-2 VA-3 VA-4 VA-5

Z *Notes:*

VA-1 Impedance is a complex quantity requiring the mathematical operation of addition and subtraction to be performed in rectangular form while multiplication and division operations are usually performed in polar form by treating the phase angle as an exponent. Impedances in rectangular form may be multiplied like other binomials, division however requires the divisor to be rationalized by multiplying the divisor and the dividend by the conjugate of the divisor (the conjugate of $R_S + X_S j = R_S - X_S j$). To eliminate this lengthy calculation, it is recommended that all multiplication and division be performed in polar form.

Input IMPEDANCE Vector Algebra — Z_i etc.

		Applicable Notes
$\mathbf{E}_g = E_g \angle 0°$ $\mathbf{Z}_i = \mathbf{Z}$ $\mathbf{I}_o = \mathbf{E}_g/\mathbf{Z}$	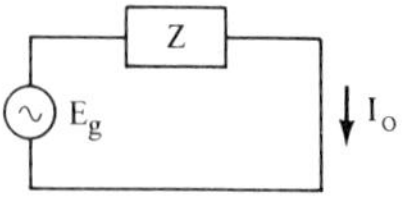	VA-1 VA-2 VA-3 VA-5
$\mathbf{I}_g = I_g \angle 0°$ $\mathbf{Z}_i = [\mathbf{Z}_1^{-1} + \mathbf{Z}_2^{-1}]^{-1}$ $\mathbf{E}_g = \mathbf{I}_g \mathbf{Z}_i$ $\mathbf{I}_o = (\mathbf{I}_g \mathbf{Z}_i)/\mathbf{Z}_1$	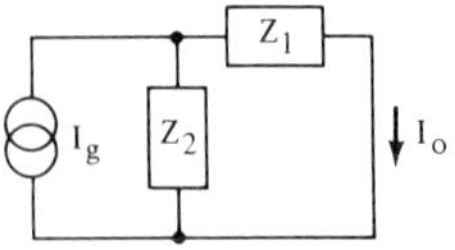	VA-1 VA-2 VA-3 VA-4 VA-5
$\mathbf{E}_g = E \angle 0°$ $\mathbf{Z}_i = \mathbf{Z}_3 + [\mathbf{Z}_2^{-1} + \mathbf{Z}_1^{-1}]^{-1}$ $\mathbf{I}_g = \mathbf{E}_g/\mathbf{Z}_i$ $\mathbf{I}_o = \mathbf{E}_g \Big/ \Big(\mathbf{Z}_i\big[(\mathbf{Z}_1/\mathbf{Z}_2) + 1\big]\Big)$	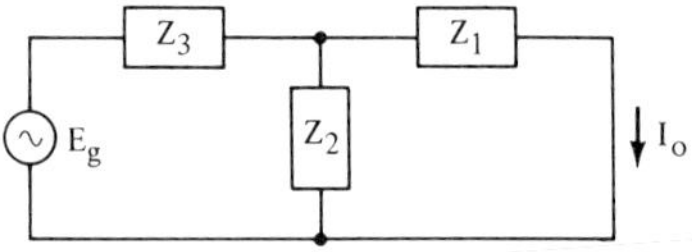	VA-1 VA-2 VA-3 VA-4 VA-5

Z *Notes:*

VA-2 E_g = Generator voltage; E_o = Output voltage
I_g = Generator current; I_o = Output current
Y_o = Output admittance; Y_t = Total admittance
Z_i = Input impedance; Z_o = Output impedance
Z_t = Total or Equivalent impedance

Input
IMPEDANCE
Vector
Algebra

Z_i etc.

Applicable Notes

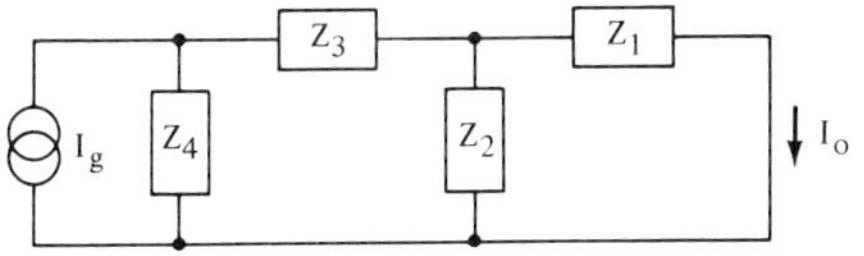

VA-1
VA-2
VA-3
VA-4
VA-5

$\mathbf{I}_g = I_g \angle 0°$

$\mathbf{Z}_i = \left(\mathbf{Z}_4^{-1} + \left[\mathbf{Z}_3 + (\mathbf{Z}_2^{-1} + \mathbf{Z}_1^{-1})^{-1}\right]^{-1}\right)^{-1}$

$\mathbf{E}_g = \mathbf{I}_g \mathbf{Z}_i$

$\mathbf{I}_o = \mathbf{I}_g \left[1 - (\mathbf{Z}_i/\mathbf{Z}_4)\right] / \left[(\mathbf{Z}_1/\mathbf{Z}_2) + 1\right]$

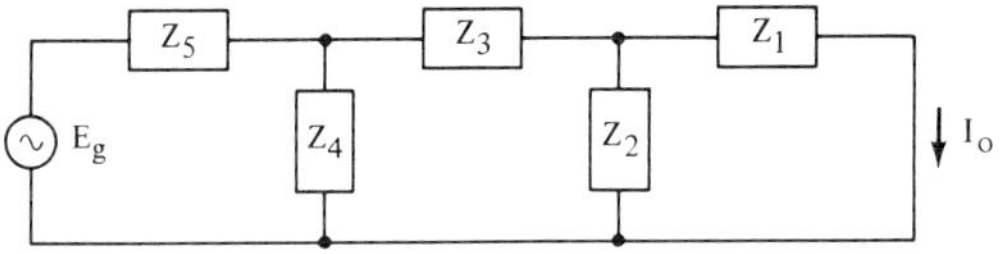

VA-1
VA-2
VA-3
VA-4
VA-5

$\mathbf{E}_g = E \angle 0°$

$\mathbf{Z}_i = \mathbf{Z}_5 + \left(\mathbf{Z}_4^{-1} + \left[\mathbf{Z}_3 + (\mathbf{Z}_2^{-1} + \mathbf{Z}_1^{-1})^{-1}\right]^{-1}\right)^{-1}$

$\mathbf{I}_g = \mathbf{E}_g/\mathbf{Z}_i$

$\mathbf{I}_o = \mathbf{I}_g \left(1 - \left[(\mathbf{Z}_i - \mathbf{Z}_5)/\mathbf{Z}_4\right]\right) / \left[(\mathbf{Z}_1/\mathbf{Z}_2) + 1\right]$

Z *Notes:*

VA-3 Impedances Z, Z_1, Z_2, Z_3, Z_4 and Z_5 may represent any resistance, reactance, series circuit, parallel circuit, unknown circuit or any circuit regardless of complexity or configuration.

VA-4 $Z^{-1} = 1/Z = Y$, $Y^{-1} = 1/Y = Z$

VA-5 Z, Z_i, Z_o, Y, Y_o, I_o and E_o all will vary with frequency except purely resistive circuits.

IMPEDANCE
Vector
Algebra

Z_i Z_o etc.

Applicable Notes

$\mathbf{I}_g = I_g \underline{/0^\circ}$

$\mathbf{Z}_i = \mathbf{Z}$

$\mathbf{Z}_o = \mathbf{Z}$

$\mathbf{E}_o = \mathbf{I}_g \mathbf{Z}$

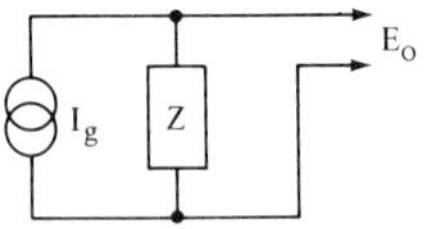

VA-1
VA-2
VA-3
VA-5

$\mathbf{E}_g = E_g \underline{/0^\circ}$

$\mathbf{Z}_i = \mathbf{Z}_1 + \mathbf{Z}_2$

$\mathbf{Z}_o = [\mathbf{Z}_1^{-1} + \mathbf{Z}_2^{-1}]^{-1}$

$\mathbf{Y}_o = \mathbf{Y}_1 + \mathbf{Y}_2$

$\mathbf{E}_o = (\mathbf{E}_g \mathbf{Z}_1)/\mathbf{Z}_i$

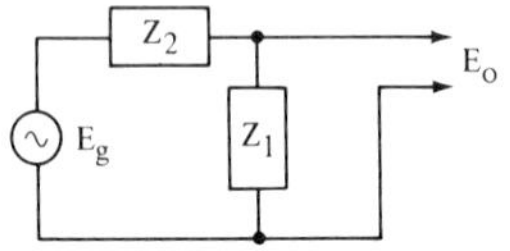

VA-1
VA-2
VA-3
VA-4
VA-5

$\mathbf{I}_g = I_g \underline{/0^\circ}$

$\mathbf{Z}_i = \left[\mathbf{Z}_3^{-1} + (\mathbf{Z}_2 + \mathbf{Z}_1)^{-1}\right]^{-1}$

$\mathbf{Z}_o = \left[\mathbf{Z}_1^{-1} + (\mathbf{Z}_2 + \mathbf{Z}_3)^{-1}\right]^{-1}$

$\mathbf{Y}_o = \mathbf{Y}_1 + (\mathbf{Y}_2^{-1} + \mathbf{Y}_3^{-1})^{-1}$

$\mathbf{E}_g = \mathbf{I}_g \mathbf{Z}_i$

$\mathbf{E}_o = \mathbf{I}_g \mathbf{Z}_1 \ [1 - (\mathbf{Z}_i/\mathbf{Z}_3)]$

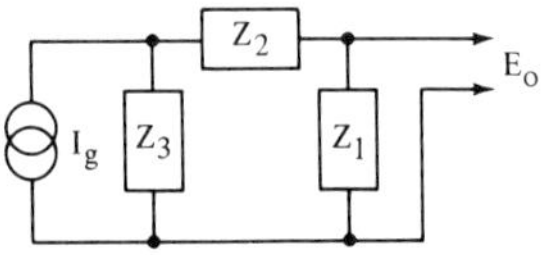

VA-1
VA-2
VA-3
VA-4
VA-5

IMPEDANCE Vector Algebra

Z_i Z_o etc.

Applicable Notes

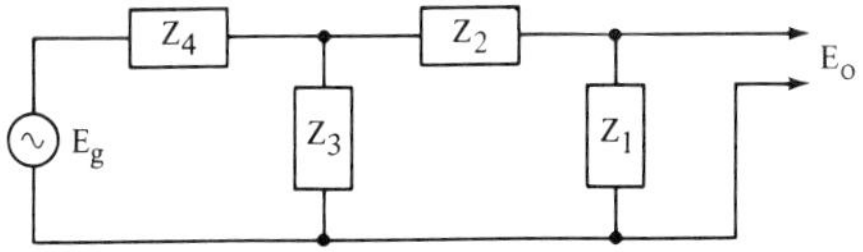

$\mathbf{E}_g = E_g \underline{/0^\circ}$

$\mathbf{Z}_i = \mathbf{Z}_4 + \left[\mathbf{Z}_3^{-1} + (\mathbf{Z}_2 + \mathbf{Z}_1)^{-1}\right]^{-1}$

$\mathbf{Z}_o = \left(\mathbf{Z}_1^{-1} + \left[\mathbf{Z}_2 + (\mathbf{Z}_3^{-1} + \mathbf{Z}_4^{-1})^{-1}\right]^{-1}\right)^{-1}$

$\mathbf{Y}_o = \mathbf{Y}_1 + \left[\mathbf{Y}_2^{-1} + (\mathbf{Y}_3 + \mathbf{Y}_4)^{-1}\right]^{-1}$

$\mathbf{I}_g = \mathbf{E}_g/\mathbf{Z}_i$

$\mathbf{E}_o = \mathbf{E}_g\left[1 - (\mathbf{Z}_4/\mathbf{Z}_i)\right]/\left[(\mathbf{Z}_2/\mathbf{Z}_1) + 1\right]$

VA-1
VA-2
VA-3
VA-4
VA-5

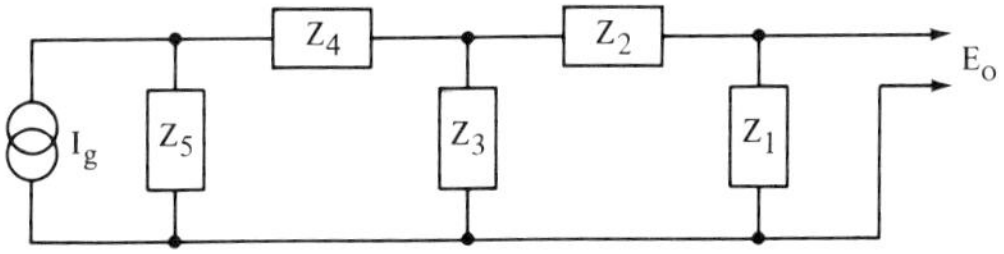

$\mathbf{I}_g = I_g \underline{/0^\circ}$ $\qquad \mathbf{E}_g = \mathbf{I}_g\mathbf{Z}_i$

$\mathbf{Z}_i = \left[\mathbf{Z}_5^{-1} + \left(\mathbf{Z}_4 + \left[\mathbf{Z}_3^{-1} + (\mathbf{Z}_2 + \mathbf{Z}_1)^{-1}\right]^{-1}\right)^{-1}\right]^{-1}$

$\mathbf{Z}_o = \left[\mathbf{Z}_1^{-1} + \left(\mathbf{Z}_2 + \left[\mathbf{Z}_3^{-1} + (\mathbf{Z}_4 + \mathbf{Z}_5)^{-1}\right]^{-1}\right)^{-1}\right]^{-1}$

$\mathbf{Y}_o = \mathbf{Y}_1 + \left(\mathbf{Y}_2^{-1} + \left[\mathbf{Y}_3 + (\mathbf{Y}_4^{-1} + \mathbf{Y}_5^{-1})^{-1}\right]^{-1}\right)^{-1}$

$\mathbf{E}_o = \left[\mathbf{I}_g(\mathbf{Z}_i - \mathbf{Z}_4)\right]/\left[(\mathbf{Z}_2/\mathbf{Z}_1) + 1\right]$

VA-1
VA-2
VA-3
VA-4
VA-5

Delta (Δ) to Wye (Y) or Reverse Conversion
Pi (π) section to Tee (T) section or reverse

Transformation

Δ or π section

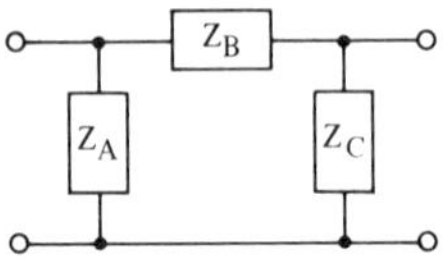

Y or T section

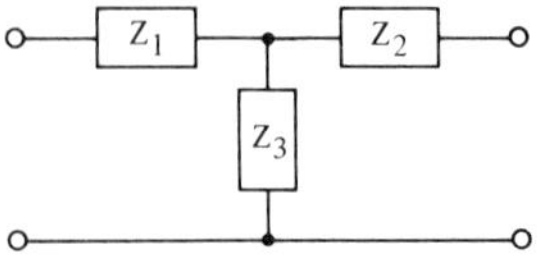

$$Z_1 = (Z_A Z_B)/[Z_A + Z_B + Z_C]$$

$$Z_2 = (Z_B Z_C)/[Z_A + Z_B + Z_C]$$

$$Z_3 = (Z_A Z_C)/[Z_A + Z_B + Z_C]$$

$$Z_A = \left[(Z_1 Z_2) + (Z_2 Z_3) + (Z_1 Z_3)\right]/Z_2$$

$$Z_B = \left[(Z_1 Z_2) + (Z_2 Z_3) + (Z_1 Z_3)\right]/Z_3$$

$$Z_C = \left[(Z_1 Z_2) + (Z_2 Z_3) + (Z_1 Z_3)\right]/Z_1$$

Page Notes:

1. Technically, delta and wye diagrams should be drawn with only three terminals.
2. Convert all impedances and intermediate solutions to both polar and rectangular form. Perform all addition in rectangular form. Perform all multiplication and division in polar form.

PASSIVE CIRCUITS

SECTION 1.2
GREEK LETTERS

α to ω Greek Alphabet

Spelling and Pronunciation	Small (Script)	Large (Capital)	Spelling and Pronunciation	Small (Script)	Large (Capital)
alpha (al′fa)	α	A	nu (nū)	ν	N
beta (bā′ta)	β	B	xi (zī)	ξ	Ξ
gamma (gam′a)	γ	Γ	omicron (om′i kron′ ō′mi kron′)	o	O
delta (del′ta)	δ	Δ	pi (pī)	π	Π
epislon (ep′sa lon′)	ϵ	E	rho (rō)	ρ	P
zeta (zā′ta)	ζ	Z	sigma (sig′ma)	σ s	Σ
eta (ā′ta)	η	H	tau (tou or taw) (ou as in out)	τ	T
theta (thā′ta)	θ	Θ	upsilon (up′sa lon′)	υ	Υ
iota (i ō′ ta)	ι	I	phi (fī or fē)	ϕ	Φ
kappa (kap′a)	κ	K	chi (kī)	χ	X
lambda (lam′da)	λ	Λ	psi (sī)	ψ	Ψ
mu (mū)	μ	M	omega (ō meg′a ō mē′ga)	ω	Ω

Note: For obvious reasons, capital greek letters A B E Z H I K M N O P T X are not used as electronic symbols.

α to η

α = Symbol for many different passive circuit quantities but no standardization has been achieved. See also–Active Circuits

β = Symbol for many different passive circuit quantities but no standardization has been achieved. See also–Active Circuits

γ = Symbol seldom used in electronics. Used for conductivity (G) in other fields.

δ = Symbol for loss angle.

δ = Ninety degrees minus the absolute value of the phase angle.

$\delta = 90° - |\theta|$

$\delta = \tan^{-1} D$

Note: The dissipation factor (D) of capacitors is specified by most USA manufacturers but the loss angle (δ) or the tangent of the loss angle ($\tan \delta$) is specified by most foreign manufacturers.

Δ = Symbol for increment or decrement. (Still used for vacuum tubes, but small signal parameters such as h_{fe} are used for semiconductors.)

ϵ = Symbol for the base of natural logarithms.

$\epsilon = 2.718281828 \cdots \quad \epsilon^{-1} = .3678794412 \cdots$

ζ = Seldom used and no standardization of meaning.

η = Efficiency. See also–Active Circuits

θ

Phase Angle Definitions

θ = Symbol for phase angle.

Note: Phi (ϕ) and other greek letters are also used as symbols for phase angle.

θ = 1. The angular difference in phase between a quantity and a reference.
2. The phase angle of voltage, current, impedance or admittance with respect to a reference.
3. The phase angle of voltage, current, impedance or admittance with respect to the phase angle of current, voltage, resistance or conductance.
4. The phase angle of voltage or impedance with respect to the phase angle of total current or with respect to 0°.
5. The phase angle of current or admittance with respect to the phase angle of total voltage or with respect to 0°.

θ = Phase angle measured and expressed in:
1. Decimal degrees
   360° = one cycle or one revolution
2. Degrees, minutes, seconds
   1° = 60′ (minutes)
   1′ = 60″ (seconds)
3. Radians
   2π radians = one cycle or one revolution
4. Grads
   400 grads = one cycle or one revolution

θ = 0° when voltage and current are in phase. 0° when circuit is or acts as a pure resistance or conductance.

θ = ±90° when circuit is or acts as a pure reactance or susceptance.

θ = +90° to −90° for all two terminal networks when θ is angle of total voltage, total current, total impedance or total admittance.

$+\theta$ = Leading phase angle. Counterclockwise rotation of a vector. Earlier in time than 0°.

$-\theta$ = Lagging phase angle. Clockwise rotation of a vector. Later in time than 0°.

θ_E = 1. The difference in phase between the total voltage and the total current when the phase angle of the total current is placed at 0°.
2. The angular difference in phase between the total voltage and the current source. ($\mathbf{I}_g = I_g \angle 0° = +I_g$ unless noted)

θ_{Eo} = Output voltage phase with respect to the phase of the voltage or current input. (Voltage or current generator E_g or $I_g = 0°$)

θ_I = 1. The angular difference in phase between the total current and the total voltage when the phase angle of the total voltage is placed at 0°.
2. The angular difference in phase between the total current and the voltage source. ($\mathbf{E}_g = E_g \angle 0° = +E_g$ unless noted)

θ_{Io} = Output current phase with respect to the phase of the voltage or current input. (Input phase = 0° unless noted)

Page Note: The phase angles of **E** and **I** may be confusing. To prevent confusion, always calculate polar impedance first and then assign zero degrees to the signal source. If the signal source is a voltage generator, $\theta_I = -\theta_Z$ and if the signal source is a current generator, $\theta_E = \theta_Z$. Use vector algebra to determine the phase angles of circuit voltages and/or currents. e.g., when a voltage source is connected to a series circuit, $\theta_I = -\theta_Z$, $\theta_{E_R} = \theta_I$, $\theta_{E_L} = \theta_I + 90°$, $\theta_{E_C} = \theta_I - 90°$, $\theta_{E_t} = 0°$.

θ_Y = 1. The angular difference between the admittance and the conductance of a circuit. (The angle of conductance G = 0°)

2. The same angle as impedance except with opposite sign.
3. The same angle as the phase angle of the total current when the phase angle of total voltage is placed at 0°.

θ_Z = 1. The angular difference between the impedance and the resistance of a circuit. (The angle of resistance R = 0°)

2. The same angle as the admittance except with opposite sign.
3. The same angle as the phase angle of the total voltage when the phase angle of total current is placed at 0°.

$\pm\theta_E = -(\pm\theta_I)$ but both may not coexist.

$\pm\theta_I = -(\pm\theta_E)$ but both may not coexist.

$\pm\theta_Y = -(\pm\theta_Z)$ $[Y\angle\pm\theta_Y = Z^{-1}\angle-(\pm\theta_Z)]$

$\pm\theta_Z = -(\pm\theta_Y)$ $[Z\angle\pm\theta_Z = Y^{-1}\angle-(\pm\theta_Y)]$

$\pm\theta_E = -(\pm\theta_I) = -(\pm\theta_Y) = \pm\theta_Z$ in all two terminal networks where the phase angle of either the total voltage or the total current is placed at 0°. (It should be understood that reactance, a component of impedance, is the cause of the difference in phase between the voltage and the current, that θ_E, θ_I, θ_Y and θ_Z are the same one and only phase angle from different reference points, that only one may be used at any one time and that if **Z** or **Y** appears as a term in a formula the other term must be $E\angle 0°$ or $I\angle 0°$.)

θ

Phase Angle Definitions

$+\theta_E$ = Inductive circuit phase angle of voltage

$-\theta_E$ = Capacitive circuit phase angle of voltage

$+\theta_I$ = Capacitive circuit phase angle of current

$-\theta_I$ = Inductive circuit phase angle of current

$+\theta_Y$ = Capacitive circuit phase angle of admittance

$-\theta_Y$ = Inductive circuit phase angle of admittance

$+\theta_Z$ = Inductive circuit phase angle of impedance

$-\theta_Z$ = Capacitive circuit phase angle of impedance

$\theta_{B_C} = +90^\circ$	$\theta_{B_L} = -90^\circ$	
$\theta_{E_C} = -90^\circ$	$\theta_{E_L} = +90^\circ$	$\theta_{E_R} = 0^\circ$
$\theta_G = 0^\circ$		
$\theta_{I_C} = +90^\circ$	$\theta_{I_L} = -90^\circ$	$\theta_{I_R} = 0^\circ$
$\theta_R = 0^\circ$		
$\theta_{X_C} = -90^\circ$	$\theta_{X_L} = +90^\circ$	

$$1\angle 0^\circ = +1 = 1 + 0_j$$

$$1\angle +90^\circ = \sqrt{-1} = 0 + 1_j$$

$$1\angle -90^\circ = -\sqrt{-1} = 0 - 1_j$$

$$1\angle \pm 180^\circ = -1 = -1 + 0_j$$

$$+270^\circ = -90^\circ, \quad -270^\circ = +90^\circ, \quad \pm 360^\circ = 0^\circ$$

Phase Angle, Series Circuits θ	**Terms**
$\lvert\theta_E\rvert = \lvert\theta_I\rvert = \lvert\theta_Y\rvert = \lvert\theta_Z\rvert = \tan^{-1}[D^{-1}]$	D_s
$\lvert\theta_E\rvert = \lvert\theta_I\rvert = \lvert\theta_Y\rvert = \lvert\theta_Z\rvert = \tan^{-1} Q$	Q_s
$\pm\theta_E = \pm\theta_Z = \tan^{-1}[\pm E_X/E_R]$	$E_R \ \pm E_X$
$\pm\theta_E = \pm\theta_Z = \tan^{-1}[\pm X/R]$	$R \ \pm X$
$\lvert\theta_E\rvert = \lvert\theta_I\rvert = \lvert\theta_Y\rvert = \lvert\theta_Z\rvert = \cos^{-1}[R/Z]$	$R \ Z$
$\pm\theta_E = \pm\theta_Z = \sin^{-1}[\pm X/Z]$	$\pm X \ Z$
$\pm\theta_E = \pm\theta_Z = \tan^{-1}\left[(E_L - E_C)/E_R\right]$	$E_R \ E_C \ E_L$
$\pm\theta_E = \pm\theta_Z = \tan^{-1}\left[(X_L - X_C)/R\right]$	$R \ X_C \ X_L$
$\pm\theta_E = \pm\theta_Z = \sin^{-1}\left[(X_L - X_C)/Z\right]$	$X_C X_L \ Z$
$(\pm\theta_Z)_t = \tan^{-1}\left[(\pm X_s)_t/(R_s)_t\right]$ $(\pm X_s)_t = \left[Z_1 \sin(\pm\theta_1)\right] + \left[Z_2 \sin(\pm\theta_2)\right]$ $(R_s)_t = (Z_1 \cos\theta_1) + (Z_2 \cos\theta_2)$	$Z_1\angle\pm\theta_1$ $Z_2\angle\pm\theta_2$

Phase Angle, Parallel Circuits θ	Terms
$\lvert\theta_E\rvert = \lvert\theta_I\rvert = \lvert\theta_Y\rvert = \lvert\theta_Z\rvert = \tan^{-1}\ [D^{-1}]$	D_p
$\lvert\theta_E\rvert = \lvert\theta_I\rvert = \lvert\theta_Y\rvert = \lvert\theta_Z\rvert = \tan^{-1} Q$	Q_p
$\pm\theta_Y = \pm\theta_I = \tan^{-1}\left[-(\pm B)/G\right]$	$\pm B\ \ G$
$\pm\theta_Y = \pm\theta_I = \sin^{-1}\left[-(\pm B)/Y\right]$	$\pm B\ \ Y$
$\pm\theta_I = \pm\theta_Y = \tan^{-1}\left[-(\pm I_X)/I_R\right]$	$\pm I_X\ \ I_R$
$\lvert\theta_E\rvert = \lvert\theta_I\rvert = \lvert\theta_Y\rvert = \lvert\theta_Z\rvert = \cos^{-1}\ [G/Y]$	$G\ \ Y$
$\pm\theta_Z = \pm\theta_E = \tan^{-1}\ [R_p/\pm X_p]$	$R\ \ \pm X$
$\lvert\theta_E\rvert = \lvert\theta_I\rvert = \lvert\theta_Y\rvert = \lvert\theta_Z\rvert = \cos^{-1}\ [Z/R_p]$	$R\ \ Z$
$\pm\theta_Z = \pm\theta_E = \sin^{-1}\ [Z/\pm X_p]$	$\pm X\ \ Z$

Page Notes: $\lvert +B\rvert = B_L = (X_L^{-1})_p$

$\lvert -B\rvert = B_C = (X_C^{-1})_p$

$\lvert +X_p\rvert = (X_L)_p = B_L^{-1}$

$\lvert -X_p\rvert = (X_C)_p = B_C^{-1}$

Phase Angle, Parallel Circuits — θ to κ	Terms
$\pm\theta_Y = \tan^{-1}\left[-(B_L - B_C)/G\right]$	B_C B_L G
$\pm\theta_Y = \sin^{-1}\left[-(B_L - B_C)/Y\right]$	B_C B_L Y
$\pm\theta_I = \tan^{-1}\left[-(I_L - I_C)/I_R\right]$	I_C I_L I_R
$\pm\theta_Z = \tan^{-1}\left[R(X_L^{-1} - X_C^{-1})\right]$	R X_C X_L
$\pm\theta_Z = \sin^{-1}\left[Z(X_L^{-1} - X_C^{-1})\right]$	X_C X_L Z
$(\pm\theta_Z)_t = \tan^{-1}\left[\pm B_t/G_t\right]$ $\pm B_t = \left[Z_1^{-1}\sin(\pm\theta_1)\right] + \left[Z_2^{-1}\sin(\pm\theta_2)\right]$ $G_t = (Z_1^{-1}\cos\theta_1) + (Z_2^{-1}\cos\theta_2)$	$Z_2\angle\pm\theta_2$ $Z_1\angle\pm\theta_1$

ι = Seldom as a symbol due to similarity to english letter i.
κ = Seldom as a symbol due to similarity to english letter k.

λ

Wavelength Definitions & Formulas

λ = Symbol for wavelength.

λ = 1. In a periodic wave, the distance between points of corresponding phase of two consecutive cycles.
2. The length of one complete cycle of a periodic wave.

λ = Wavelength measured and expressed in various units of distance such as inches, feet, centimeters or meters.

$\lambda = v/f$ where v is the velocity of the wave in the medium through which it is traveling. f = frequency of wave.
Note: In physics, the symbol c is used for the velocity of light.

Wavelength of Sound in Air

$\lambda \approx 1136/f$ feet @ 25° C

$\lambda \approx 346.3/f$ meters @ 25° C

$\lambda \simeq (1051 + 1.1\ T_F)/f$ feet @ std pressure

$\lambda \simeq (331.3 + .6\ T_C)/f$ meters @ std pressure

Wavelength of Electromagnetic Waves

$\lambda \approx (9.8 \cdot 10^8)/f$ feet

$\lambda \approx (3 \cdot 10^8)/f$ meters

$\lambda = (2.997\,93 \cdot 10^8)/f$ meters (in vacuum)

μ

Micro,
Mu Factor,
Permeability

μ = Symbol for micro.

μ = Prefix meaning 1/1,000,000. 10^{-6} multiplier prefix for most basic units. $\mu V = 10^{-6}$ Volts, $\mu A = 10^{-6}$ Amperes, $\mu F = 10^{-6}$ Farads, $\mu H = 10^{-6}$ Henries.

μ = Symbol for mu factor.

μ = Amplification factor (voltage) in vacuum tubes.

$\mu = \Delta E_p / \Delta E_g$ (with I_p constant)

$\mu = g_m r_p$ E_g = grid voltage, E_p = plate voltage
g_m = mutual conductance (transconductance)
r_p = dynamic plate resistance.

μ = Term not used with semiconductors.

See–Active Circuits A_v

μ = Symbol for magnetic permeability.

μ = The magnetic equivalent of electrical conductivity. The magnetic conductivity of a material compared to air or vacuum.

μ = B/H
B = Flux density in gauss
H = Magnetizing force in oersteds

ϕ

Total Magnetic Flux, Complementary Angles

ϕ = Symbol for total magnetic flux.

ϕ = The total measure of the magnetized condition of a magnetic circuit when acted upon by a magnetomotive force.

ϕ = The total number of lines of magnetic flux.

ϕ = Maxwells or lines of flux units (CGS).

ϕ = Weber units (SI).

1 maxwell (Mx) = 1 line of flux

1 maxwell (Mx) = 10^{-8} Weber (Wb)

$\phi = F/\mathfrak{R}$ where F = magnetomotive force

$\phi = F\mathcal{P}$ $\mathfrak{R}$ = reluctance

$\mathcal{P}$ = permeance

$\phi = BA$ where B = flux density in gauss

A = cross sectional area of magnetic path in cm^2.

$\phi = BA$ where B = flux density in lines per in^2

A = cross sectional area of magnetic path in in^2

ϕ = Alternate symbol for phase angle.

ϕ = Symbol for complementary angle.

ϕ = Complement of phase angle θ.

$\phi = \pm 90^\circ - \theta \quad (\phi + \theta = \pm 90^\circ)$

ω

Angular Velocity Definitions & Formulas

ω = Symbol for angular velocity or angular frequency.

ω = The rate at which an angle changes in radians per second. The angular change in a uniformly rotating system measured in radians per second. (2π radians = 360°, 2π radians per second = rps = r/s = Hz)

	Terms
$\omega = 2\pi f = 6.2831853 \cdots f$	f
$\omega = B_C/C$	B_C C
$\omega = (B_L L)^{-1}$	BL L
$\omega = (X_C C)^{-1}$	C X_C
$\omega = X_L/L$	L X_L

Resonant Angular Velocity

$$\omega_r = \sqrt{(LC)^{-1}}$$

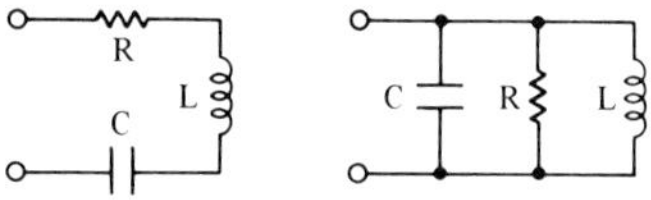

f_r Definition 1

$$\omega_r = \left[(LC) - (L/R)^2\right]^{-\frac{1}{2}}$$

$(-x)^{-\frac{1}{2}}$ exception

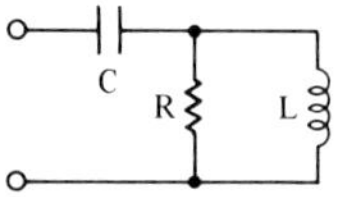

$$\omega_r \approx \sqrt{(LC)^{-1}}$$

ω_r

Resonant Angular Velocity

$\omega_r = \sqrt{(LC)^{-1} - (CR)^{-2}}$ $\sqrt{-x}$ exception

f_r definition 1

$\omega_r \approx \sqrt{(LC)^{-1}}$

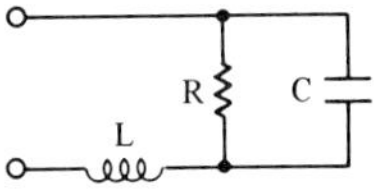

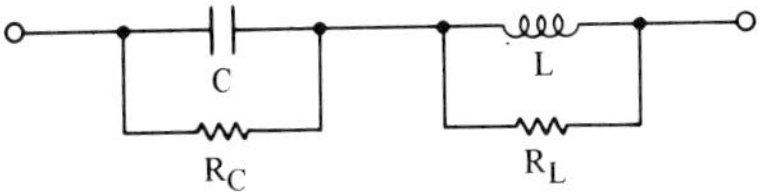

$\omega_r = \sqrt{\left[(R_C^2C)^{-1} - L^{-1}\right] / \left[(L/R_L^2) - C\right]}$ $\sqrt{-x}$ exception

f_r definition 1

$\omega_r \approx \sqrt{(LC)^{-1}}$

$\omega_r = \sqrt{(LC)^{-1} - (R/L)^2}$ $\sqrt{-x}$ exception

f_r definition 1

$\omega_r \approx \sqrt{(LC)^{-1}}$

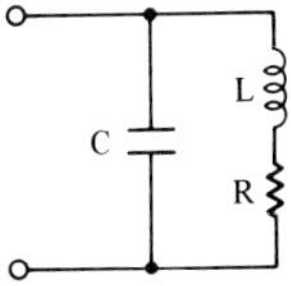

$\omega_r = \left[(LC) - (CR)^2\right]^{-\frac{1}{2}}$ $(-x)^{-\frac{1}{2}}$ exception

f_r definition 1

$\omega_r \approx \sqrt{(LC)^{-1}}$

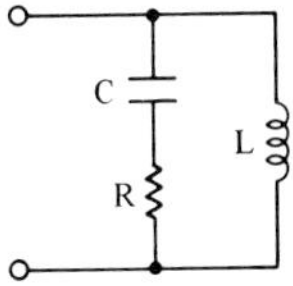

$\omega_r = \sqrt{\left[C^{-1} - (R_L^2/L)\right] / \left[L - R_C^2C\right]}$

f_r definition 1 $\sqrt{-x}$ exception

$\omega_r \approx \sqrt{(LC)^{-1}}$

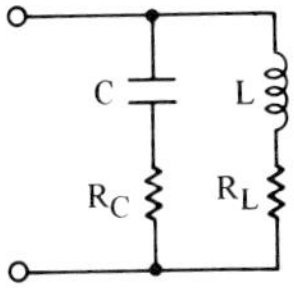

Ω = Symbol for ohm.

Ω = 1. The basic unit of resistance, reactance and impedance.

2. That resistance which will develop a current of one ampere from an applied potential of one volt.

3. That reactance or impedance which will develop a steady state rms current of one ampere from an applied sinewave potential of one volt rms.

4. The resistance of a uniform column of mercury 106.3 cm long weighing 14.4521 g at a temperature of 0°C.

Ω = Unit often used with multiplier prefixes.

$\mu\Omega = 10^{-6}$ ohms

$m\Omega = 10^{-3}$ ohms

$k\Omega = 10^{3}$ ohms

$M\Omega = 10^{6}$ ohms

$G\Omega = 10^{9}$ ohms

$T\Omega = 10^{12}$ ohms

Note: kΩ is frequently contracted to K
MΩ is frequently contracted to M
Megohm is frequently contracted to Meg

Ω = A real (positive or 0°) quantity when a unit of resistance.

Ω = A magnitude or a complex quantity when a unit of reactance or impedance.

SECTION TWO
TRANSISTORS

2.1 STATIC (DC) CONDITIONS

A to E DC Transistor Symbol Definitions

$\bar{a}$ – See–h_{FB} or $\bar{\alpha}$

A_I = Static current amplification (seldom used)

A_V = Static voltage amplification (seldom used)

BV_{CBO} – See–$V_{(BR)CBO}$

BV_{CEO} – See–$V_{CEO(SUS)}$

BV_{CER} – See–$V_{CER(SUS)}$

BV_{CES} – See–$V_{CES(SUS)}$

BV_{CEV} – See–$V_{CEV(SUS)}$

BV_{CEX} – See–$V_{CEX(SUS)}$

BV_{EBO} – See–$V_{(BR)EBO}$

E – See–V for dc transistor voltages
See also–V, Opamp
See also–E, Passive Circuits

E = The original symbol for the electric force originally known as electromotive force. This force is now known as voltage, potential or potential difference. The voltage symbol E has been superseded by V for dc transistor voltages and for all operational amplifier voltages.

$E_{S/b}$ – (second breakdown energy) See–$I_{S/b}$

Note: The term second breakdown energy ($E_{S/b}$) has never been appropriate for static transistor conditions since continuous power at any level converts to infinite energy.

h

Static (DC) Hybrid Parameters

h_{FB} = Seldom used common-base static forward-current transfer ratio.

h_{FB} = DC alpha ($\overline{\alpha}$)

$h_{FB} = I_C/I_E$

$h_{FB} = h_{FE}/(h_{FE} + 1)$

h_{FE} = Common-emitter static forward-current transfer ratio at a specified collector current, collector voltage and junction temperature.

h_{FE} = DC beta ($\overline{\beta}$)

$h_{FE} = I_C/I_B$

$h_{FE} = (I_E/I_B) - 1$

$h_{FE} = \left[(I_E/I_C) - 1\right]^{-1}$

$h_{FE(INV)}$ = Seldom used h_{FE} when collector and emitter leads are interchanged.

h_{IE} = Seldom used common-emitter static input resistance.

h *Notes:*

The DC counterparts of h_{fc}, h_{ib}, h_{ob}, h_{oc}, h_{oe}, h_{rb}, h_{rc}, and h_{re} are very seldom used.

h_{FE} usually has a different value than h_{fe} measured under the same conditions.

h_{IE} and h_{IB} will have a much higher value than their small signal counterparts measured under the same conditions.

I_B = DC base current.

I_C = DC collector current. See–I Note ①

I_E = DC emitter current.

I_{BB} = Base supply current.

I_{CC} = Collector supply current.

I_{CO} – See–I_{CBO}

I_{EE} = Emitter supply current.

I_{EO} – See–I_{EBO}

I_{CBO} = DC collector to base leakage current at a specified voltage and temperature with emitter open. ② ③

I_{CEO} = DC collector to emitter leakage current at a specified voltage and temperature with base open. ② ③

I_{CER} = DC collector to emitter leakage current at a specified voltage and temperature with a specified base to emitter resistance. See–I Notes ② ③

I_{CES} = DC collector to emitter leakage current at a specified voltage and temperature with the base and emitter shorted. See–I Notes ② ③

I_{CEV} = DC collector to emitter leakage current at a specified voltage and temperature with a specified base to emitter reverse bias voltage. See–I Notes ② ③

I_{CEX} = DC collector to emitter leakage current at a specified voltage and temperature with a specified base to emitter circuit. See–I Notes ② ③

I_{EBO} = DC base to emitter reverse bias leakage current at a specified voltage and temperature with open collector. See–I Notes ② ③

I_F = Forward bias dc current.

I_N = Noise current. See–I_N, Passive Circuits
See also–NF

I_O = DC output current. See also–Passive Circuits

I_R – See–I, Passive Circuits

$I_{S/b}$ = Symbol for second breakdown current.

$I_{S/b}$ = The collector current at which second breakdown occurs at a specified collector voltage, case or junction temperature and pulse duration.

Second breakdown occurs when the combination of voltage, current, temperature, time and a current constriction within the transistor produces spot heating sufficient to thermally maintain or increase the collector current regardless of base bias. In the usual transistor circuit, if second breakdown has been allowed to occur, transistor failure will also occur due to excessive spot junction temperature.

Second breakdown is not the same as thermal failure where failure may be predicted from low voltage thermal resistance calculations. Second breakdown may occur at positive, zero, or negative base bias.

Circuit design should be such that the manufacturers second breakdown specifications are not exceeded under worst case conditions. Alternately, the second breakdown characteristics of transistors may be measured with special non-destructive procedures.

Static (DC) Transistor Currents $I_B \; I_C \; I_E$	Applicable Notes
$I_B = I_C/h_{FE}$	④
$I_B = I_E/(h_{FE} + 1)$	⑤
$I_B = V_{BE}/h_{IE}$	⑥
$I_B \approx \left[\log^{-1} (V_{BE}/.06)\right]/(10^{13} h_{FE})$	⑦
$I_C = h_{FE} I_B$	④
$I_C = I_E - I_B$	⑤
$I_C = (h_{FE} V_{BE})/h_{IE}$	⑥
$I_C \approx \left[\log^{-1} (V_{BE}/.06)\right] (5 \cdot 10^{-16} h_{FE})$	⑦
$I_C \approx 10^{-13} \left[\log^{-1} (V_{BE}/.06)\right]$	
$I_E = I_C + I_B$	④
$I_E = (h_{FE} + 1) I_B$	⑤
$I_E = \left[V_{BE}(h_{FE} + 1)\right]/h_{IE}$	⑥
$I_E \approx (5 \cdot 10^{-16}) (h_{FE} + 1) \left[\log^{-1} (V_{BE}/.06)\right]$	⑦

I *Notes:*

① The subscript of I_C is a capital letter for DC. It is often difficult to distinguish between a capital and a lower case C subscript. I_c (lower case) is rms collector current and i_C (upper case) is instantaneous total collector current.

Static (DC) Transistor Currents	$I_B \quad I_C \quad I_E$	Applicable Notes
$I_C \approx 10^{-13}\left[\log^{-1}(V_{BB}/.06)\right]$ $I_B = I_C/h_{FE}$ $I_E = I_C + I_B$	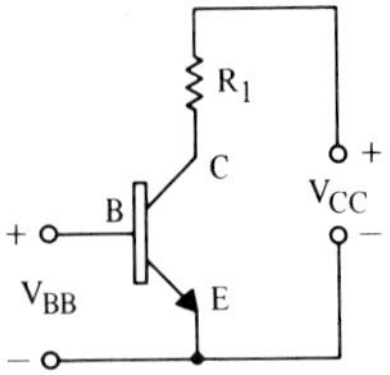	① ② ③ ④ ⑤ ⑥
$I_C = \left[h_{FE}(V_{BB} - V_{BE})\right]/R_2$ $V_{BE} \approx .06\left[\log(10^{13}\, I_C)\right]$ $V_{BE} \approx .6$ $I_B = I_C/h_{FE}$ $I_E = I_C + I_B$	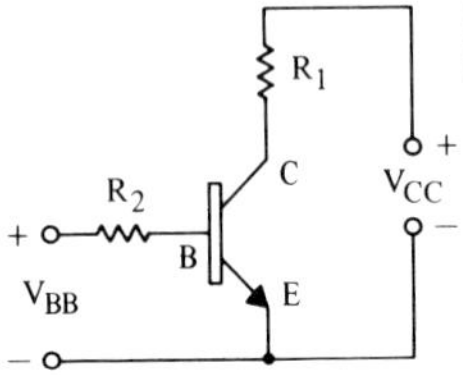	① ② ③ ④ ⑤ ⑥ ⑦
$I_C = \left[h_{FE}(V_{CC} - V_{BE})\right]/R_2$ $V_{BE} \approx .06\left[\log(10^{13}\, I_C)\right]$ $V_{BE} \approx .6$ $I_B = I_C/h_{FE}$ $I_E = I_C + I_B$	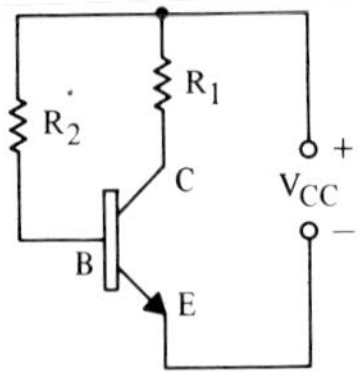	① ② ③ ④ ⑤ ⑥ ⑦

I *Notes:*

② The standard specified temperature is 25°C

③ Transistor leakage currents have a temperature dependent component and a voltage dependent component.

④ h_{FE}, V_{BE} and h_{IE} are temperature, current and voltage dependent.

⑤ $\log x = \log_{10} x$, $\log^{-1} x = \text{antilog } x = 10^x$

Static (DC) Transistor Currents $I_B \quad I_C \quad I_E$

Applicable Notes

$$I_E = (V_{CC} - V_{BE}) \Big/ \Big(R_1 + \big[R_3/(h_{FE} + 1)\big]\Big)$$

$$V_{BE} \approx .06 \left[\log \left(10^{13} I_C\right)\right] \approx .6$$

$$I_B = I_E/(h_{FE} + 1)$$

$$I_C = I_E - I_B$$

$$I_{CC} = I_E$$

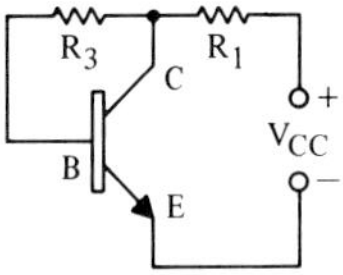

① ② ③ ④ ⑤ ⑥ ⑦

$$I_E = (V_{CC} - V_{BE}) \Big/ \Big(R_1 + R_5 + \big[R_3/(h_{FE} + 1)\big]\Big)$$

$$V_{BE} \approx .06 \left[\log \left(10^{13} I_C\right)\right] \approx .6$$

$$I_B = I_E/(h_{FE} + 1)$$

$$I_C = I_E - I_B$$

$$I_{CC} = I_E$$

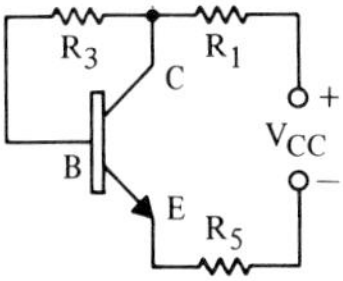

① ② ③ ④ ⑤ ⑥ ⑦

$$I_C = \big[V_{CC} - (V_{BE} R_X)\big] \Big/ \Big(R_1 + \big[(R_1 + R_3)/h_{FE}\big]\Big)$$

$$V_{BE} \approx .06 \left[\log \left(10^{13} I_C\right)\right] \approx .6$$

$$R_X = \big[(R_1 + R_3)/R_4\big] + 1$$

$$I_B = I_C/h_{FE}$$

$$I_E = I_C + I_B$$

$$I_{CC} = I_E + (V_{BE}/R_4)$$

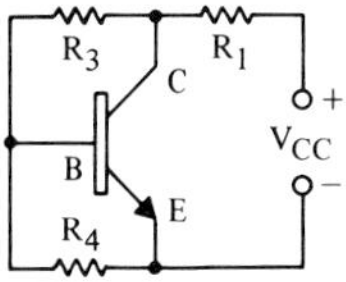

① ② ③ ④ ⑤ ⑥ ⑦

Static (DC) Transistor Currents | I_B I_C I_E | **Applicable Notes**

$$I_C = h_{FE}\left(\left[(V_{CC} - V_{BE})/R_2\right] - \left[V_{BE}/R_4\right]\right)$$

$$V_{BE} \approx .06\left[\log\left(10^{13} I_C\right)\right]$$

$$V_{BE} \approx .6$$

$$I_B = I_C/h_{FE}$$

$$I_E = I_C + I_B$$

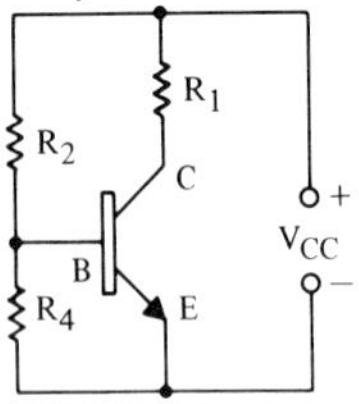

Applicable Notes: ① ② ③ ④ ⑤ ⑥ ⑦

$$I_C = \left[h_{FE}(V_{BB} - V_{BE})\right] \Big/ \left(R_2 + \left[R_5(h_{FE} + 1)\right]\right)$$

$$V_{BE} \approx .06\left[\log\left(10^{13} I_C\right)\right]$$

$$V_{BE} \approx .6$$

$$I_B = I_C/h_{FE}$$

$$I_E = I_C + I_B$$

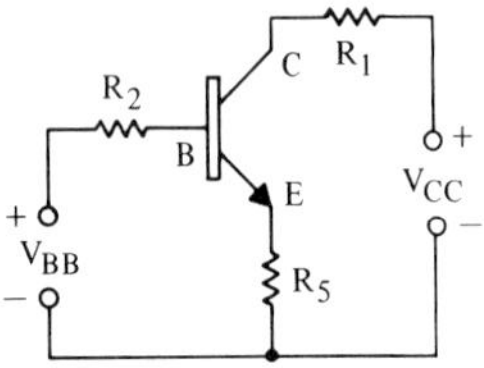

Applicable Notes: ① ② ③ ④ ⑤ ⑥ ⑦

$$I_C = \left[h_{FE}(V_X - V_{BE})\right] \Big/ \left(R_X + \left[R_5(h_{FE} + 1)\right]\right)$$

$$V_{BE} \approx .06\left[\log\left(10^{13} I_C\right)\right]$$

$$V_{BE} \approx .6$$

$$V_X = V_{CC}\Big/\left[(R_2/R_4) + 1\right]$$

$$R_X = \left(R_2^{-1} + R_4^{-1}\right)^{-1}$$

$$I_B = I_C/h_{FE}$$

$$I_E = I_C + I_B$$

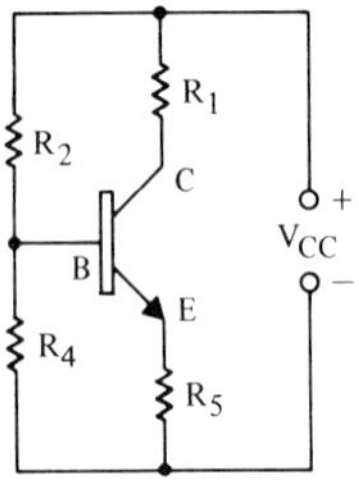

Applicable Notes: ① ② ③ ④ ⑤ ⑥ ⑦

Static (DC) Transistor Currents $I_B \quad I_C \quad I_E$	Applicable Notes
$$I_C = \frac{[V_{CC} - (V_{BE}R_X)]}{\left(R_1 + R_6 + [(R_1 + R_3 + R_6)/h_{FE}]\right)}$$ $$V_{BE} \approx .06\,[\log(10^{13}I_C)] \approx .6$$ $$R_X = [(R_1 + R_3 + R_6)/R_4] + 1$$ $$I_B = I_C/h_{FE}$$ $$I_E = I_C + I_B$$ $$I_{CC} = I_E + (V_{BE}/R_4)$$ 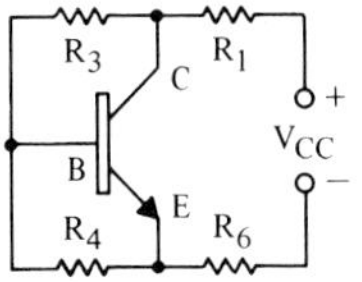	① ② ③ ④ ⑤ ⑥ ⑦
$$I_C = [V_{CC} - (V_{BE}R_{X1})]/[R_{X2} + (R_{X3}/h_{FE})]$$ $$V_{BE} \approx .06\,[\log(10^{13}I_C)] \approx .6$$ $$R_{X1} = [(R_1 + R_3)/R_4] + 1$$ $$R_{X2} = R_1 + R_5 + [(R_1R_5 + R_3R_5)/R_4]$$ $$R_{X3} = R_{X2} + R_3$$ $$I_B = I_C/h_{FE}$$ $$I_E = I_C + I_B$$ $$I_{CC} = I_E + [(I_ER_5 + V_{BE})/R_4]$$ 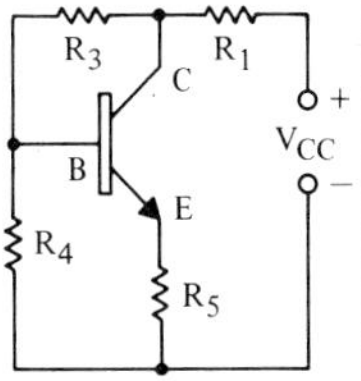	① ② ③ ④ ⑤ ⑥ ⑦

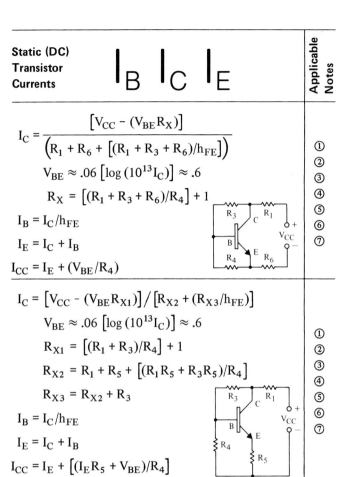

I *Notes:*

⑥ "Exact formulas" apply to silicon, germanium, npn, pnp, small signal and power transistors. (Exact formulas are not really exact since h_{FE} will vary somewhat with collector current, collector voltage and temperature.)

⑦ The V_{BE} of silicon transistors varies with temperature at the rate of approximately –2.2 mV per °C.

Static (DC) Transistor Currents

I_B I_C I_E

$$I_C = [V_{CC} - (V_{BE} R_{X1})] / [R_{X2} + (R_{X3}/h_{FE})]$$

$$V_{BE} \approx .06\,[\log(10^{13} I_C)] \approx .6$$

$$R_{X1} = (R_1 G_4) + (R_3 G_4) + 1$$

$$R_{X2} = (R_1 R_5 G_4) + (R_3 R_5 G_4) + R_1 + R_5$$

$$R_{X3} = R_{X2} + R_3$$

$$R_1 = R_{1A} + R_{1B}$$

$$G_4 = 1/R_4$$

$$I_B = I_C / h_{FE}$$

$$I_E = I_C + I_B$$

$$I_{CC} = I_E + [(I_E R_5 + V_{BE})/R_4]$$

$$V_C = V_{CC} - (I_{CC} R_{1A}) - (I_C R_{11})$$

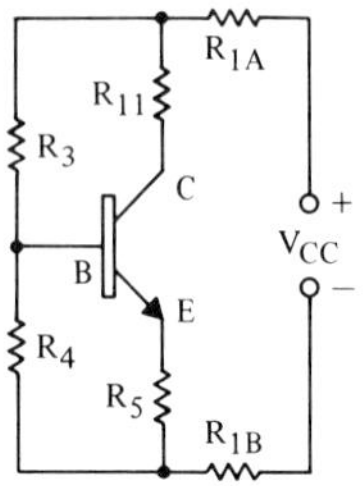

Page Notes:

1. R_{1A}, R_{1B}, G_4, R_5 and/or R_{11} may equal zero.
2. R_4 must be manually converted to G_4 since conventional mathematics and calculators will not allow division by zero or infinity.
3. R_4 may equal infinity. When $R_4 = \infty$, $G_4 = 0$.
4. Reverse power supply polarity and emitter arrow for pnp transistors.
5. The effect of varying collector voltage upon collector current has been assumed to be negligible.

LV_{CEO} – See–$V_{CEO(SUS)}$

LV_{CER} – See–$V_{CER(SUS)}$

LV_{CES} – See–$V_{CES(SUS)}$

LV_{CEV} – See–$V_{CEV(SUS)}$

LV_{CEX} – See–$V_{CEX(SUS)}$

n = Region of transistor where electrons are the majority carriers.

npn = Transistor type having two n regions and one p region. (positive polarity V_{CC} and V_{BB})

p = Region of transistor where holes are the majority carriers.

pnp = Transistor type having two p regions and one n region. (negative polarity V_{CC} and V_{BB})

P_C = Collector power dissipation

$P_C = V_{CE}I_C$

P_D = Device power dissipation. See–P_T

P_T = Total power dissipation of transistor.

$P_T = (V_{CE}I_C) + (V_{BE}I_B)$

r_B = T equivalent static internal series base resistance.

r_C = T equivalent static internal series collector resistance.

r_E = T equivalent static internal series emitter resistance.

$r_{CE(SAT)}$ = Collector to emitter saturation resistance.

R_θ

Thermal Resistance

R_θ = Symbol for thermal resistance.
(old symbol was θ)

R_θ = 1. The opposition to the transfer of thermal energy which develops an increase in temperature at the thermal energy source.
2. The ratio of temperature rise in degrees Celsius to the power dissipated in watts.

$R_{\theta CA}$ = Case to ambient (usually air) thermal resistance.
(formerly θ_{C-A} or θ_{CA})

$R_{\theta CS}$ = Case to (heat) sink thermal resistance.
(formerly θ_{C-S} or θ_{CS})

$R_{\theta JA}$ = Junction to ambient (usually air) thermal resistance.
(formerly θ_{J-A} or θ_{JA})

$R_{\theta JC}$ = Junction to case thermal resistance.
(formerly θ_{J-C} or θ_{JC})

$R_{\theta JT}$ = Junction to tab thermal resistance.
(formerly θ_{J-T} or θ_{JT})

$R_{\theta SA}$ = (Heat) sink to ambient (usually air) thermal resistance.
(formerly θ_{S-A} or θ_{SA})

$R_{\theta TS}$ = Tab to (heat) sink thermal resistance.
(formerly θ_{T-S} or θ_{TS})

R_θ = Thermal resistance expressed in °C per watt.

$R_{\theta xy} = (T_y - T_x)/P$

T = Temperature in °C
P = Power in watts

$R_{\theta JA} \approx (T_J - T_A)/(V_{CE} I_C)$

$R_{\theta JA} = (T_J - T_A)/\left[(V_{CE} I_C) + (V_{BE} I_B)\right]$

$R_{\theta JA} = R_{\theta JC} + R_{\theta CS} + R_{\theta SA}$

R to T

DC or Static Definitions

R_B = External series base resistance.

R_C = External series collector resistance.

R_E = External series emitter resistance.

R_L = Load resistance.

R_S = Source resistance.

R_{BC} – See–R_{CB}

R_{BE} = External base to emitter resistance.

R_{CB} = External collector to base resistance.

R_{CE} = External collector to emitter resistance.

R_{EB} – See–R_{BE}

R_{EC} – See–R_{CE}

T_A = Ambient temperature.

T_C = Case temperature. (T_C meaning "temperature in °C" is not used for semiconductors since temperature is given in °C unless noted.)

T_J = Junction temperature.

$T_J = T_A + (P_t R_{\theta JA})$

$T_J = T_A + \left[P_t(R_{\theta SA} + R_{\theta CS} + R_{\theta JC})\right]$

T_L = Lead temperature.

T_S = (Heat) sink temperature.

T_T = Tab temperature.

T_{STG} = Storage temperature.

V

DC Transistor Voltage Symbol Definitions

V_B = Base voltage.

V_{BB} = Base supply voltage.

V_{BC} – See–V_{CB}

V_{BE} = Base to emitter forward bias voltage

$V_{BE(ON)}$ = Base to emitter forward bias voltage with normal collector to base reverse bias voltage.

$$V_{BE(ON)} \approx \left(.06\left[\log\left(10^{13} I_C\right)\right]\right) - \left[.0022\left(T_J - 27\right)\right]$$

$V_{BE(SAT)}$ = Base to emitter forward bias voltage with collector in saturation. (typically, saturation occurs when the collector to base junction becomes forward biased)

$V_{(BR)CBO}$ = Collector to base breakdown voltage with emitter open-circuited.

$V_{(BR)CEO}$ – See–$V_{CEO(SUS)}$

$V_{(BR)CER}$ – See–$V_{CER(SUS)}$

$V_{(BR)CES}$ – See–$V_{CES(SUS)}$

$V_{(BR)CEV}$ – See–$V_{CEV(SUS)}$

$V_{(BR)CEX}$ – See–$V_{CEX(SUS)}$

$V_{(BR)EBO}$ = Emitter to base breakdown voltage with collector open-circuited.

V_C = Collector voltage.

V_{CB} = Collector to base voltage.

V_{CBO} = See–$V_{(BR)CBO}$

V_{CC} = Collector supply voltage.

V_{CE} = Collector to emitter voltage.

V

DC Transistor Voltage Symbol Definitions

V_{CEO} – See–$V_{CEO(SUS)}$

$V_{CEO(SUS)}$ = Collector to emitter sustaining voltage with base open.

V_{CER} – See–$V_{CER(SUS)}$

$V_{CER(SUS)}$ = Collector to emitter sustaining voltage with specified base to emitter resistance.

V_{CES} – See–$V_{CES(SUS)}$

$V_{CES(SUS)}$ = Collector to emitter sustaining voltage with base to emitter short-circuit.

V_{CEV} – See–$V_{CEV(SUS)}$

$V_{CEV(SUS)}$ = Collector to emitter sustaining voltage with specified base to emitter voltage.

V_{CEX} – See–$V_{CEX(SUS)}$

$V_{CEX(SUS)}$ = Collector to emitter sustaining voltage with specified base to emitter circuit.

V_E = Emitter voltage.

V_{EB} = Emitter to base reverse bias voltage.

V_{EBO} – See–$V_{(BR)EBO}$

V_{EE} = Emitter supply voltage.

V_{RT} = Reach through voltage (certain old transistors only).

Note:

Collector to emitter breakdown voltage of almost all present production transistors is measured at a current above the negative resistance region where the voltage is sustained over a wide range of current and is therefore called sustaining voltage.

DC Transistor Voltages V_B V_C V_E

Applicable Notes

$V_E = 0$

$V_B = V_{BE} \approx .6$

$V_{BE} \approx .06 \left[\log(10^{13} I_C)\right]$

$V_C = V_{CC} - \left[h_{FE}(V_{CC} - V_{BE})(R_1/R_2)\right]$

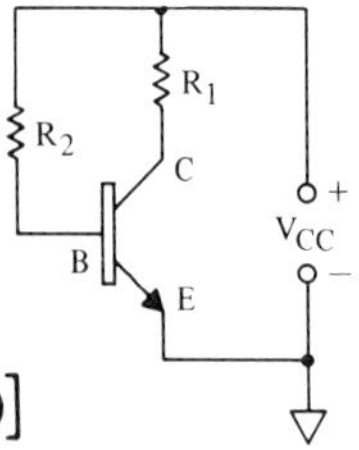

I-① I-② I-③ I-④ I-⑤ I-⑥ I-⑦

$V_E = 0$

$V_B = V_{BE} \approx .6$

$V_{BE} \approx .06 \left[\log(10^{13} I_C)\right]$

$V_C \approx V_{CC} \Big/ \left(\left[h_{FE}(R_1/R_3)\right] + 1\right)$

$V_C = \left[(V_{CC} - V_{BE}) \Big/ \left(\left[(h_{FE} + 1)(R_1/R_3)\right] + 1\right)\right] + V_{BE}$

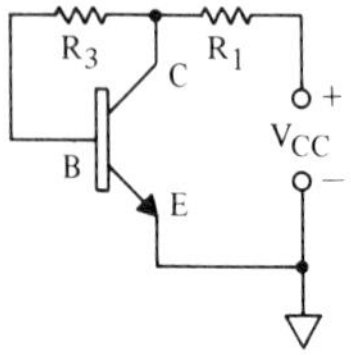

I-① I-② I-③ I-④ I-⑤ I-⑥ I-⑦

$V_E = [V_{CC} - V_{BE}] \Big/ \left[\left(R_2/\left[R_5(h_{FE} + 1)\right]\right) + 1\right]$

$V_C = V_{CC} - R_1 h_{FE}\left(\left[V_{CC} - V_{BE}\right]/\left[R_2 + R_5(h_{FE} + 1)\right]\right)$

$V_B = V_E + V_{BE}$

$V_{BE} \approx .06 \left[\log(10^{13} I_c)\right]$

$V_{BE} \approx .6$

$V_C \approx V_{CC} - \left(\left[R_1 V_{CC}\right]/\left[R_5 + (R_1/h_{FE})\right]\right)$

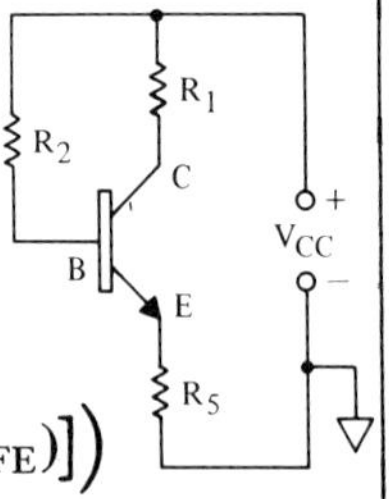

I-① I-② I-③ I-④ I-⑤ I-⑥ I-⑦

DC Transistor Voltages V_B V_C V_E	Applicable Notes

$$V_C = V_{CC} - \left(\left[R_1 h_{FE}(V_X - V_{BE})\right] / \left[R_X + R_5(h_{FE} + 1)\right]\right)$$

$$V_E = [V_X - V_{BE}] / \left(R_X / \left[R_5(h_{FE} + 1)\right]\right)$$

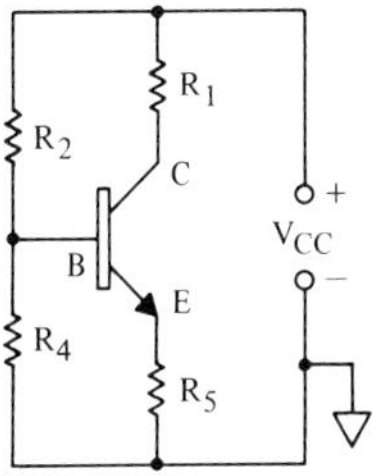

$$V_B = V_E + V_{BE} \approx V_E + .6$$

$$V_{BE} \approx .06\left[\log\left(10^{13} I_C\right)\right] \approx .6$$

$$V_X = V_{CC} / \left[(R_2/R_4) + 1\right]$$

$$R_X = \left[R_2^{-1} + R_4^{-1}\right]^{-1}$$

Applicable Notes: I-① I-② I-③ I-④ I-⑤ I-⑥ I-⑦

$$V_C = V_{CC} - (I_{CC} R_1)$$

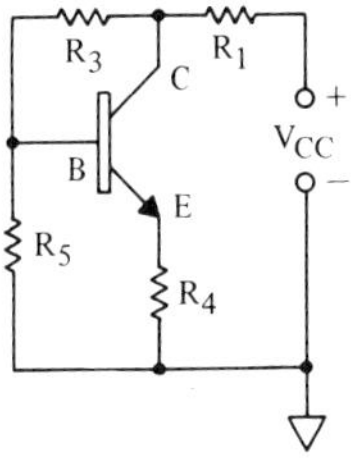

$$V_E = I_E R_5$$

$$V_B = V_E + V_{BE} \approx V_E + .6$$

$$V_{BE} \approx .06\left[\log\left(10^{13} I_C\right)\right] \approx .6$$

$$I_B = I_C / h_{FE}$$

$$I_E = \left[I_C(h_{FE} + 1)\right] / h_{FE} \approx I_C$$

$$I_{CC} = I_E + \left[(I_E R_5 + V_{BE})/R_4\right]$$

$$I_C = \left[V_{CC} - (V_{BE} R_{X1})\right] / \left[R_{X2} + (R_{X3}/h_{FE})\right]$$

$$R_{X1} = \left[(R_1 + R_3)/R_4\right] + 1$$

$$R_{X2} = R_1 + R_5 + \left[(R_1 R_5 + R_3 R_5)/R_4\right]$$

$$R_{X3} = R_{X2} + R_3$$

Applicable Notes: I-① I-② I-③ I-④ I-⑤ I-⑥ I-⑦

α to θ

Static (DC) Definitions & Formulas

α = Greek script letter alpha.

$\overline{\alpha}$ = Static (DC) alpha.

Note: Although "DC alpha" is still verbalized, the equivalent hybrid parameter symbol h_{FB} has almost completely superceeded $\overline{\alpha}$ as the accepted written symbol. See–h_{FB}

$\overline{\alpha} = h_{FB}$

$\overline{\alpha}$ = Common base static forward current transfer ratio. See–h_{FB}

Note: $\overline{\alpha}$ and h_{FB} are seldom used with modern transistors since specifications are in the common emitter form h_{FE}.

$\overline{\alpha} = I_C/I_E = h_{FE}/(h_{FE} + 1)$

$\overline{\alpha} = (h_{FE} I_B)/I_E = I_C/\left[I_B(h_{FE} + 1)\right]$

β = Greek script letter beta.

$\overline{\beta}$ = Static (DC) beta.

Note: DC beta is often verbalized, but the equivalent hybrid parameter symbol h_{FE} is used on all specifications and most other written or printed usage. See–h_{FE}

$\overline{\beta} = h_{FE}$

$\overline{\beta}$ = Common emitter static forward current transfer ratio at specified I_C, V_{CE} and T_J. See–h_{FE}

$\overline{\beta} = I_C/I_B = (I_E/I_B) - 1 = \left[(I_E/I_C) - 1\right]^{-1}$

$\overline{\beta} = [\overline{\alpha}^{-1} - 1]^{-1} = [h_{FB}^{-1} - 1]^{-1}$

θ = Greek letter theta = Obsolete symbol for thermal resistance. See–R_θ

TRANSISTORS

SECTION 2.2

Small Signal Conditions

a to C Small-Signal Low Frequency Definitions

a = Substitute symbol for α (not recommended)

A_i = Small-signal current amplification. (small-signal current gain)

A_i = The ratio of output current to input current

$A_i = \alpha = h_{fb}$ when circuit is common base with output ac shorted.

$A_i = \beta = h_{fe}$ when circuit is common emitter with output ac shorted.

A_v = Small-signal voltage amplification. (small-signal voltage gain)

A_v = The ratio of output voltage to input voltage

C_c = Collector to case capacitance.

$C_{b'c}$ = Collector to base feedback capacitance.

C_{cb} = Collector to base feedback capacitance.

C_{ob} – See–C_{obo}

C_{oe} – See–C_{oeo}

C_{ibo} = Common base open-circuit input capacitance.

C_{ieo} = Common emitter open-circuit input capacitance.

C_{obo} = Common base open-circuit output capacitance.

C_{oeo} = Common emitter open-circuit output capacitance.

Small-Signal Low-Frequency Common Base

A_i Current Amplification

Applicable Notes

$A_i = i_o/i_g$

$A_i = i_c/i_e$

$A_i = \alpha$ (alpha)

$A_i = h_{fb}$

$A_i = h_{fe}/(h_{fe} + 1)$

$A_i \simeq 1$

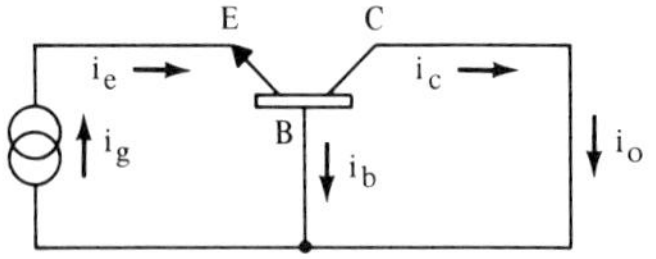

Applicable Notes: ① ③ ④ ⑤ ⑥

$A_i = i_o/i_g$

$A_i = i_c/i_e$

$A_i \approx 1$

$A_i \simeq \left[h_{fe}^{-1}(h_{oe}R_L + 1) + 1\right]^{-1}$

(accuracy typically > 4 digits)

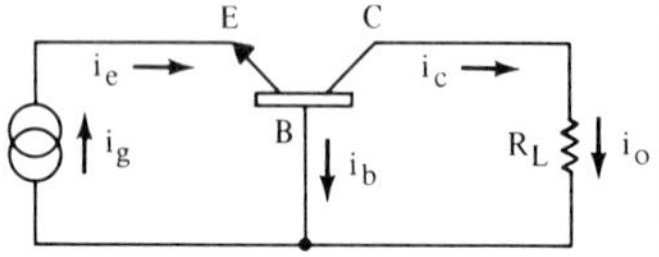

Applicable Notes: ① ③ ④ ⑤ ⑥

$A_i = i_o/i_g$

$A_i = i_c/i_e$

$A_i \approx 1$

$A_i \simeq \left[h_{fc}^{-1}(h_{oe}R_L + 1) + 1\right]^{-1}$

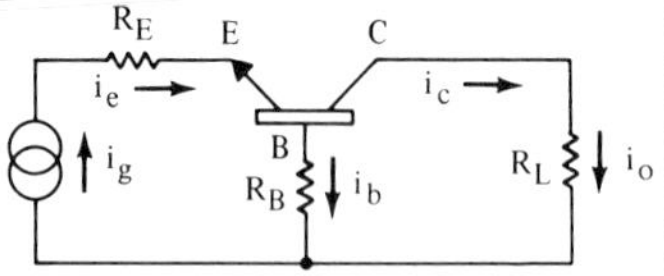

Applicable Notes: ① ③ ④ ⑤ ⑥

A *Notes:*

① —ↀ— is the graphic symbol for an alternating current generator (infinite impedance) or any very high impedance signal source.

Small-Signal Low Frequency Common Collector

A_i Current Amplification

Applicable Notes

$A_i = i_o/i_g = i_e/i_b$

$A_i = h_{fc}$

$A_i = h_{fe} + 1$

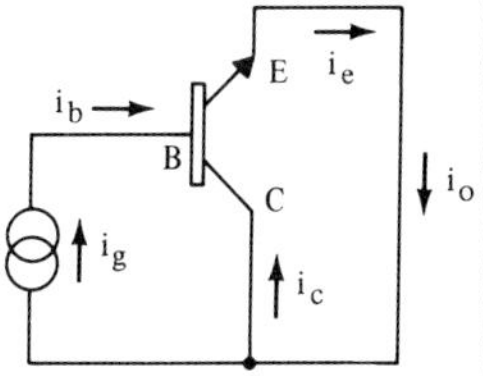

Applicable Notes: ① ③ ④ ⑤ ⑥

$A_i = i_o/i_g = i_e/i_b$

$A_i \approx h_{fe}$

$A_i = \left[h_{fe} + 1\right] / \left[(h_{oe}R_L) + 1\right]$

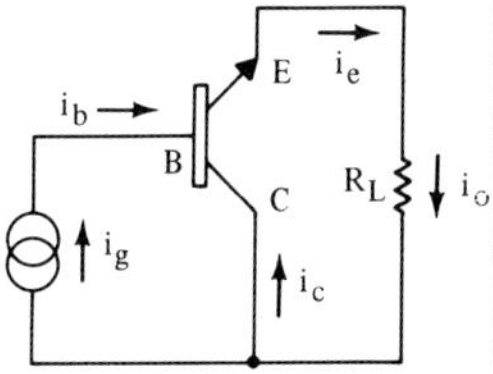

Applicable Notes: ① ③ ④ ⑤ ⑥

$A_i = i_o/i_g = i_e/i_b$

$A_i \approx h_{fe}$

$A_i \simeq h_{fe}\left[h_{oe}(R_L + R_C) + 1\right]$

$A_i = \left(\left[h_{fe} - (h_{oe}R_L)\right] / \left[h_{oe}(R_L + R_C) + 1\right]\right) + 1$

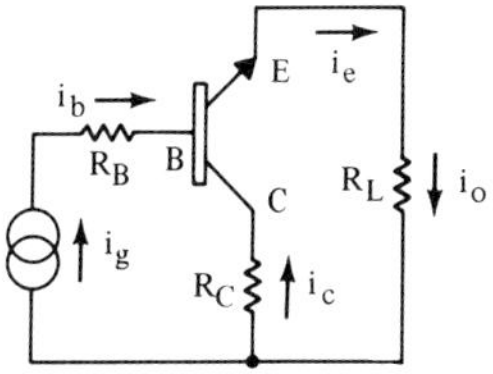

Applicable Notes: ① ③ ④ ⑤ ⑥

A *Notes:*

② —⊖— is the graphic symbol for an ac voltage generator (zero impedance) or any very low impedance signal source.

③ Formulas apply to silicon, germanium, npn and pnp bipolar transistors. Emitter arrows and the power supply polarity (if shown) must be reversed for pnp transistors.

④ Small-signal parameters will vary with temperature as well as with dc bias currents and voltages.

⑤ Small-signal parameters if specified by the manufacturer seldom have maximum or minimum limits and may vary widely. The relationships of parameters, however, will hold very closely to the formulas.

Small-Signal Low Frequency Common Emitter

A_i Current Amplification

Applicable Notes

$$A_i = i_o/i_g = i_c/i_b$$

$$A_i = \beta \text{ (beta)}$$

$$A_i = h_{fe}$$

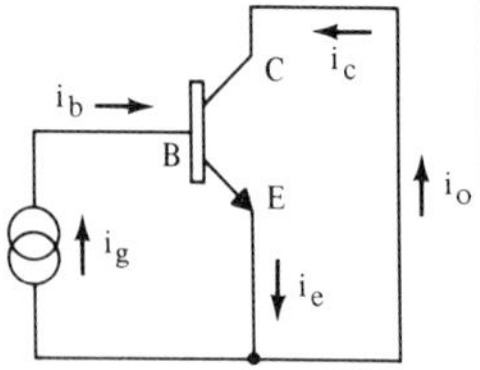

Applicable Notes: ① ③ ④ ⑤ ⑥

$$A_i = i_o/i_g = i_c/i_b$$

$$A_i \approx h_{fe}$$

$$A_i = h_{fe}/\left[h_{oe}R_L + 1\right]$$

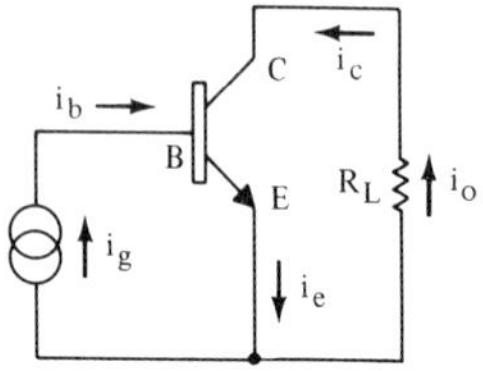

Applicable Notes: ① ③ ④ ⑤ ⑥

$$A_i = i_o/i_g = i_c/i_b$$

$$A_i \approx h_{fe}$$

$$A_i \simeq h_{fe}/\left[h_{oe}(R_L + R_E) + 1\right]$$

$$A_i = \left[h_{fe} - h_{oe}R_E\right]/\left[h_{oe}(R_L + R_E) + 1\right]$$

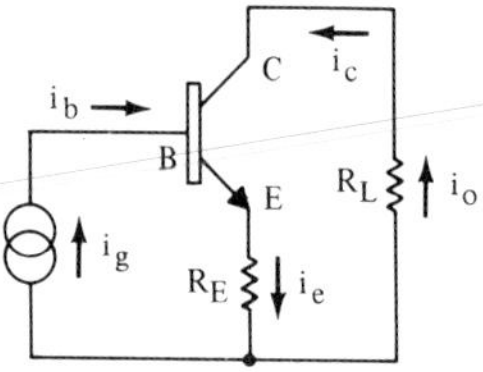

Applicable Notes: ① ③ ④ ⑤ ⑥

$$A_i = i_o/i_g$$

$$A_i = i_c/i_b$$

$$A_i \approx \left[(R_L/R_F) + h_{fe}^{-1}\right]^{-1}$$

$$A_i \simeq h_{fe}\left[R_L(R_F^{-1} h_{fe} + R_F^{-1} + h_{oe}) + 1\right]^{-1}$$

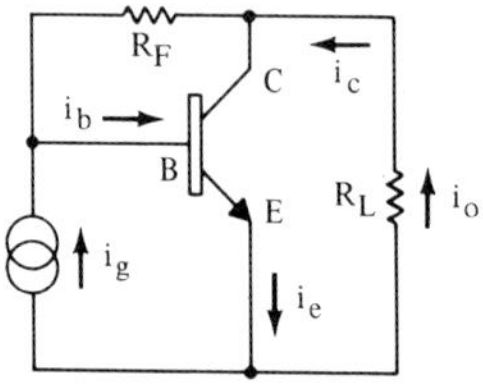

Applicable Notes: ① ③ ④ ⑤ ⑥

Small-Signal Low Frequency Common Base | A_V | Voltage Amplification | Applicable Notes

$A_v = e_o/e_g = e_c/e_e$

$A_v \approx 37 I_C R_L$

$A_v \approx (h_{fe} R_L)/h_{ie}$

$A_v \simeq \left[h_{ie} h_{fe}^{-1}(R_L^{-1} + h_{oe}) - h_{re}\right]^{-1}$

$$A_v = \frac{\left[h_{ie} h_{fe}^{-1} h_{oe} - h_{re} + 1\right]}{\left[h_{ie} h_{fe}^{-1}(R_L^{-1} + h_{oe}) - h_{re}\right]}$$

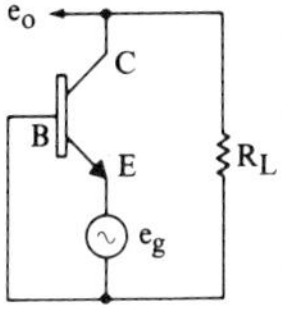

Applicable Notes: ② ③ ④ ⑤ ⑥

$A_v = e_o/e_g$

$A_v = e_c/e_e$

$A_v \approx (h_{fe} R_L)/(h_{ie} + R_B)$

$A_v \simeq \left[h_{fe}^{-1}(h_{ie} + R_B)(R_L^{-1} + h_{oe}) - h_{re}\right]^{-1}$

$$A_v = \frac{\left[(h_{ie} + R_B) h_{fe}^{-1} h_{oe} - h_{re} + 1\right]}{\left[h_{fe}^{-1}(h_{ie} + R_B)(R_L^{-1} + h_{oe}) - h_{re}\right]}$$

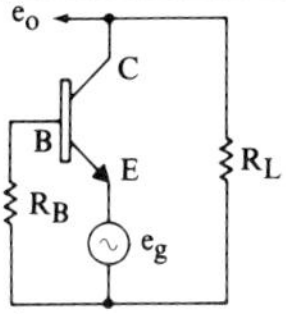

Applicable Notes: ② ③ ④ ⑤ ⑥

$A_v = e_o/e_g$

$A_v \approx R_L/\left[R_E + h_{fe}^{-1}(h_{ie} + R_B)\right]$

$A_v \simeq \left[h_{fe}^{-1}(h_{ie} + R_B + h_{fe} R_E)(R_L^{-1} + h_{oe}) - h_{re}\right]^{-1}$

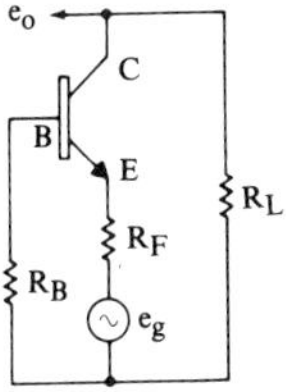

Applicable Notes: ② ③ ④ ⑤ ⑥

Small-Signal Low Frequency Common Collector — A_V — **Voltage Amplification** — **Applicable Notes**

$A_v = e_o/e_g$

$A_v = e_e/e_b$

$A_v \simeq 1$

$A_v = \left[h_{ie}(h_{oe} + R_L^{-1})(h_{fe} + 1)^{-1} - h_{re} + 1\right]^{-1}$

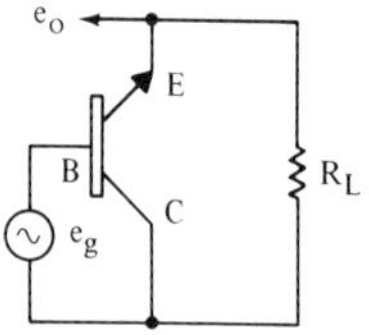

Applicable Notes: ② ③ ④ ⑤ ⑥

$A_v = e_o/e_g$

$A_v \approx 1$

$A_v \simeq 1$
when $R_g << (h_{fe} R_L)$
and $R_L >> (h_{re}/h_{oe})$

$A_v = \left(\left[(h_{ie} + R_g)(h_{oe} + R_L^{-1})(h_{fe} + 1)^{-1}\right] - h_{re} + 1\right)^{-1}$

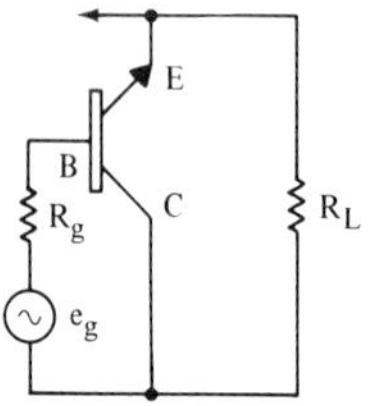

Applicable Notes: ② ③ ④ ⑤ ⑥

$A_v = e_o/e_g$

$A_v \approx 1$

$A_v = (\text{Common emitter } A_v)$

$\cdot\, R_L(h_{fe}^{-1} + 1)\, R_{L2}^{-1}$

See–Common emitter formulas

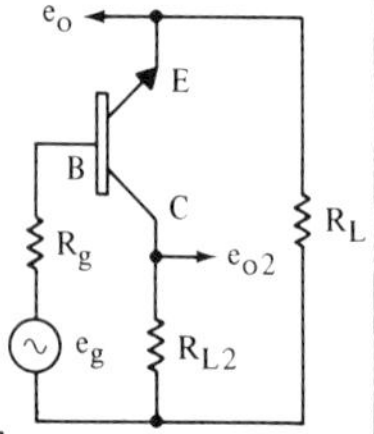

Applicable Notes: ② ③ ④ ⑤ ⑥

A *Notes:*

⑥ Most small-signal parameters are drastically different from the dc parameters due to the nonlinear nature of transistors. A diode or transistor junction which develops .6 volts from 1 mA forward current has a dc resistance of 600 ohms according to ohms law,

Small-Signal Low Frequency Common Emitter — A_V — **Voltage Amplification** — **Applicable Notes**

$$A_v = e_o/e_g$$
$$A_v \approx 37 I_C R_L$$
$$A_v \simeq (h_{fe} R_L)/h_{ie}$$
$$A_v = \left[h_{ie} h_{fe}^{-1} (R_L^{-1} + h_{oe}) - h_{re}\right]^{-1}$$

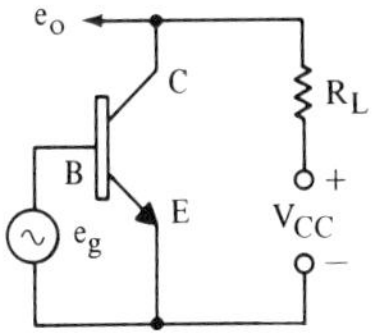

Applicable Notes: ② ③ ④ ⑤ ⑥

$$A_v = e_o/e_g$$
$$A_v \approx R_L\,[.027 I_C^{-1} + R_g h_{fe}^{-1}]^{-1}$$
$$A_v \simeq (h_{fe} R_L)/(h_{ie} + R_g)$$
$$A_v = \left[h_{fe}^{-1} (R_g + h_{ie})(R_L^{-1} + h_{oe}) - h_{re}\right]^{-1}$$

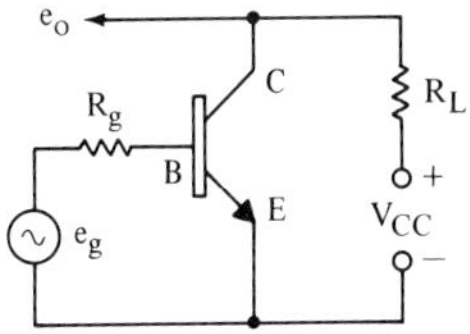

Applicable Notes: ② ③ ④ ⑤ ⑥

$$A_v = e_o/e_g$$
$$A_v \approx R_L/R_E$$
$$A_v \simeq (h_{fe} R_L)/\left[h_{ie} + (h_{fe} R_E)\right]$$
$$A_v = \left[h_{fe}^{-1} h_{ie} (R_L^{-1} + h'_{oe}) + R_E R_L^{-1} (h_{fe} + 1) + h_{fe}^{-1} h'_{oe} - h_{re}\right]^{-1}$$
$$h'_{oe} = (R_E + h_{oe}^{-1})^{-1}$$

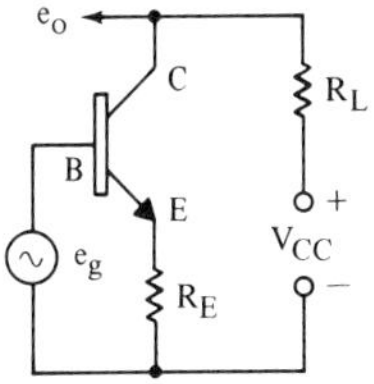

Applicable Notes: ② ③ ④ ⑤ ⑥

A *Notes:*

⑥ Cont. but if a small ac signal is superimposed upon the dc and measured, the ac resistance will be found to be about 26 ohms. This small-signal resistance (r) is often verbally expressed as impedance (z), but admittance (y) is used at frequencies where internal capacitances are significant.

Small-Signal Low Frequency Common Emitter

A_V Voltage Amplification

Applicable Notes

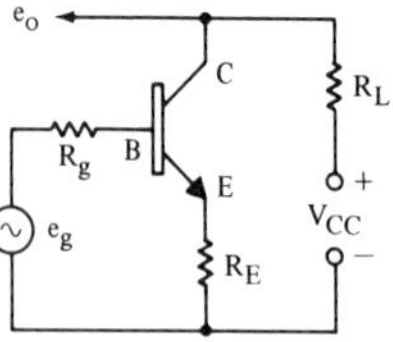

$$A_v = e_o/e_g$$

$$A_v \approx R_L \left[R_E(R_g h_{ie}^{-1} + 1)\right]^{-1}$$

$$A_v \simeq h_{fe} R_L \left[h_{ie} + R_g + h_{fe} R_E\right]^{-1}$$

$$A_v = \Big(h_{fe}^{-1} R_{BX}(R_L^{-1} + h'_{oe}) + h_{fe}^{-1} R_{EX}\left[R_L^{-1}(h_{fe} + 1) + h'_{oe}\right]\Big)^{-1}$$

$$R_{BX} = R_g + \left[h_{ie} - h_{re} h_{oe}^{-1}(h_{fe} + 1)\right]$$

$$R_{EX} = R_E + h_{re} h_{oe}^{-1}$$

$$h'_{oe} = (h_{oe}^{-1} + R_E)^{-1}$$

Applicable Notes: ② ③ ④ ⑤ ⑥

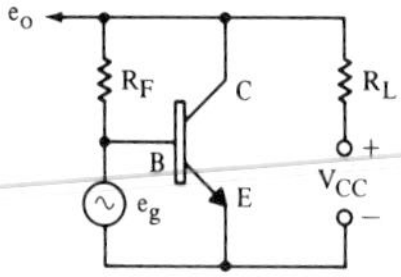

$$A_v = e_o/e_g$$

$$A_v \approx 38 I_C (R_L^{-1} + R_F^{-1})^{-1}$$

$$A_v \simeq h_{ie}^{-1}\left[h_{fe}(R_L^{-1} + R_F^{-1})^{-1}\right]$$

$$A_v = \Big(h_{ie} h_{fe}^{-1}\left[R_L^{-1} + R_F^{-1} + h_{oe}\right] - h_{re}\Big)^{-1}$$

Applicable Notes: ② ③ ④ ⑤ ⑥

$$A_v = e_o/e_g$$

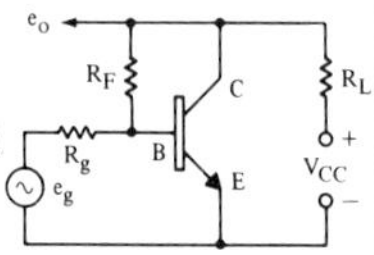

$$A_v \approx R_F/R_g$$

$$A_v = A_{vx}\Big(\left[R_g R_F^{-1}(A_{vx} + 1)\right] + 1\Big)^{-1}$$

$$A_{vx} \simeq 38 I_C (R_L^{-1} + R_F^{-1})^{-1}$$

$$A_{vx} = \Big(h_{ie} h_{fe}^{-1}\left[R_L^{-1} + R_F^{-1} + h_{oe}\right] - h_{re}\Big)^{-1}$$

Applicable Notes: ② ③ ④ ⑤ ⑥

Small-Signal Low Frequency Common Emitter	A_V	Voltage Amplification	Applicable Notes

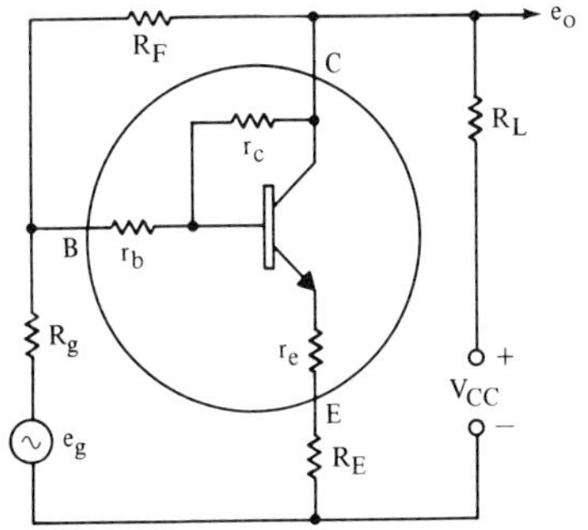

$$A_v = e_o/e_g$$

$$A_v = \Big(\big[R_{BX}h_{fe}^{-1}(R_L^{-1} + G_{CX})\big] + R_{EX}h_{fe}^{-1}\big[G_{CX} + R_L^{-1}(h_{fe} + 1)\big]\Big)^{-1}$$

$$r_e = h_{re}h_{oe}^{-1}$$

$$r_c = h_{oe}^{-1}(h_{fe} + 1)$$

$$r_b = h_{ie} - \big[h_{re}h_{oe}^{-1}(h_{fe} + 1)\big]$$

$$R_{BX} = R_g + \big[(R_F r_b)(R_F + r_b + r_c)^{-1}\big]$$

$$R_{EX} = R_E + r_e + \frac{\big[(r_b + r_c)(R_F + r_b + r_c)^{-1}\big]}{(h_{fe} + 1)}$$

$$R_{CX} = (R_F r_c)(R_F + r_b + r_c)^{-1}$$

$$G_{CX} = \Big(\big[R_{CX}(h_{fe} + 1)^{-1}\big] + R_{EX}\Big)^{-1}$$

Applicable Notes: ② ③ ④ ⑤ ⑥

e

Small-Signal Voltage Definitions

e = Symbol for emitter. (small signal subscript)

e = Small-signal voltage. (rms or instantaneous)

e_b = Small-signal base voltage.

e_c = Small-signal collector voltage.

e_e = Small-signal emitter voltage.

e_g = Generator voltage.

e_i = Input voltage.

e_N = Noise voltage (rms).

e_N = Thermal noise voltage or equivalent input total transistor noise voltage.

$e_{N(\sqrt{Hz})}$ = Noise voltage per root hertz. (BW = 1 Hz or $e_N/\sqrt{BW}$ for white noise)

$e_{N(s)}$ = Transistor shot noise (white noise) voltage.

$e_{N(th)}$ = Thermal noise voltage. (white noise voltage of an ideal resistance at a specified temperature)

$e_{N(TR)}$ = Transistor noise voltage.

$e_{N(1/f)}$ = 1/f noise voltage of a transistor. (Resistor 1/f noise is known as excess or current noise)

e_o = Output voltage.

e_p = Peak voltage.

e_s – See–$e_{N(s)}$.

e_t = Total or equivalent voltage.

$e_{1/f}$ – See–$e_{N(1/f)}$.

e_b e_c Small-Signal Voltages		Applicable Notes
$e_b = i_g h_{ie}$ $e_c = 0$	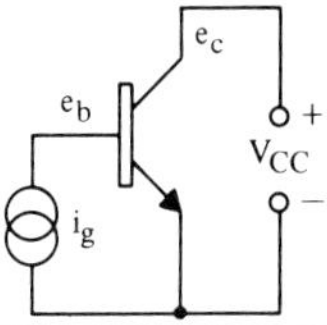	① ② ③ ④ ⑤
$e_b \simeq i_g h_{ie}$ when $A_v < 50$ $e_b = i_g Z_i$ See—Z_i $e_c \simeq -h_{fe} i_g R_L$ $e_c = -A_v i_g Z_i$ See—A_v, Z_i	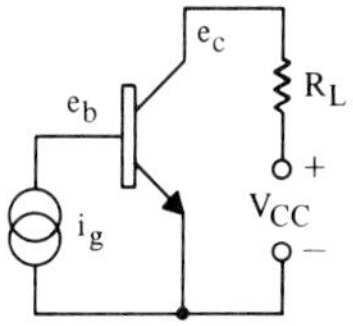	① ② ③ ④ ⑤
$e_b = e_g$ $e_c \simeq -37 I_C e_g R_L$ when $A_v < 50$ $e_c = -e_g A_v$ See—A_v	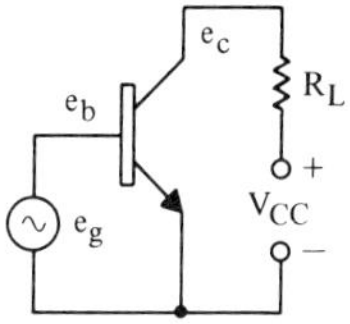	② ③ ④ ⑤ ⑥
$e_b \simeq e_g / [(R_S/h_{ie}) + 1]$ $e_b = e_g / [(R_S/Z_i) + 1]$ $e_c \simeq -(e_g h_{fe} R_L)/(R_S + h_{ie}^{-1})$ when $A_v < 50$ $e_c = -(e_g A_i R_L)/(R_S + Z_i^{-1})$ See—A_i, Z_i	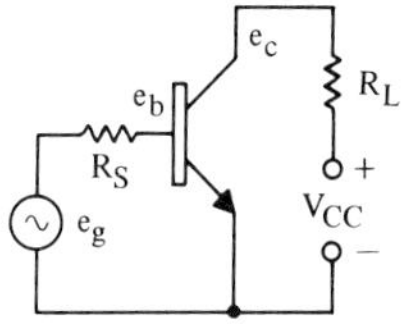	② ③ ④ ⑤

e *Notes:*

① —ꝏ— is the graphic symbol for an alternating current generator. (an infinite impedance signal source)

② Transistors must be biased into an active region.

③ Approximations apply to high beta silicon small-signal transistors while exact formulas apply to all bipolar transistors.

e_b e_c e_e Small-Signal Voltages

Applicable Notes

$e_b = e_g$

$e_c \simeq -(e_g R_L)/R_E$ when $R_E >> (37 I_C)^{-1}$

$e_c = -e_g A_v$ See—A_v

$e_e \simeq e_b$ when $R_E >> (37 I_C)^{-1}$

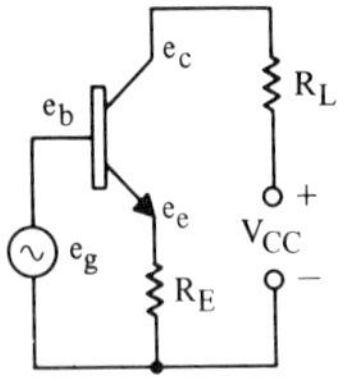

Applicable Notes: ② ③ ④ ⑤ ⑥

$e_b \simeq e_g / \left[R_S(R_B^{-1} + h_{ie}^{-1}) + 1\right]$

$e_b = e_g / \left[R_S(R_B^{-1} + Z_i^{-1}) + 1\right]$

$e_c \simeq -37 I_C\, e_b R_L$

$e_c = -e_b A_v$ See—A_v, Z_i

$e_e = 0$

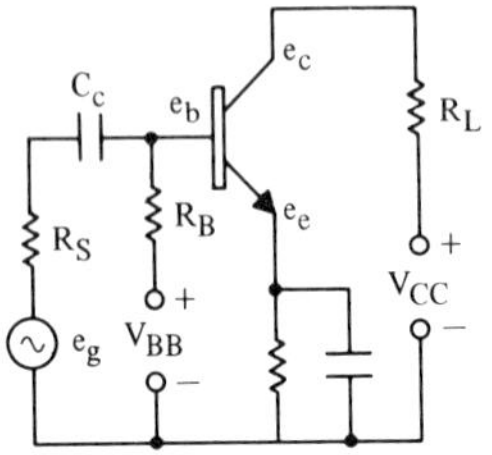

Applicable Notes: ② ③ ④ ⑤

$e_b = e_g$

$e_c \simeq -(e_g h_{fe}) \left[h_{ie}(R_C^{-1} + R_L^{-1})\right]^{-1}$

$e_c = -(e_g A_i) \left[Z_i(R_C^{-1} + R_L^{-1})\right]^{-1}$

$e_e = 0$

See—A_i, Z_i

$e_o = e_c$

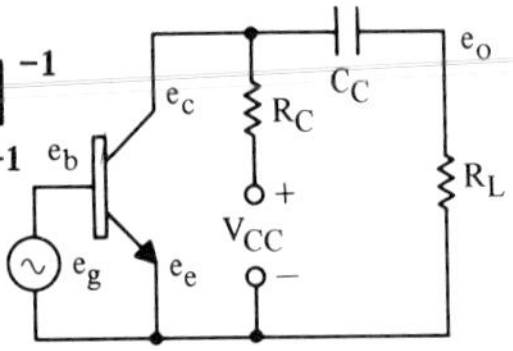

Applicable Notes: ② ③ ④ ⑤ ⑥

e *Notes:*

④ Formulas apply to pnp transistors as well as the npn transistors shown. Reverse emitter arrow and power supply polarity for pnp transistors.

⑤ Transistor parameters will vary with collector voltage and temperature as well as collector current.

⑥ The resistance of base bias resistors must be included in all calculations where the generator source resistance is of significance.

e_N Noise Voltage

$$e_{N(th)} = \sqrt{4K_B T_K R_S \overline{BW}}$$ Thermal noise voltage

$$\left(e_{N(\sqrt{Hz})}\right)_{th} = \sqrt{4K_B T_K R_S}$$ Thermal noise voltage per root hertz

$$e_{N(s)} = r_e\sqrt{2qI_B\overline{BW}}$$ Transistor shot noise (white noise)

$$\left(e_{N(\sqrt{Hz})}\right)_s = r_e\sqrt{2qI_B}$$ Transistor shot noise per root hertz

$$e_{N(1/f)} = \sqrt{(e_N)^2_{TR} - (e_N)^2_s}$$ Both same bandwidth

$$(e_N)_{TR} = \sqrt{(e_N)^2_s + (e_N)^2_{1/f}}$$ Both same bandwidth

Total Equivalent Input Noise Voltage

$$(e_N)_t = \left(\left[R_S(i_N)_{TR}\right]^2 + (e_N)^2_{TR} + (e_N)^2_{th}\right)^{\frac{1}{2}}$$

(all same BW)

Wideband Total Noise Voltage Output

$$e_{N(OUT)} = A_v\left(\left[R_S(i_N)_{TR}\right]^2 + (e_N)^2_{TR} + (e_N)^2_{th}\right)^{\frac{1}{2}}$$

(all same BW)

Total Spot Noise Voltage Output

$$e_{N(OUT)} = A_v\left(\left[R_S(i_N)_{TR}\right]^2 + (e_N)^2_{TR} + 4K_S T_K R_S\right)^{\frac{1}{2}}$$

all noise terms are for 1 Hz BW at same frequency

e_N *Notes:* K_B = Boltzmann constant ($1.38 \cdot 10^{-23}$ J/K); T_K = Kelvin temperature (°C + 273.15); q = Charge of electron ($1.6 \cdot 10^{-19}$); BW = Bandwidth, See–Opamp, BW_{NOISE}; r_e = Internal transistor dynamic emitter resistance; $r_e \approx .027/I_C$; $(i_N)_{TR}$ = Transistor noise current equivalent input); A_v = stage voltage amplification; I_B, I_C = dc base and collector currents.

F to g_{me}

Small-Signal Definitions

F – See–NF, See also–F, Passive Circuits

F_N – See–NF

f_c = Symbol for cutoff frequency. (The half power or 3 dB down frequency)

See–f_c, Opamps

f_T = Gain-bandwidth product. The frequency at which the common emitter small-signal forward current transfer ratio falls to unity.
(dc biased for large signal)

f_t = Same as f_T except biased for small-signal.

$f_{\alpha b}$ – See–f_{hfb}

$f_{\alpha e}$ – See–f_{hfe}

f_{hfb} = Common base small-signal forward current transfer ratio cutoff frequency with output ac shorted.

f_{hfe} = Common emitter small-signal forward current transfer ratio cutoff frequency with output ac shorted.

G_{pb} = Common base small-signal average power gain.

G_{pe} = Common emitter small-signal average power gain.

$G_{pe(conv)}$ = Common emitter conversion gain.

g_{me} = Common emitter small-signal transconductance.

h_{fb} h_{fc} h_{fe} Small-Signal Forward Current Ratios

h_{fb} = Common base small-signal forward current transfer ratio with output ac shorted.

h_{fb} = Small signal alpha (α)

$h_{fb} = i_c/i_e$

$h_{fb} = h_{fe}/(h_{fe} + 1)$

---

h_{fc} = Common collector (emitter follower) small-signal forward current transfer ratio with output ac shorted.

$h_{fc} = i_e/i_b$

$h_{fc} = h_{fe} + 1$

---

h_{fe} = Common emitter small-signal forward current transfer ratio with output ac shorted.

h_{fe} = Small-signal beta (β).

$h_{fe} = i_c/i_b$

$h_{fe} = \alpha/(1 - \alpha)$

$h_{fe} = h_{fb}/(1 - h_{fb})$

$h_{fe} = h_{fc} - 1$

$h_{fe} = h_{fb}h_{fc}$

h_{fe} = Common emitter current gain when output is ac shorted.

$h_{fe} = (i_c h_{ie})/e_{be}$ when output is ac shorted.

h_{ib} h_{ic} h_{ie} Small-Signal Input Impedance

h_{ib} = Common base small-signal input impedance with output ac shorted.

$h_{ib} = e_e/i_e$

$h_{ib} = h_{ie}/(h_{fe} + 1)$

$h_{ib} = r_e + [r_b/(h_{fe} + 1)]$

$h_{ib} \approx 1/(37I_C)$

h_{ic} = Common collector (emitter follower) small-signal input impedance with output ac shorted.

$h_{ic} = h_{ie}$ (since emitter is ac shorted)

h_{ie} = Common emitter small-signal input impedance with output ac shorted.

$h_{ie} = e_b/i_b$

$h_{ie} = h_{ic}$

$h_{ie} = h_{ib}(h_{fe} + 1)$

$h_{ie} = r_b + r_e(h_{fe} + 1)$

$h_{ie} \approx h_{fe}/(37I_C)$

$h_{ie} \simeq (26.7h_{fe})/(1000I_C)^{.78}$

Approximations apply to small-signal silicon transistors.

h_{ob} h_{oc} h_{oe} Small-Signal Output Admittance

h_{ob} = Common base small-signal output admittance with input ac open-circuited.

$h_{ob} = i_c/e_c$ when emitter is ac open-circuited (constant current emitter supply)

$h_{ob} = y_{ob}$ (y_{ob} is generally used at high frequencies)

$h_{ob} \simeq h_{oe}/(h_{fe} + 1)$

$h_{ob} = r_c + r_b$

$h_{ob} = \left(\left[h_{oe}^{-1}(h_{fe} + 1)\right] + \left[h_{ie} - h_{re}h_{oe}^{-1}(h_{fe} + 1)\right]\right)^{-1}$

---

h_{oc} = Common collector (emitter follower) small-signal output admittance with input ac open-circuited.

$h_{oc} = h_{oe}$ (since input is open-circuited)

---

h_{oe} = Common emitter small-signal output admittance with input ac open-circuited.

$h_{oe} = i_c/e_c$ when base is ac open-circuited (constant current base supply)

$h_{oe} = y_{oe}$ (y_{oe} is generally used at high frequencies)

$h_{oe} = h_{oc}$

$h_{oe} = (h_{fe} + 1)/r_c$

$h_{oe} \approx 20\ \mu S$ $(50\ k\Omega)^{-1}$ when $I_C \approx 1$ mA, $V_{CE} > 5$ V, $T \approx 25°C$

h_{rb} h_{rc} h_{re} Small-Signal Reverse Voltage Ratios

h_{rb} = Common base small-signal reverse voltage transfer ratio with input ac open-circuited.

$h_{rb} = e_e/e_c$ when emitter is ac open-circuited (constant current emitter supply)

$$h_{rb} = r_b/(r_b + r_c)$$

$$h_{rb} \simeq \left[h_{ie}h_{oe}(h_{fe} + 1)^{-1}\right] - h_{re}$$

$$h_{rb} = \left[\left(\left[h_{ie}h_{oe}(h_{fe} + 1)^{-1}\right] - h_{re}\right)^{-1} + 1\right]^{-1}$$

---

h_{rc} = Common collector small-signal reverse voltage transfer ratio with input ac open-circuited.

$$h_{rc} = 1 - h_{re}$$

---

h_{re} = Common emitter small-signal reverse voltage transfer ratio with input ac open-circuited.

$h_{re} = e_b/e_c$ when base is ac open-circuited (constant current base supply)

$$h_{re} = r_e h_{oe}$$

$$h_{re} = \left[r_e(h_{fe} + 1)\right]/r_c$$

$h_{re} \approx 1.33 \cdot 10^{-6}\, h_{fe}$ when $I_C \approx 1$ mA and $V_{CE} > 5$ V

Note: h_{re} is very V_{CE} sensitive at low voltage. h_{re} typically is very nonlinear over large variations of I_C.

i

Small-Signal Current Definitions

I – See–I, Static (DC) Symbols
See also–I, Passive Circuits

i = Small-signal current.

i_b = Small-signal base current.

i_c = Small-signal collector current.

i_e = Small-signal emitter current.

i_g = Small-signal generator (source) current.

i_{in} = Small-signal input current

i_N = Noise current.

$i_{N(TRANSISTOR)}$ = That portion of the input equivalent internal transistor noise which is proportional to the external resistance in shunt with the input (source resistance)

i_N *Notes:*

① $i_{N(TRANSISTOR)}$ does not include the thermal noise or the excess noise currents of the effective external source resistance.

② $i_{N(TRANSISTOR)}$ may be included in the $e_{N(TRANSISTOR)(TOTAL)}$ if a source resistance (R_S) has been specified.

③ Much of the confusion regarding noise voltages and noise currents results from the difficulty in proper identification of the symbols for the various noise voltages and noise currents.

④ See also–e_N, NF; See also–V_n, i_n, Opamps; See also–E_N, I_N, N_{th}, NI, Passive Circuits

i_o = Small-signal output current.

i_p = Small-signal peak current.

i_b i_c i_e Small-Signal Transistor Currents

Applicable Notes

$i_b = i_g$

$i_c = i_g h_{fe}$

$i_e = i_g(h_{fe} + 1)$

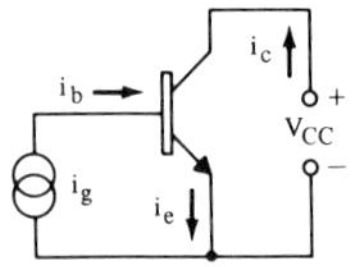

Applicable Notes: ① ② ③ ④ ⑤

$i_b = i_g$

$i_c \simeq i_g h_{fe}$

$i_c = i_g A_i$ See–A_i

$i_e \simeq i_g(h_{fe} + 1)$

$i_e = i_g(A_i + 1)$ See–A_i

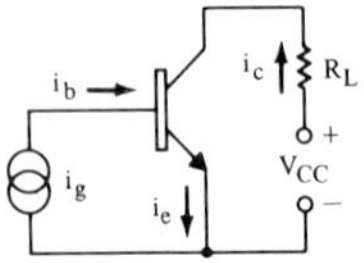

Applicable Notes: ① ② ③ ④ ⑤

$i_b \simeq e_g/h_{ie}$

$i_b = e_g/Z_i$ See–Z_i

$i_c \simeq (e_g h_{fe})/h_{ie}$

$i_c = (e_g A_v)/R_L$ See–A_v

$i_e \simeq i_c$

$i_e = \left[e_g(A_v + 1)\right]/R_L$
See–A_v

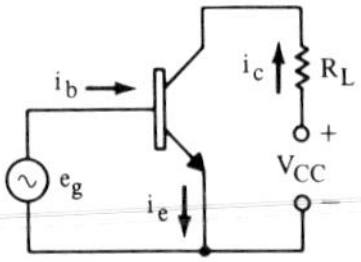

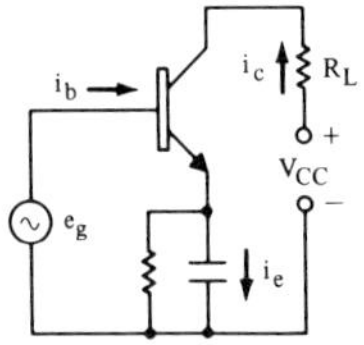

Applicable Notes: ② ③ ④ ⑤ ⑥

i *Notes:*

① –ꝏ– is the graphic symbol for an infinite impedance alternating current generator. (any very high impedance source)

② Transistors must be biased into an active region.

③ Approximations apply to high beta small-signal silicon transistors. Exact formulas apply to all bipolar transistors.

i_b i_c i_e i_o Small-Signal Currents

Applicable Notes

$i_b \simeq e_g/[(h_{fe}R_E) + h_{ie}]$

$i_b = e_g/Z_i$ See—Z_i

$i_c \simeq e_g/R_E$ when $R_E >> (37I_C)^{-1}$

$i_c = (e_g A_v)/R_L$ See—A_v

$i_e \simeq i_c$ when $h_{fe} > 100$

$i_e = [e_g(A_v + 1)]/R_L$ See—A_v

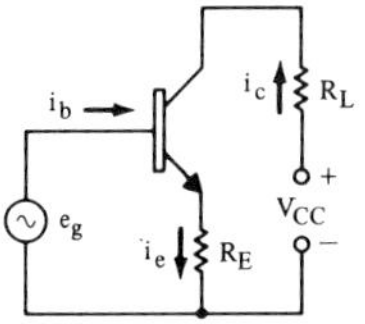

Applicable Notes: ② ③ ④ ⑤ ⑥

$i_b \simeq e_g/(R_S + h_{ie})$

$i_b = e_g/(R_S + Z_i)$

$i_c \simeq (e_g h_{fe})/(R_S + h_{ie})$

$i_c = (e_g A_i)/(R_S + Z_i)$ See—A_i, Z_i

$i_e \simeq i_c$ when $h_{fe} > 100$

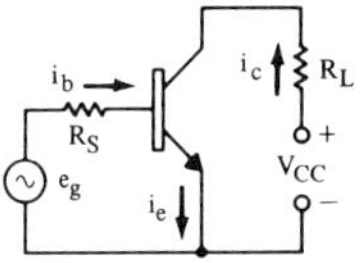

Applicable Notes: ② ③ ④ ⑤

$i_c \simeq e_g/R_E$ when $R_E >> (37I_C)^{-1}$

$i_o \simeq (e_g/R_E)\,[(R_L/R_C) + 1]^{-1}$

$i_o = i_c/[(R_L/R_C) + 1]$

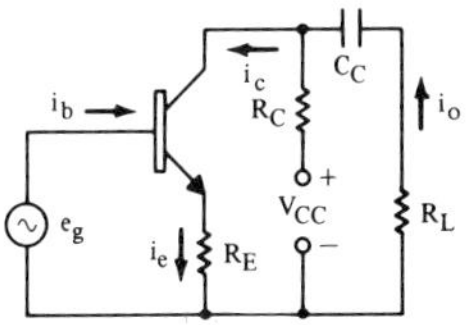

Applicable Notes: ② ③ ④ ⑤ ⑥

---

i *Notes:*

④ All formulas apply to pnp transistors as well as the npn transistors shown. Reverse emitter arrow and power supply polarity for pnp transistors.

⑤ Transistor parameters will vary with collector voltage and temperature as well as with collector current.

⑥ The resistance of base bias resistors must be included properly in all calculations where the signal source resistance is of significance.

NF

Noise Figure

NF = Symbol for noise figure (also known as noise factor) (other symbols include F and F_N)

NF = 1. The ratio (usually in decibels) of the output signal-to-noise power to the input signal-to-noise power.
2. The ratio in decibels of the total output noise to that portion of the output noise generated thermally by the input termination resistance.

$$NF = 20\left[\log\left(e_{ni}/e_{nR'}\right)\right]$$

$$NF = 20\left(\log\left[e_{no}/(e_{nR'}A_v)\right]\right)$$

where $e_{nR'} = e_{nR} \Big/ \left[(R_S/r_i) + 1\right]$

and $e_{nR} = \sqrt{4K_B T_K \overline{BW}}$

See also – $e_{N(out)}$ and BW_{NOISE}, Opamp

NF Notes:

① A_v = Voltage amplification, $\overline{BW}$ = Average bandwidth (rms bandwidth), e_{ni} = Input equivalent total noise voltage, e_{no} = Output noise voltage, e_{nR} = Thermal noise voltage of source resistance, K_B = Boltzmann constant ($1.38 \cdot 10^{-23}$ J/°K), r_i = Transistor input resistance, R_S = Source resistance (the total effective resistance presented to the transistor input), T_K = Kelvin temperature (°C + 273.15), log = Base 10 logarithm.

② The standard noise temperature (T_N) of the source resistance is 290 K (16.85°C) if unspecified.

See also – e_N Notes

V v Definitions

V = The unit symbol for volt.
See—V, Passive Circuits

V = The quantity symbol for voltage.
See—V, Static (dc) Parameters
See also—V, Opamp

Note: A definite trend exists towards the elimination of E and e as symbols for voltage. At present, E and e predominate in passive circuits, V and v predominate in operational amplifiers, V has superseded E in dc transistor parameters and e predominates for ac transistor parameters.

---

v = Symbol for small signal voltage.
See—e See also—V, Opamp

v_b – See—e_b

v_c – See—e_c

v_e – See—e_e

v_g – See—e_g

v_i – See—e_i

v_N – See—e_N (V_N—See—V, Opamp)

v_o – See—e_o

v_p – See—e_p

v_s – See—$e_{N(s)}$

v_t – See—e_t

$v_{1/f}$ – See—$e_{N(1/f)}$

Small Signal
Low Frequency
Common Base

Z_i

Input Impedance

Applicable Notes

$Z_i \approx 1/(37I_C)$

$Z_i = r_e + r_b(h_{fe} + 1)^{-1}$

$Z_i = h_{ib}$

$Z_i = h_{ie}/(h_{fe} + 1)$

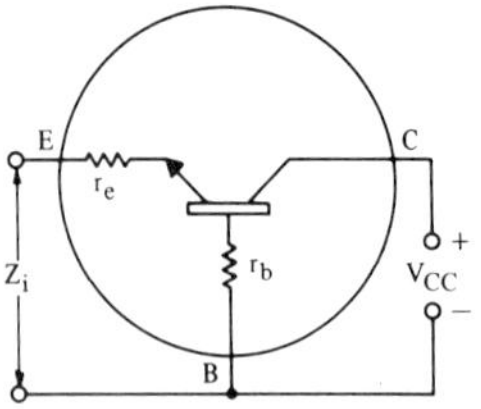

Applicable Notes: ③ ④ ⑤ ⑥ ⑨

$Z_i \approx 1/(37I_C)$
(when $A_v < 50$)

$Z_i \simeq h_{ie}/(h_{fe} + 1)$
(when $A_v > 50$)

$Z_i = (h_{oe}R_L + 1)/(h_{oe}R_L + 1 + h_{fe})$

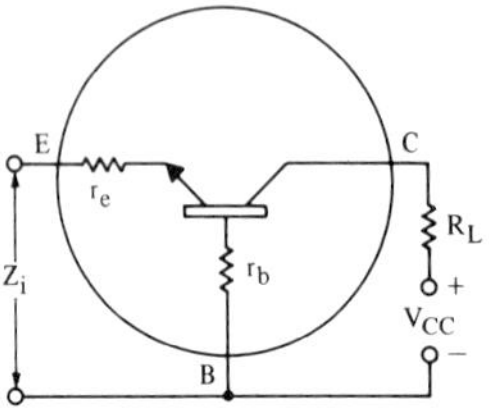

Applicable Notes: ③ ④ ⑤ ⑥ ⑧ ⑨

Small Signal
Low Frequency
Common Collector

Z_i

Input Impedance

Applicable Notes

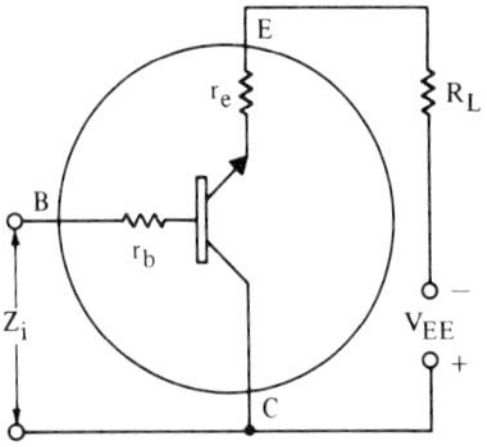

$Z_i \approx h_{fe}R_L$

$Z_i \simeq h_{ie} + R_L(h_{fe} + 1)$

$Z_i = h_{ie} + \left[(h_{fe} + 1)/(h_{oe} + R_L^{-1})\right]$

Applicable Notes: ③ ④ ⑤ ⑥ ⑧ ⑨

Z NOTES

Z Notes:

① –○○– is the graphic symbol for an infinite impedance alternating current generator. (an ac current source) In practice, any very high impedance source of current may be substituted.

② –(~)– is the graphic symbol for a zero impedance signal generator. (an ac voltage source) In practice, any very low impedance signal source may be substituted.

③ Approximations apply to high beta, small signal, silicon transistors. Exact formulas apply to all bipolar transistors.

④ Formulas apply to pnp as well as the npn transistors shown. Reverse emitter arrow and power supply polarity for pnp transistors.

⑤ All internal dynamic resistances (r_b, r_c, r_e) vary with operating conditions. Primarily, r_e varies with emitter current while r_c varies primarily with temperature and collector voltage. Usually, r_b is assumed to be a non-varying resistance.

⑥ All biasing resistors connected in shunt with an input are effectively in parallel with the input impedance. The equivalent resistance of all parallel quantities must be used in all calculations where the source resistance becomes significant. $Z_i' = (Z_{i(R)}^{-1} + R_1^{-1} + R_2^{-1})^{-1}$

⑦ $x^{-1} = 1/x$

⑧ In the usual circuit where the collector is capacitor coupled to a load, the series collector resistor and the load resistance are effectively in parallel and the net parallel resistance should be used in all ac calculations. $R_L = (R_1^{-1} + R_2^{-1})^{-1}$

⑨ Base biasing is not shown but transistors must be biased into an active region.

⑩ Collector bias and base bias circuits are not shown, however the transistors must be biased into an active region.

Small-Signal
Low Frequency
Common Emitter

Z_i

Input
Impedance

$Z_i \approx h_{fe}/(37 I_C)$

$Z_i = r_b + r_c(h_{fe} + 1)$

$Z_i = h_{ie}$

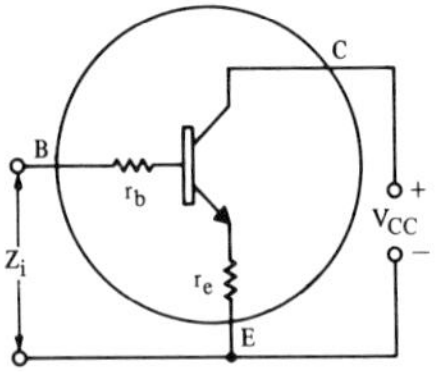

③ ④ ⑤ ⑥ ⑨

$Z_i \approx h_{fe}/(37 I_C)$
(when $A_v < 50$)

$Z_i \simeq h_{ie}$
(when $A_v < 50$)

$Z_i = h_{ie} + h_{re} h_{fe} h_{oe}^{-1} \left[(R_L h_{oe}^{-1} + 1)^{-1} - 1\right]$

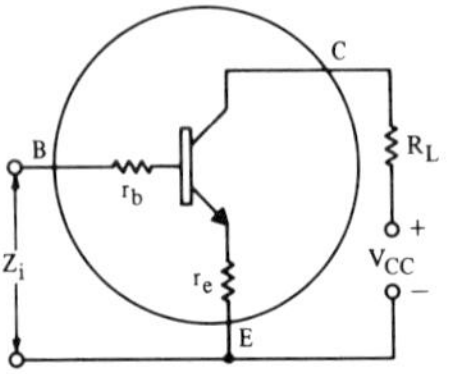

③ ④ ⑤ ⑥ ⑦ ⑧ ⑨

$Z_i \approx h_{fe}[(37 I_C)^{-1} + R_E]$
(when $A_v < 50$)

$Z_i \simeq h_{ie} + h_{fe} R_E$

$$Z_i = \frac{\left[(h_{fe} + 1)(R_E + r_e)\right]}{\left(\left[h_{fe}^{-1}(r_c R_L^{-1} + 1) + 1\right]^{-1} + 1\right)}$$

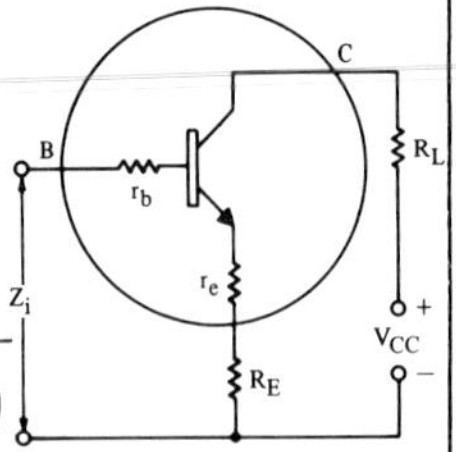

$r_b = h_{ie} - h_{re} h_{oe}^{-1}(h_{fe} + 1)$

$r_c = h_{oe}^{-1}(h_{fe} + 1)$

$r_e = h_{re} h_{oe}^{-1}$

③ ④ ⑤ ⑥ ⑦ ⑧ ⑨

Small-Signal Low Frequency Common Base	Z_O	Output Impedance	Applicable Notes

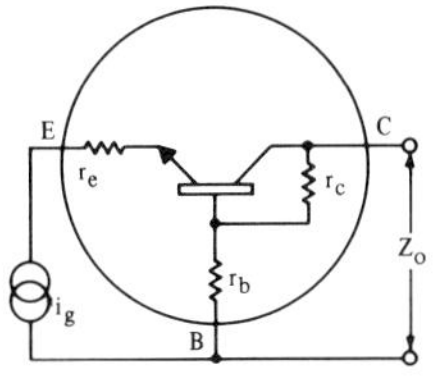

$Z_o \simeq (h_{fe} + 1)/h_{oe}$

$Z_o = r_c + r_b$

$Z_o = 1/h_{ob}$

$Z_o = \left(\left[h_{oe}^{-1}(h_{fe} + 1)\right] + \left[h_{ie} - h_{re}h_{oe}^{-1}(h_{fe} + 1)\right]\right)^{-1}$

Applicable Notes: ① ③ ④ ⑤ ⑥ ⑦ ⑩

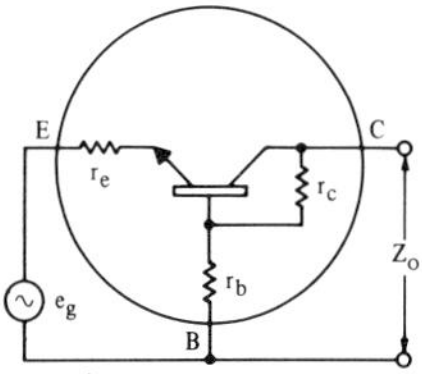

$Z_o \approx 200\ \text{k}\Omega$
(when $I_C \approx 1$ mA)

$Z_o \simeq h_{oe}^{-1} + h_{oc}^{-1}(50 I_C h_{ie} h_{fe}^{-1} - 1)^{-1}$

$Z_o = h_{oe}^{-1}\left(\left[(h_{ie}h_{oe}/h_{fe}h_{re}) - 1\right]^{-1} + 1\right)$

Applicable Notes: ② ③ ④ ⑤ ⑥ ⑦ ⑩

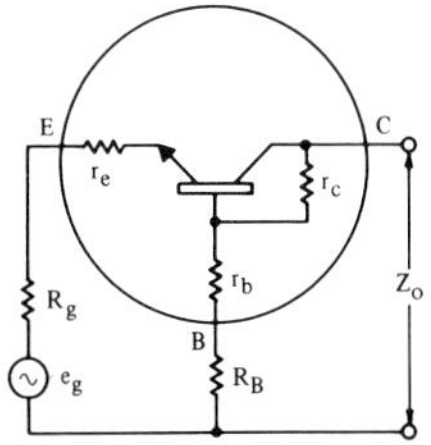

$Z_o < R_B + (h_{fe}/h_{oe})$

$Z_o = h_{oe}^{-1} + h_{fe}h_{oe}^{-1}\left(\left[(R_B + r_b)/(R_g + r_e)\right] + 1\right)^{-1}$

$r_e = h_{re}h_{oe}^{-1}$

$r_b = h_{ie} - r_e(h_{fe} + 1)$

Applicable Notes: ② ③ ④ ⑤ ⑥ ⑦ ⑩

Small-Signal Low Frequency Common Collector

Z_o

Output Impedance

Applicable Notes

$Z_o \approx 50\ \text{k}\Omega$
(when $I_C \approx 1$ mA, $V_{CE} > 5$)

$Z_o = 1/h_{oc}$

$Z_o = 1/h_{oe}$

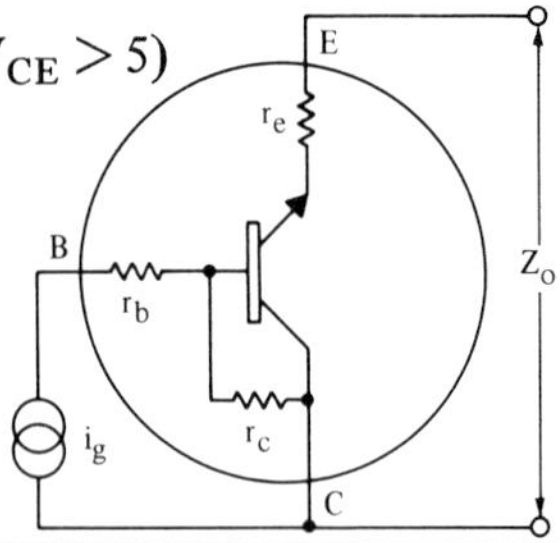

Applicable Notes: ① ③ ④ ⑤ ⑥ ⑩

$Z_o \approx 1/(37I_C)$

$Z_o \simeq h_{ie}/h_{fe}$

$Z_o = r_e + r_b(h_{fe} + 1)^{-1}$

$Z_o = h_{ie}/(h_{fe} + 1)$

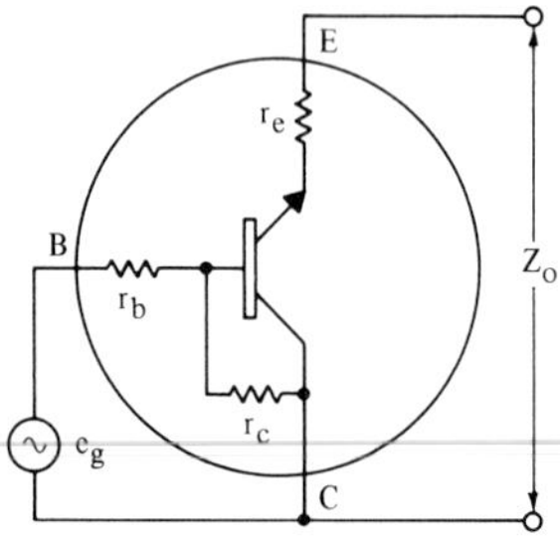

Applicable Notes: ② ③ ④ ⑤ ⑥ ⑦ ⑩

$Z_o \approx (37I_C)^{-1} + (R_g/h_{fe})$

$Z_o \simeq (R_g + h_{ie})/h_{fe}$

$Z_o = r_e + \left[(R_g + r_b)/(h_{fe} + 1)\right]$

$Z_o = (h_{ie} + R_g)/(h_{fe} + 1)$

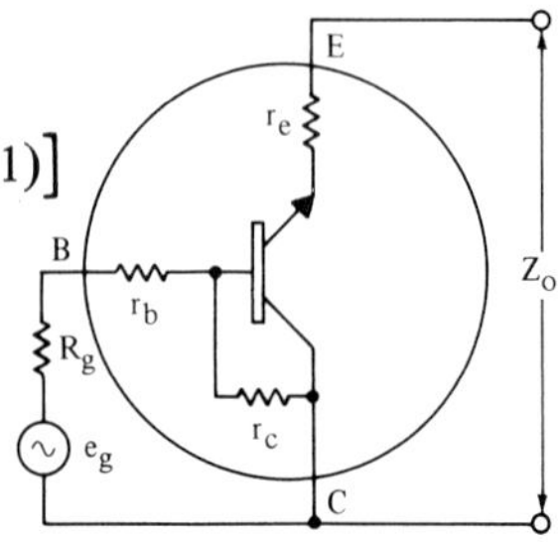

Applicable Notes: ② ③ ④ ⑤ ⑥ ⑦ ⑩

Small-Signal Low Frequency Common Emitter

Z_O

Output Impedance

Applicable Notes

$Z_o \approx 50\ \mathrm{k\Omega}$
(when $I_C \approx 1$ mA

$Z_o \simeq r_c/(h_{fe} + 1)$

$Z_o = r_c(h_{fe} + 1)^{-1} + r_e$

$Z_o = 1/h_{oe}$

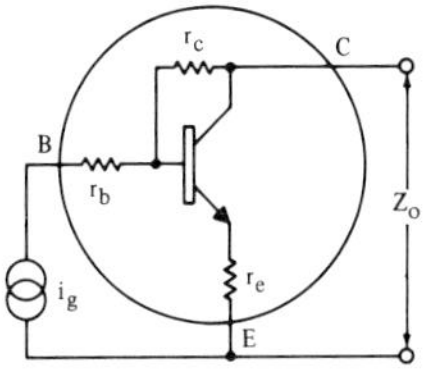

Applicable Notes: ① ③ ④ ⑤ ⑥ ⑦ ⑩

$Z_o \approx 200\ \mathrm{k\Omega}$
(when $I_C \approx 1$ mA)

$Z_o \simeq h_{oe}^{-1} + h_{oe}^{-1}(50 I_C h_{ie} h_{fe}^{-1} - 1)^{-1}$

$Z_o = h_{oe}^{-1}\left(\left[(h_{ie}h_{oe}/h_{fe}h_{re}) - 1\right]^{-1} + 1\right)$

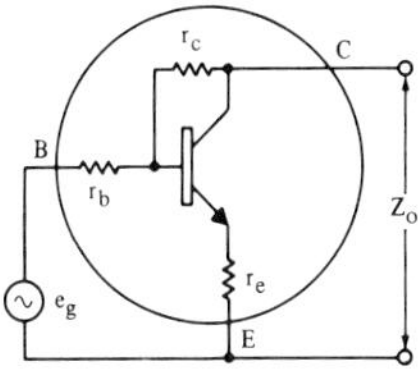

Applicable Notes: ② ③ ④ ⑤ ⑥ ⑦ ⑩

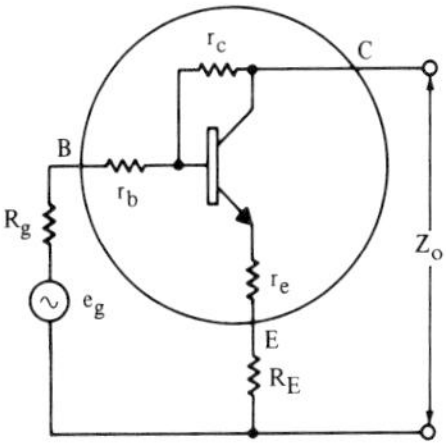

$Z_o < R_g + (h_{fe}/h_{oe})$

$Z_o = (h_{oe}^{-1} + h_{fe}h_{oe}^{-1})\Big/\left(\left[(R_g + r_b)/(R_E + r_e)\right] + 1\right)$

$r_e = h_{re}h_{oe}^{-1}, \quad r_b = h_{ie} - r_e(h_{fe} + 1)$

Applicable Notes: ② ③ ④ ⑤ ⑥ ⑦ ⑩

α β Small-Signal Current Ratios

α = Greek script letter alpha.

α = Symbol for small signal common base forward current transfer ratio with output ac shorted.

Note: Although alpha predominates as the "oral symbol," the equivalent hybrid parameter h_{fb} has almost completed superseded α as the accepted written symbol. See—h_{fb}

$\alpha = h_{fb}$

$\alpha = h_{fe}/(h_{fe} + 1)$ or $(h_{fe}^{-1} + 1)^{-1}$

$\alpha = i_c/i_e$ when e_c and $e_b = 0$

$\alpha \simeq 1$

$\alpha < 1$ exception—very early point contact transistors

β = Greek script letter beta.

β = Symbol for small signal common emitter forward current transfer ratio with output ac shorted.

Note: Although beta predominates as the "oral symbol," the equivalent hybrid parameter h_{fe} has almost completely superseded β as the accepted written symbol. See—h_{fe}

$\beta = h_{fe}$

$\beta = i_c/i_b$ when e_c and $e_e = 0$

$\beta = (i_c h_{ie})/e_b$ when e_c and $e_e = 0$

$\beta = h_{fb}/(h_{fb} - 1)$ or $(h_{fb}^{-1} - 1)^{-1}$

$\beta = (i_e/i_b) - 1 = \left[(i_e/i_c) - 1\right]^{-1}$

SECTION THREE

OPERATIONAL AMPLIFIERS

3.1 DEFINITIONS

A

Opamp Symbol Definitions

A – See–A_V

a – See–α (alpha)

A_{CL} – See–A_{VCL}

A_{DIFF} – See–A_{VD}

$A_{(fo)}$ – See–A_{vo}

A_I = Large signal current amplification (gain). Also dc current gain in direct coupled circuits.

A_i = Small signal current amplification (gain).

A_{IAC} = Alternating current amplification (gain).

A_{IDC} = Direct current amplification (gain).

A_m = Gain margin. The reciprocal of the open-loop voltage amplification at the lowest frequency at which the open-loop phase shift is such that the output is in phase with the inverting input. See also–θ_m

A_o – See–A_{vo}

A_{OL} – See–A_{VOL}

A_V = Large signal voltage amplification (gain). Also dc voltage gain in direct coupled circuits.

A_v = Small signal voltage amplification (gain).

A_{VAC} = AC voltage amplification (gain).

A_{VD} = Large signal differential voltage amplification (gain).

A_{VDC} = DC voltage amplification (gain).

A_{VCL} = Large signal closed-loop voltage amplification. The large signal voltage gain of an opamp stage with inverse feedback. Applies also to dc voltage gain in direct coupled circuits. This symbol is used in place of A_V only when the meaning would otherwise be confusing. See also–A_V

A_{vcl} = Small-signal closed-loop voltage amplification. The small-signal voltage gain of an operational amplifier stage with inverse feedback applied. See also–A_v

A_{vo} = Midband voltage amplification. The voltage amplification at the midband or reference frequency (f_o).

A_{VOL} = Large-signal open-loop voltage amplification. The large-signal voltage gain of an operational amplifier before application of inverse feedback.

A_{vol} = Small-signal open-loop voltage amplification. The small-signal voltage gain of an operational amplifier (opamp) before application of inverse feedback.

B = See–BW, See also–B, Passive Circuits.

B_1 = See–$BW_{(A_v = 1)}$

B_{OM} = Maximum output swing bandwidth.

BW = Bandwidth

$BW_{(-3\,dB)}$ = Half power or 3 dB down bandwidth

$BW_{(-3\,dB)}$ = f_o/Q (bandpass filters)

$BW_{(A_v = 1)}$ = Unity gain bandwidth. The range of frequencies within which the open-loop voltage amplification is greater than unity. Unity gain bandwidth is also known as gain-bandwidth product but, is only approximately equal to actual gain-bandwidth product. (See–GBW, f_T). The unity-gain bandwidth is equal to the product of the small-signal closed-loop voltage amplification (A_{vcl}) and the closed-loop flat-response bandwidth only when the open-loop voltage gain is inversely proportional to frequency in the frequency range between the top bandpass frequency and the unity-gain frequency.

BW_{NOISE} Noise Bandwidth

BW_{NOISE} = Bandwidth used to compute noise output. (other symbols include: $\overline{B}$, B, BW, BW_n)

BW_{NOISE} = Noise bandwidth with zero noise contribution from frequencies above or below bandwidth limits

BW_{NOISE} = Noise bandwidth measured with filters having nearly rectangular response curves. ("cliff" or "brick wall" filters)

Effective Noise Bandwidth from zero to the 3 dB Down Frequency using Butterworth Filters

$BW_{NOISE} = 1.57\ BW_{-3\,dB}$	6 dB per octave filter
$BW_{NOISE} = 1.11\ BW_{-3\,dB}$	12 dB per octave filter
$BW_{NOISE} = 1.05\ BW_{-3\,dB}$	18 dB per octave filter
$BW_{NOISE} = 1.025\ BW_{-3\,dB}$	24 dB per octave filter
$BW_{NOISE} = BW_{-3\,dB}$	∞ dB per octave filter

Notes:

6 dB per octave = 20 dB per decade
(first order filter)

12 dB per octave = 40 dB per decade
(second order filter)

dB per decade = 3.333 (dB per octave)
dB per octave = .3 (dB per decade)

$BW_{(A_v = 1)} = [A_{vcl}]/[BW_{cl}]$

BW_{cl} = Small signal flat response bandwidth.

$BW_{cl} \simeq [BW_{(A_v = 1)}]/A_{vcl}$

BW_{NOISE} – See preceding page

BW_p = Power bandwidth. See also–PBW

$BW_p = SR/[\pi V_{opp}]$

C_B = Bypass capacitor. Bootstrap capacitor.

C_C = Coupling capacitor.

C_I, C_i, C_{IN}, C_{in} = Input capacitance.

CMRR = Common mode rejection ratio. The ratio of differential voltage gain to common mode voltage gain.

C_O, C_o, C_{out} = Output capacitance.

C_p = Parallel capacitance.

C_T, C_t = Total capacitance.

D – See–THD

d = Damping factor. (other symbols include α and δ) The reciprocal of the Q factor in most applications. A symbol used in high and low pass filter formulas where the 3 dB down definition of Q factor is not applicable. Note: Nearly everyone understands the meaning of Q factor regardless of the difficulty with an all encompassing definition. See–Q

$d = 1/Q$

dB

Editorial

The decibel (dB) has a long history of use as a voltage or current ratio regardless of the original, and still the only completely unqualified, definition of the term as a power ratio.

The lack of an official sanction for the use of the decibel as a voltage or current ratio has created the ridiculous situation where, input and output resistances are "assumed" to be equal when it is known that large differences exist, specifications use mV/V, mV/V, V/V, V/mV and dBV for voltage ratios, and the same engineer who uses the decibel for voltage ratios on a daily basis, switches to a "more technically correct" form when writing a specification or a paper. (Some opamp manufacturers do specify certain voltage ratios in decibels)

The time has arrived for the official definition of the decibel to be expanded to include voltage and current ratios. There can be absolutely no misunderstanding of the expanded definition as long as the decibel is stated as a power, voltage or current ratio, so you may start using this very useful tool now, without waiting for an official definition.

There will be a few ultra conservatives who will insist that the decibel can only be used correctly as a power ratio but they will be a very small minority. Some of this same group will insist that the word gain means only power and although completely understandable the use of the term "voltage gain" is technically incorrect.

The use of any of the words gain, amplification or decibels without reference to power, voltage or current is not recommended since all three would have to be interpreted as power in the absence of an adjective.

dB = Decibel. A logarithmic ratio of power, voltage or current. See—dB editorial on preceding page. See also—dB, Passive Circuits

$$dB = 10[\log(P_o/P_i)]$$

$$dB = 20[\log(V_o/V_i)]$$

$$dB = 20[\log(I_o/I_{in})]$$

dBf = Power in decibels referenced to one femtowatt. ($fW = 10^{-15}$ W)

dBm = Power in decibels referenced to one milliwatt.

dB_{re} – See—dB_{REF}

dB_{REF} = Reference level in decibels.

dBV = 1. Voltage in decibels referenced to one volt.
2. Voltage ratio in decibels. (not recommended)

E – See—V See also—E, Passive Circuits

e – See—V See also—e, Transistors and e, Passive Circuits

e_g, e_i, e_{in} – See—V_g, V_i, V_s

E_N, E_n, e_N, e_n etc – See—V_n
– See also—e_N, NF, Transistors
– See also—E_N, NI, Passive Circuits

Note:

The transition from E to V as the quantity symbol for voltage is complete in this opamp section. The symbol E was used exclusively in the passive circuit section while the transistor section used V for dc voltages only. It is expected that eventually the symbol V will replace E for all electronic usage.

F G

F = Noise factor. Noise factor is also known as noise figure (NF). F may represent the average or the spot noise factor. See–NF, Transistors

$\overline{F}$ = Average noise factor.

f_1 – See–B1, $BW_{(A_V = 1)}$

$F(f)$ = Spot noise factor.

f_c = Cutoff frequency. The frequency at which the output falls to one-half power or 3 dB down from maximum.

f_{IN}, f_{in} = Input frequency

f_o = Reference, center, midband, resonant, oscillation or output frequency.

f_p = Frequency of pole. (poles and zeros)

f_r = Resonant frequency.

f_T, f_t = Unity gain frequency. The frequency at which the open-loop voltage gain falls to unity. Has exactly the same meaning as $BW_{(A_V = 1)}$ in all integrated circuit opamps. See–$BW_{(A_V = 1)}$

f_z = Frequency of zero. (poles and zeros)

G = Conductance See–G, Passive Circuits

GBW = Gain-bandwidth product. The product of the small signal voltage amplification (A_V) and the bandwidth (BW). See–$BW_{(A_V = 1)}$

GBW $\simeq$ or = $BW_{(A_V = 1)}$ or f_T

Depending upon the exact definition.

G H I

Opamp Symbol Definitions

G_m = Large-signal forward transconductance.

g_m = Small-signal forward transconductance.

G_P = Large-signal power gain.

G_p = Small-signal power gain.

H_o = Function at reference frequency. (f_o)

$I+$ = Positive dc supply current

$I-$ = Negative dc supply current

I_A = Amplifier dc supply current

I_{ABC} = Amplifier bias current.

I_B = Bias current

I_{CC} = Positive dc supply current

I_D = Device dc supply current

I_{D+} = Device positive dc supply current

I_{D-} = Device negative dc supply current.

I_{DG} = Non-inverting input grounded current.

I_{DO} = Non-inverting input open current.

I_{EE} = Device negative dc supply current.

I_g = Small-signal generator (source) current.

I_{IB} = Input bias current

I_{IN}, I_{in} = Input signal current

I_{IO} = Input offset current. The difference between the bias currents into the two input terminals of an opamp with the output at zero volts.

$|I_{IO}|$ = The magnitude of input offset current. See also–I_{IO}

I_n = Device equivalent-input noise current. That component of device total equivalent-input noise which varies with· the external source resistance and therefore is properly represented by an infinite impedance current source in parallel with the input terminals.

$I_n = \sqrt{I_{ns}^2 + I_{nf}^2}$

i_n – See–I_n

I_{nf} = Device equivalent-input 1/f noise current. That part of I_n which has a spectral density which is inversely proportional to frequency.

I_{nR} = Thermal (white) noise current of resistance See–I_N, Passive Circuits

I_{ns} = Device equivalent-input shot (white) noise current.

I_O = Large signal output current.

I_o = Small signal output current.

I_{O+} = Large signal positive swing output current.

I_{O-} = Large signal negative swing output current.

I_{OPP} = Peak to peak output current.

I_{OS} = Short-circuit output current. The maximum output current available from the device with the output shorted to ground or either supply.

I_P = Large signal peak current

I_p, I_{pk}, I_{peak} = Small signal peak current

I_s = 1. Source current. See–I_g, I_{in}
2. Shot noise current. See–I_{ns}

I_{SC} – See I_{OS}

I_T, I_t, i_T = Total current

I_{TH} = Threshold current.

J, j – See–J, j, Passive Circuits

K = 1. Kelvin temperature. (°C + 273.15)
2. Voltage gain See–A_v
3. Any constant.

k = 1. Any constant
2. Boltzmann constant.

k_B = Boltzmann constant. ($1.38 \cdot 10^{-23}$ J/°K)

k_{CMR} = Common mode rejection ratio. See–CMRR

k_{SVR} = Supply voltage rejection ratio. The absolute value of the ratio of change in supply voltage to the change in input offset voltage. The reciprocal of PSRR or PSS. See also–PSRR, PSS, k_{SVS}

$k_{SVR} = |\Delta V_{CC}/\Delta V_{IO}|$

k_{SVS} = Supply voltage sensitivity. The absolute value of the ratio of change in input offset voltage to the change in supply voltages producing it. The reciprocal of k_{SVR}. See Also–PSRR, PSS, k_{SVR}

$k_{SVS} = |\Delta V_{IO}/\Delta V_{CC}|$

L M N

Opamp Symbol Definitions

L = 1. Inductance. See–L, Passive Circuits

2. Level. Signal level in decibels with respect to a noted reference level.

---

mAdc = Direct current milliampere.

MAG = Maximum available (power) gain.

MUF = Maximum usable frequency.

mW/°C = Milliwatt per degree Celsius.

MΩ, M = Megohm

N = 1. Noise. See also–V_n, I_n.

2. Noise power. See–P_N, Passive Circuits

3. Number. A pure number or a ratio.

NF – See–F, See also–NF, Transistors

NI – See–NI, $E_{N(EX)}$, Passive Circuits.

N_P – See–P_N, Passive Circuits

N_{th} – See–P_N, Passive Circuits, See also–V_{nR}

$nV/\sqrt{Hz}$, $nV/(Hz)^{\frac{1}{2}}$, $nV/\sqrt{\sim}$ =

Nanovolts per hertz or nanovolts per root hertz or nanovolts per root cycle. The spot noise voltage in nanovolts. The noise voltage in nanovolts for a bandwidth of one hertz at a specified frequency.

$nV/\sqrt{Hz} = (V_{n(nV)})/\sqrt{BW}$

only when the noise voltage has constant spectral density. (only when the noise voltage is white noise)

os, OS = Overshoot

P_C = 1. (Device) power consumption.
2. Collector power dissipation. See—P_C, Transistors

P_D = 1. Device power dissipation
2. Power dissipation.

PF, p.f. = Power factor. See—pf, Passive Circuits

pF = Picofarad. (10^{-12} farad)

P_i, P_{IN}, P_{in} = Input power.

P_N = Noise power. See—P_N, Passive Circuits.

P_o = Output power.

PSRR = Power supply rejection ratio. The absolute value of the ratio of the change in input offset voltage to the change in power supply voltage producing it. This ratio is usually in μV/V or in dB. When all are given in decibels and disregarding the sign of the decibel ratio, K_{SVR}, K_{SVS}, PSS, PSRR, VSRR, $|\Delta V_{CC}/\Delta V_{IO}|$ and $|\Delta V_{IO}/\Delta V_{CC}|$ are all equal. It is hoped that the industry will soon standardize on only one of these symbols.

P_{SRR} = See—PSRR

PSS = Power supply sensitivity. See—PSRR

PSS± – See—PSS

PSS+ = Positive power supply sensitivity. See—PSRR

PSS- = Negative power supply sensitivity. See—PSRR

P_T, P_t, P_{tot} = Total power.

Q = Q factor. In simple bandpass filters, the ratio of the resonant frequency to the 3 dB down bandwidth. In highpass or lowpass filters where the 3 dB down definition is not applicable, the reciprocal of the damping factor (d). See also—Q, Passive Circuits.

Note: The Q factor is also known as the merit, quality, storage, magnification and energy factor. There is no known simple definition of Q which will encompass all of the applications. The general meaning of the term appears to be understood but the exact meaning, except in a few applications, is open to interpretation.

$Q = 1/d$

$Q = f_o/BW_{(-3\,dB)}$

$Q = f_r/BW_{(-3\,dB)}$

Q_L = Loaded Q factor.

Q_o = Q factor at center or reference frequency (f_o).

Q_u = Unloaded Q factor.

R = Resistance See—R, Passive Circuits

r = Small signal (dynamic) resistance. Any resistance of a semiconductor device which may be non-linear and therefore produce a different value between dc and small signal measurements.

R_F = Feedback Resistor

R_g = Generator resistance. See—R_S

R_I = 1. Input resistor. (Not recommended)
2. Large signal input resistance. (Not recommended)

R_i = Small signal input resistance. See also–Z_i

r_i = Device small signal input resistance.

r_{id} = Device differential input resistance.

R_{IN} = Large signal input resistance.

R_{in} – See–R_i

r_{in} – See–r_i

R_L = Load resistance.

R_O = Large signal output resistance.

R_o = Small signal output resistance. See also–Z_o

r_o = Device small signal output resistance.

R_{OPT} = Optimum resistance. e.g. $R_{s(OPT)} = V_n/I_n$

R_{OUT}, R_{out} – See–R_O, R_o

R_P, R_p = Parallel resistance.

r_p = Dynamic plate resistance (vacuum tube) (anode resistance (r_a) is also used).

R_S = Source resistance.

R_s = Series resistance.

R_T, R_t = Total resistance.

R_{th} – See–R_θ, Transistors

R_θ – See–R_θ, Transistors

S = 1. Sensitivity.
2. Signal. See–sig

s = Laplace transform function.

S+ – See–PSS+

S- – See–PSS-

S± – See–k_{SVS}, PSRR, PSS

sig = Signal. Any electrical, visual, audible or other indication used to convey information.

S/N = Signal to noise ratio.

SR = Slew rate. The closed-loop average-time rate-of-change of output voltage for a step-signal input. A specification used to determine the maximum combination of frequency and peak-to-peak output signal without the distortion associated with rise and fall time.

$SR = \pi \, PBW \, V_{OPP}$

$SR_{(A_V = 1)}$ = Slew rate when closed-loop voltage amplification is unity.

T = 1. Temperature. (°C unless noted)
2. Time constant. See–T, Passive Circuits
3. Time. See–t
4. Loop gain. (A_{VOL}/A_{VCL})

t = 1. Time. Time or period in seconds
2. Temperature. See–T

T_A = Ambient temperature. The average temperature of the air in the immediate vicinity of the device.

TC = Temperature coefficient.

T_C = Case temperature.

TC_{IIO} = Temperature coefficient of input offset current. The ratio of the change in input offset voltage to the change in free-air temperature when averaged over a specified temperature range.

$TC_{IIO} = \left| [(I_{IO})_1 - (I_{IO})_2] / [(T_A)_1 - (T_A)_2] \right|$

TC_{VIO} = Temperature coefficient of input offset voltage. The ratio of the change in input offset voltage to the change in free-air temperature when averaged over a specified temperature range.

$TC_{VIO} = \left| [(V_{IO})_1 - (V_{IO})_2] / [(T_A)_1 - (T_A)_2] \right|$

t_f = Fall time. The time required for the trailing edge of an output pulse to fall from 90% to 10% of the final voltage in response to a step function pulse at the input.

THD = Total harmonic distortion.

THD = $\sqrt{V_2^2 + V_3^2 \cdots + V_n^2}/V_1$ where V_1 is a sine-wave input signal (fundamental) and V_2 through V_n are the 2nd through nth harmonic respectively.

T_{high} = High temperature.

T_K = Kelvin temperature. (°C + 273.15)

T_L = Lead temperature.

T_{low} = Low temperature.

t_{os} = Time of output short-circuit.

t_p, t_{pd} = Pulse duration

t_{PLH} – See–t_r

TUV

Opamp Symbol Definitions

t_r = Rise time. The time required for an output voltage step to rise from 10% to 90% of the final value.

t_{setlg} = Settling time. See–t_{tot}

T_{stg} = Storage temperature.

t_{THL} – See–t_f

t_{tot} = Total response time. (Settling time) The time between a step-function change of the input signal level and the instant at which the magnitude of the output signal reaches for the last time a specified level range.

U = Teletypewriter or computer printer substitute for greek letter mu (μ).

u = Typewriter substitute for greek letter mu (μ).

V = Symbol for the voltage quantity as well as for the volt unit.

V_A = DC or rms large signal voltage.

V_a = Small signal rms signal voltage.

v_A = Instantaneous large signal voltage.

v_a = Instantaneous small signal voltage.

+V = Any positive dc voltage.

-V = Any negative dc voltage.

V+ = Positive polarity power supply voltage.

$V-$ = Negative polarity power supply voltage.

VAC – See–V_{AC} (V AC or V ac = unit-symbol)

V_{AC} = Alternating current voltage.

V_{BB} = Base power supply voltage or base bias voltage.

V_{CC} = Collector supply voltage. (positive polarity in all present IC opamps)

$+V_{CC}$ = Positive polarity collector supply voltage.

V_{CM} = Common mode voltage.

VDC – See–V_{DC} (V DC or V dc = unit symbol)

V_{DC} = Direct current voltage.

V_{EE} = Emitter supply voltage. (negative polarity in all present IC opamps)

$-V_{EE}$ = Negative polarity emitter supply voltage.

V_g = Generator (signal) rms voltage.

V_I = Input voltage range.

V_i = Input (signal) rms voltage.

v_i = Instantaneous input voltage.

V_{IC} – See–V_{ICM}

V_{ICM} = Common mode input voltage.

V_{ICR} = Common mode input voltage range.

V_{ID} = Differential input voltage.

V

V_{IDR} = Differential input voltage range.

V_{IN} = Large signal input voltage.

V_{in} = Small signal input voltage.

V_{IO} = Input offset voltage. The dc voltage that must be applied between the input terminals to force the quiescent dc output to zero.

$|V_{IO}|$ = The magnitude of V_{IO}. See–V_{IO}

V_{IOR} = Input offset voltage adjustment range.

V_{IR} = Input voltage range. See also–V_I

V_n – See following page

V_O = Large signal output voltage.

V_o = Small signal output voltage.

v_O = Instantaneous large signal output voltage.

v_o = Instantaneous small signal output voltage.

$V_{O(CM)}$ = Common mode output voltage.

V_{OM} = Maximum output voltage.

$V_{OM}+$, V_{OM+} = Maximum positive output voltage.

$V_{OM}-$, V_{OM-} = Maximum negative output voltage.

V_{OO} = Output offset voltage.

V_{OOS} – See–V_{OO}

V_{OPP} = Peak to peak output voltage.

$V_{OP\text{-}P}$, $V_{O(p\text{-}p)}$ – See–V_{OPP}

V_n = 1. Any rms noise voltage
2. The equivalent-input rms noise voltage of that part of the device total noise which is independent of source resistance.

Notes:

1. The other parts of total equivalent-input noise voltage (V_{ni}) are the voltages developed by device noise current through the source resistance and that developed thermally by the source resistance.

2. Noise voltages vary with bandwidth. Wide band noise may be any bandwidth but is usually specified for a 10.7 kHz bandwidth. Narrow-band noise voltages are for a bandwidth of 1 Hz and usually are specified in nV. ($nV/\sqrt{Hz}$)

$\overline{v_n^2}$ = The mean square noise voltage

V_{nf} = 1/f rms noise voltage

V_{ng} = Generator (noise generator) rms noise voltage

V_{ni} = The total equivalent input rms noise voltage

$V_{ni} = V_{no}/A_V$

$V_{ni} = \sqrt{BW(V_n^2 + I_nR_S + 4K_BT_KR_S)}$ See–BW_{NOISE}

V_{no} = The total output rms noise voltage

$V_{no} = A_VV_{ni}$

V_{nR} = Source resistance (R_S) rms thermal noise voltage.

V_{ns} = 1. Device rms shot noise voltage
2. See–V_{nR}

V_{nt} = 1. Any rms thermal noise voltage
2. See–V_{nR}

V_{nT} = Device total equivalent-input rms noise voltage including V_n and (I_nR_S)

V_{OR} = Output voltage range.

V_{OUT} – See–V_O

V_p, V_{pk}, V_{peak} = Peak voltage.

V_{p-p} = Peak to peak voltage.

V_{PS} = Power supply voltage.

V_Q = Quiescent voltage.

V_S = 1. Signal voltage
2. Source voltage
3. Supply voltage

$+V_S$, V_S+ = Positive polarity supply voltage.

$-V_S$, V_S- = Negative polarity supply voltage.

Z_i = Small signal closed-loop input impedance.

z_i = Device small signal open-loop input impedance.

Z_{i+} = Small signal closed-loop non-inverting input impedance.

$Z_{i+} = (r_i A_{vol})/A_{vcl}$

Z_{i-} = Small signal closed-loop inverting input impedance.

$Z_{i-} \simeq$ Series input resistor R.

$Z_{i-} = R + \left[R_F/(A_{vol} + 1)\right]$

z_{ic} = Device common mode input impedance. The parallel sum of the small signal open-loop impedance between each input terminal and ground.

Z to Ω

Opamp Symbol Definitions

z_{id} = Device differential input impedance.

Z_o = Small signal closed-loop output impedance.

$Z_o = z_o / [(A_{VOL}/A_{VCL}) + 1]$

z_o = Device small signal output impedance.

z_{od} = Differential output impedance. (opamps with differential output)

α – See–d etc.

α_{IIO} – See–TC_{IIO}

α_{VIO} – See–TC_{VIO}

$\Delta I_{IO}/\Delta T$ – See–TC_{IIO}

$\Delta V_{CC}/\Delta V_{IO}$ – See–k_{SVR} etc.

$\Delta V_{IO}/\Delta T$ – See–TC_{VIO}

$\Delta V_{IO}/\Delta V_{CC}$ – See–k_{SVS} etc.

δ – See–d etc.

θ_m = Phase margin. The absolute value of the open-loop phase shift between the output and the inverting input at the frequency at which the modulus of the open-loop amplification is unity.

ϕ_m – See–θ_m

ω_c = Cutoff (–3dB) angular velocity (angular frequency).

ω_o = Reference angular velocity (angular frequency).

ω_r = Resonant angular velocity (angular frequency).

OPERATIONAL AMPLIFIERS

SECTION 3.2
FORMULAS
AND
CIRCUITS

DC or Low Frequency, Large or Small Signal	A_I Current Amplification	Applicable Notes
$A_I = I_O/I_g$ $A_I = -R_F/R_L$ $A_i = i_o/i_g$ $A_i = -R_F/R_L$	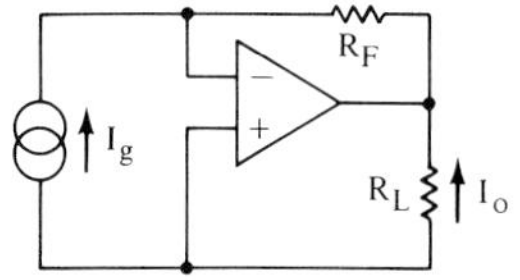	① ② ④ ⑫ ⑮
$A_I = I_O/I_g$ $A_I = R_1/R_L$ $A_i = i_o/i_g$ $A_i = R_1/R_L$	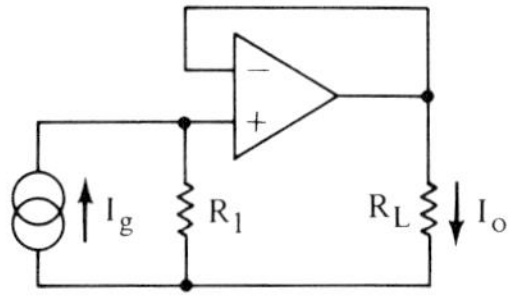	① ② ⑫ ⑮
$A_I = I_O/I_g$ $A_I = I_{RL}/I_{R1}$ $A_I = (V_O R_1)/(V_i R_L)$ $A_I = (R_1/R_L)\left[(R_F/R_2) + 1\right]$	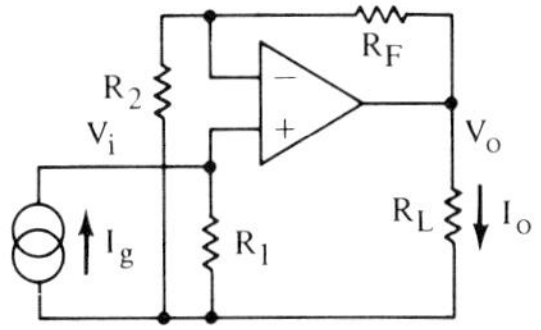	① ② ⑫ ⑮
Precision Voltage-to-Current Converter (Bilateral) 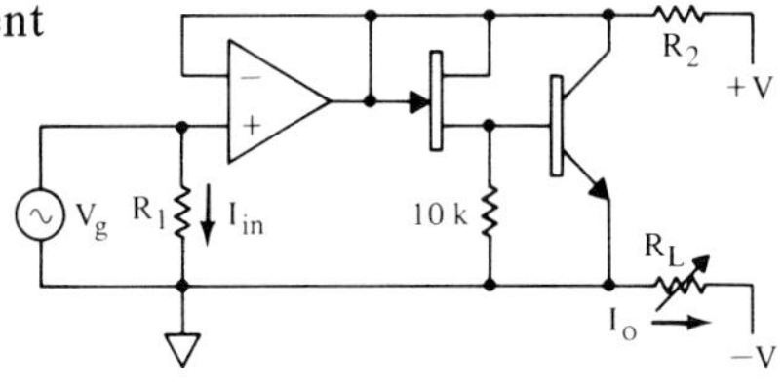 $A_I = I_O/I_{in}$ $I_O = V_g/R_2$ $I_{in} = V_g/R_1$ $A_I = R_1/R_2$		① ③ ⑪ ⑫ ⑮

DC or Low Frequency, Large or Small Signal	A_V Voltage Amplification	Applicable Notes

$A_V = V_O/V_{IN}$

$A_V = -R_F/R_B$

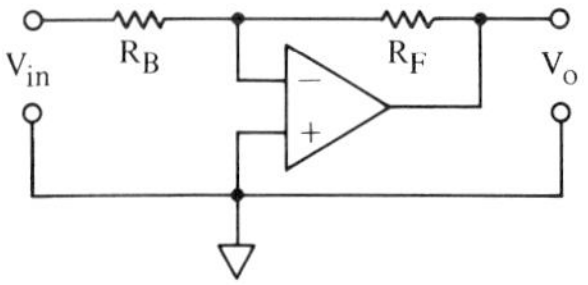

Applicable Notes: ① ④ ⑪ ⑫ ⑮

$A_V = V_O/V_{IN}$

$A_V = (R_F/R_B) + 1$

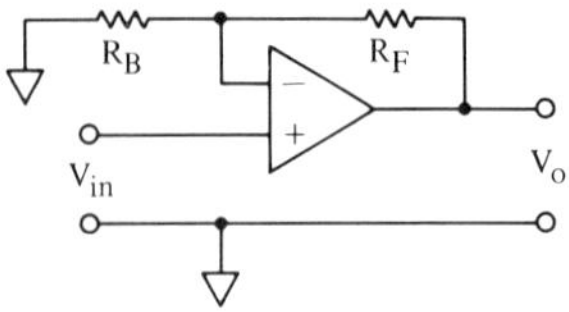

Applicable Notes: ① ⑫ ⑮

$A_V = V_O/V_S$

$A_V = -R_F/(R_B + R_S)$

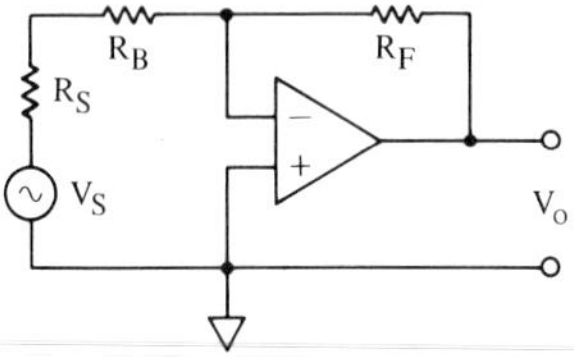

Applicable Notes: ① ③ ④ ⑫ ⑮

$A_V = V_O/V_S$

$A_V \simeq (R_F/R_B) + 1$

error typically
extremely small
at low frequencies

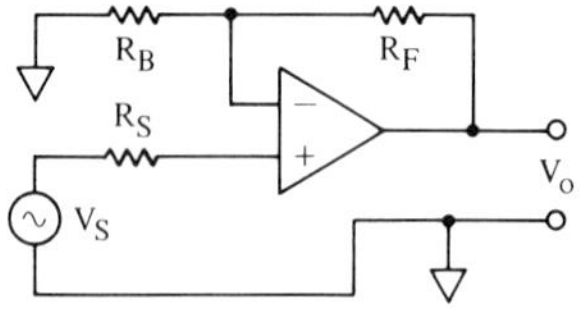

Applicable Notes: ① ③ ⑫ ⑮

$$A_V = \left[(R_F/R_B) + 1\right] / \left[(R_S/Z_i) + 1\right]$$

$$Z_i = r_i\left[(A_{VOL}/A_{VCL}) + 1\right]$$

Small Signal Voltage Gain At Any Frequency

A_V

Voltage Amplification

Applicable Notes

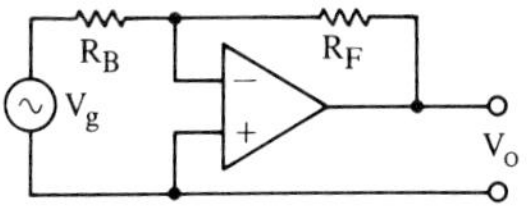

$A_v = V_o/V_g$

$A_v = -R_F/R_B$

when $(A_v f_{in}) < .1\ BW_{(Av=1)}$

or when $(A_{vol} @ f_{in}) > 100\ A_v$

$A_v = -\left[R_F^{-1}(R_B + R_B k^{-1}) + k^{-1}\right]^{-1}$

where $k = A_{vol} @ f_{in}$

Applicable Notes: ① ③ ④ ⑩ ⑪ ⑫ ⑮

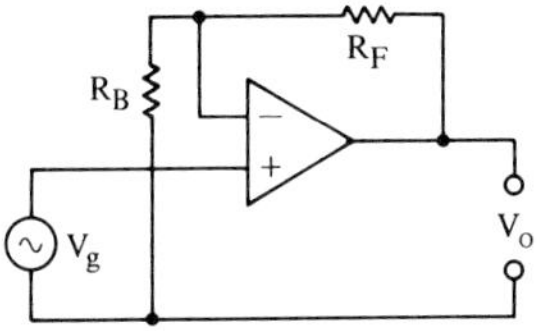

$A_v = V_o/V_g$

$A_v = (R_F/R_B) + 1$

when $(A_v f_{in}) < .1\ BW_{(Av=1)}$

or when $(A_{vol} @ f_{in}) > 100\ A_v$

$A_v = \left[R_F^{-1}(R_B + R_B k^{-1}) + k^{-1}\right]^{-1} + 1$

where $k = A_{vol} @ f_{in}$

Applicable Notes: ① ③ ⑩ ⑫ ⑮

Note: Above formulas are for small-signal output only. The output may be limited by device slew rate at frequencies above 5 kHz and at any output level greater than small-signal.

See—SR, PBW, V_{OPP}

DC or Low Frequency, RMS or Instantaneous, Small or Large Signal

A_V Voltage Amplification

Applicable Notes

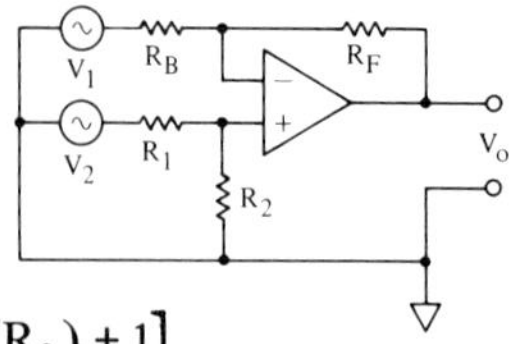

$A_{V(-)} = -V_O/V_1$

$A_{V(+)} = V_O/V_2$

$A_{V(-)} = -R_F/R_B$

$A_{V(+)} = \left[(R_F/R_B) + 1\right] / \left[(R_1/R_2) + 1\right]$

$$v_O = \left(v_2\left[(R_F/R_B) + 1\right] / \left[(R_1/R_2) + 1\right]\right) - \left[(v_1 R_F)/R_B\right]$$

v_O = Instantaneous output voltage

$A_{V(-)} = A_{V(+)}$ when $R_2/R_1 = R_F/R_B$

When $A_{V(-)} = A_{V(+)}$

$v_O = (v_2 - v_1)(R_F/R_B)$

$V_{O(rms)} = (V_{2(rms)} - V_{1(rms)})(R_F/R_B)$
$(\theta_{V2} = \theta_{V1})$

$V_{O(rms)} = (V_{2(rms)} + V_{1(rms)})(R_F/R_B)$
$(\theta_{V2} \pm \theta_{V1} = 180°)$

$V_{O(DC)} = (V_{2(DC)} - V_{1(DC)})(R_F/R_B)$
(assuming zero offset)

Applicable Notes: ① ③ ④ ⑪ ⑫ ⑮

Opamp *Notes:*

① 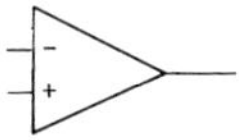 is the graphic symbol for an operational amplifier (opamp) with differential inputs. The minus input (the inverting input) develops an inverted (180°) output while the plus input (the non-inverting input) develops a non-inverted (0°) output.

DC or Low Frequency, RMS or Instantaneous, Small or Large Signal	A_V	Voltage Amplification	Applicable Notes

Low Gain from Inverting Input and High Gain from Non-inverting Input.

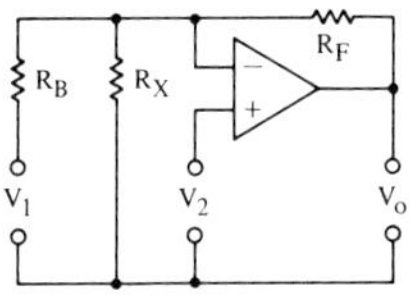

$A_{V1} = V_O/V_1$

$A_{V2} = V_O/V_2$

$A_{V1} = -R_F/R_B$

$A_{V2} = \left[R_F(R_B^{-1} + R_X^{-1})\right] + 1$

$R_X = \left[R_F^{-1}(A_{V2} - 1) - R_B^{-1}\right]^{-1}$

$v_O = (v_2 A_{V2}) - (v_1 A_{V1})$

v = instantaneous voltage

Applicable notes: ① ④ ⑩ ⑪ ⑫ ⑮

360° Phase Shifter with Flat Freq. Response (0° midband, ±180° @ DC, ±180° @ ∞ freq.)

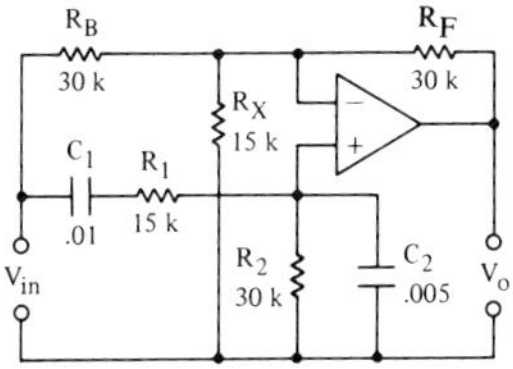

$A_V = |V_O/V_{IN}|$

$A_V = |R_F/R_B|$

when

$R_X = \left[R_F^{-1}(2R_1R_2^{-1} + 2) - R_B^{-1}\right]^{-1}$

and $X_{C1} = R_1$ @ f_o and $X_{C2} = R_2$ @ f_o

(f_o = midband and 0° phase shift freq.)

Note: Two of above circuits with different capacitor values will maintain a 90° phase differential ±1.5° over > 20 to 1 frequency range. (e.g. Resistors per above, $(C_1)_1 = .03\ \mu F$, $(C_2)_1 = .015\ \mu F$, $(C_1)_2 = .0068\ \mu F$, $(C_2)_2 = .0033\ \mu F$)

Applicable notes: ① ⑧ ⑩ ⑪ ⑫ ⑮

DC or Low Frequency, Large or Small Signal	A_V	Voltage Amplification	Applicable Notes

Positive and Negative Feedback
(decreases input impedance)

$A_V = -V_O/V_{IN}$

$$A_V = \frac{1 - (R_1/R_2)}{\left[(R_B R_1)/(R_2 R_F)\right] - 1}$$

when divisor is positive

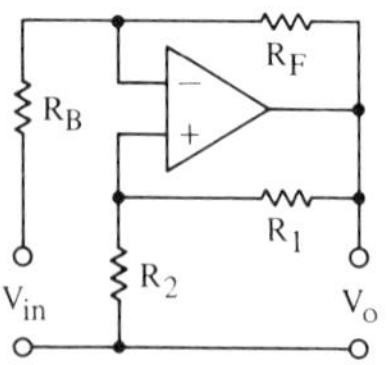

Applicable notes: ① ④ ⑪ ⑫ ⑮

Positive and Negative Feedback
[negative input impedance (negative immittance)]
(input voltage produces reverse
current through source)

$A_V = V_O/V_S$

$$A_V = 1 + \frac{1 + (R_1/R_2)}{\left[(R_B R_1)/(R_S R_F)\right] - 1}$$

when divisor is positive

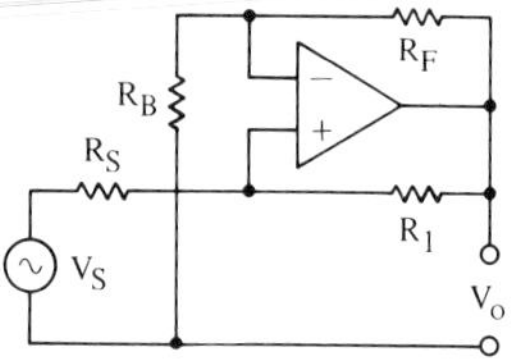

R_S must be resistive over entire unity gain bandwidth of op-amp and divisor must be positive to prevent oscillation or latch-up.

Applicable notes: ① ③ ⑫ ⑮

DC or Low Frequency, Large or Small Signal

A_V

Voltage Amplification

Applicable Notes

$A_V = -V_O/V_S$

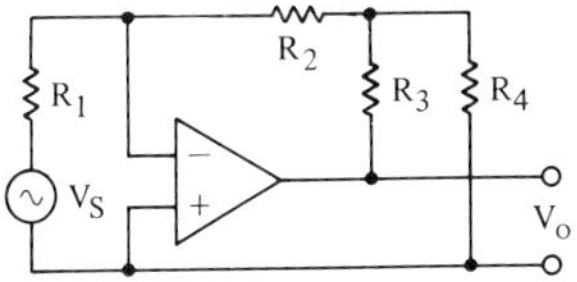

$A_V = -\left[R_2 + R_3 + (R_2 R_3/R_4)\right]/R_1$

Applicable Notes: ① ③ ④ ⑪ ⑫ ⑮

$A_V = V_O/V_S$

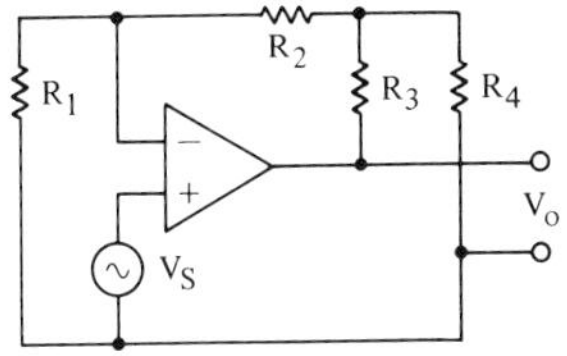

$A_V = \left(\left[R_2 + R_3 + (R_2 R_3/R_4)\right]/R_1\right) + 1$

Applicable Notes: ① ③ ⑫ ⑮

Opamp *Notes:*

②  is the graphic symbol for an infinite impedance alternating current generator or any very high impedance signal source. If the signal source resistance is too low to be ignored, consider the source resistance to be in parallel with an ideal alternating current generator. (*Note:* The graphic symbol for a direct current source is )

③  is the graphic symbol for a zero impedance signal generator or any very low impedance signal source. If the signal source resistance is too high to be ignored, consider the source resistance to be in series with an ideal voltage generator.

Opamp notes continued at V_{OPP}.

Required Passband Gain for VCVS 12 dB per Octave Low Pass Filter

$A_V = V_O/V_S$

$A_{VO} = (R_F/R_B) + 1$

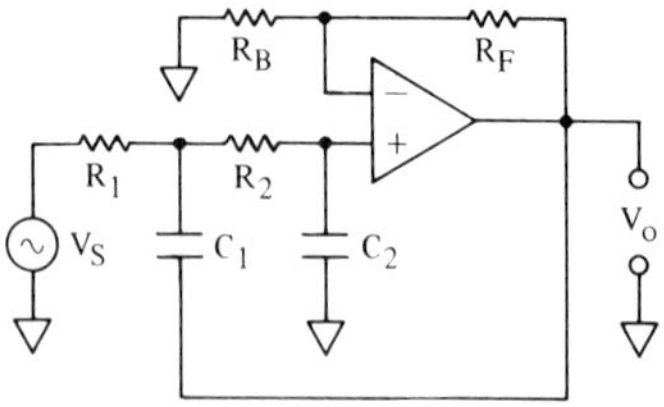

$A_{V(RQD)} = 1 + \left[C_2C_1^{-1}(R_2R_1^{-1} + 1)\right] - \left[d\sqrt{R_2R_1^{-1}C_2C_1^{-1}}\right]$

$A_{V(RQD)} = 2 + (R_2/R_1) - \left(d\sqrt{R_2/R_1}\right)$ when $C_1 = C_2$

$A_{V(RQD)} = 1 + (2C_2/C_1) - \left(d\sqrt{C_2/C_1}\right)$ when $R_1 = R_2$

$A_{V(RQD)} = 3 - d$ when $C_1 = C_2$ and $R_1 = R_2$

$d = \sqrt{2}$ for flattest response (Butterworth)
$d = 2$ for isolated two section RC response
$d = 1.731$ for best transient response (Bessell)
$d = 1.045$ for 1 dB hump (1 dB ripple Chebyshev)
$d = .895$ for 2 dB hump (2 dB ripple Chebyshev)
$d = .767$ for 3 dB hump (3 dB ripple Chebyshev)

$f_o = \left[2\pi\sqrt{R_1R_2C_1C_2}\right]^{-1}$

$f_c = f_o$ when $d = \sqrt{2}$

$f_c = f_o\left[k + (k^2 + 1)^{\frac{1}{2}}\right]^{\frac{1}{2}} \quad k = 1 - 2/d^2$

① ③ ⑤ ⑥ ⑦ ⑩ ⑪ ⑫ ⑬ ⑭ ⑮

Required Passband Gain for VCVS 12 dB per Octave High Pass Filter

$A_V = V_O/V_S$

$A_{VO} = (R_F/R_B) + 1$

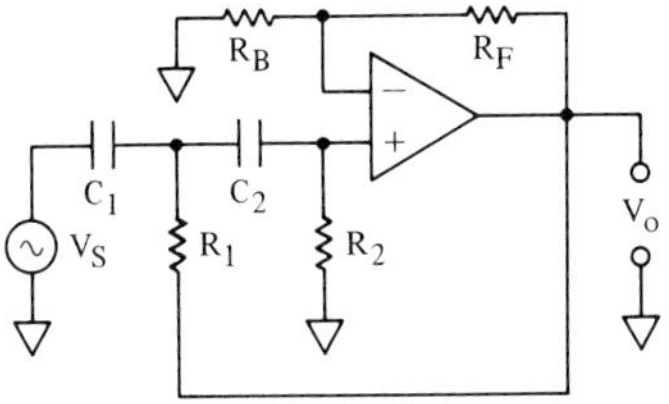

$$A_{V(RQD)} = 1 + \left[R_1R_2^{-1}(C_1C_2^{-1} + 1)\right] - \left[d\sqrt{C_1C_2^{-1}R_1R_2^{-1}}\right]$$

$$A_{V(RQD)} = 1 + (2R_1/R_2) - \left(d\sqrt{R_1/R_2}\right) \text{ when } C_1 = C_2$$

$$A_{V(RQD)} = 2 + (C_1/C_2) - \left(d\sqrt{C_1/C_2}\right) \text{ when } R_1 = R_2$$

$$A_{V(RQD)} = 3 - d \text{ when } C_1 = C_2 \text{ and } R_1 = R_2$$

$d = \sqrt{2}$ for flattest response (Butterworth)
$d = 2$ for isolated two section RC response
$d = 1.731$ for best transient response (Bessell)
$d = 1.045$ for 1 dB hump (1 dB ripple Chebyshev)
$d = .895$ for 2 dB hump (2 dB ripple Chebyshev)
$d = .767$ for 3 dB hump (3 dB ripple Chebyshev)

$$f_o = \left[2\pi\sqrt{R_1R_2C_1C_2}\right]^{-1}$$

$$f_c = f_o \text{ when } d = \sqrt{2}$$

$$f_c = f_o/\left[k + (k^2 + 1)^{\frac{1}{2}}\right]^{\frac{1}{2}} \quad k = 1 - 2/d^2$$

Applicable Notes: ① ③ ⑤ ⑥ ⑦ ⑩ ⑫ ⑬ ⑭ ⑮

Multiple Feedback Bandpass Filter

$A_V = V_O/V_S$

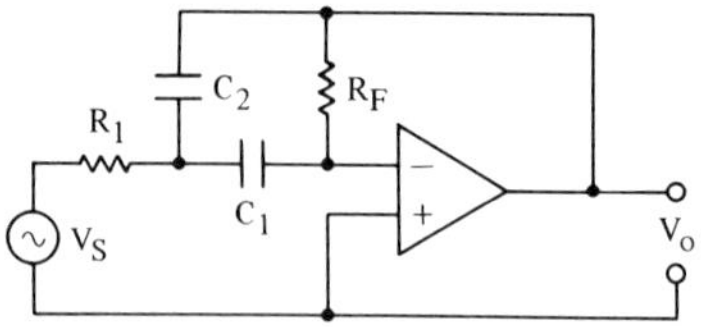

Let $C_1 = C_2$

$A_{VO} = R_F/(2R_1)$

$Q = \sqrt{.5\ A_{VO}}$

$f_r = \left[2\pi C_1 \sqrt{R_F R_1}\right]^{-1}$

$f_r = 500/\left[\pi C_1 \sqrt{R_F R_1}\right] (C_{(\mu F)}, R_{(k\Omega)})$

Applicable Notes: ① ③ ⑩ ⑪ ⑫ ⑮

Higher Input Impedance, Lower Gain Version

$A_V = V_O/V_S$

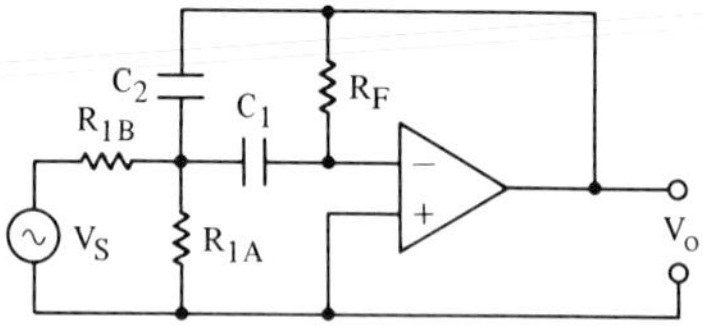

Let $C_1 = C_2$, Let $R_{1B} = 10 R_{1A}$

$A_{VO} = R_F/(2R_{1B})$

$Q = \sqrt{11 R_2/4R_{1A}}$

$f_r = \left[2\pi C_1 \sqrt{R_F R_{1B}/11}\right]^{-1}$

$f_r = 166.9/\left[C_1 \sqrt{R_F R_{1A}}\right] (C_{(\mu F)}, R_{(k\Omega)})$

Applicable Notes: ① ③ ⑩ ⑪ ⑫ ⑮

BW_{-3dB} 3 dB Down Bandwidth

Applicable Notes

$$BW_{(-3\,dB)} \simeq \frac{\sqrt{2}\ BW_{(A_V=1)}}{A_{vcl}+1}$$

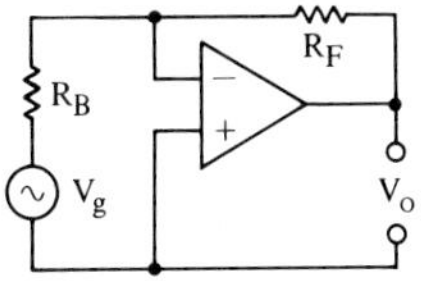

$$BW_{(-3\,dB)} \simeq \frac{\sqrt{2}\ BW_{(A_V=1)}}{(R_F/R_B)+1}$$

See also—PBW for large signal

Applicable Notes: (1) (3) (11) (12) (15)

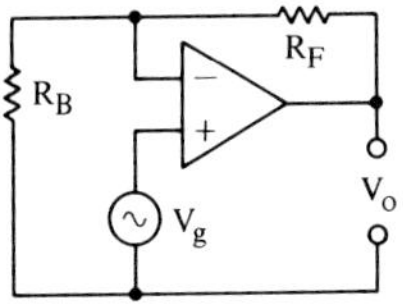

$$BW_{(-3\,dB)} \simeq \left[\sqrt{2}\ BW_{(A_V=1)}\right]/A_{vcl}$$

$$BW_{(-3\,dB)} \simeq \left(R_B/R_F\right)\ \sqrt{2}\ BW_{(A_V=1)}$$

See also—PBW for large signal

Applicable Notes: (1) (3) (12) (15)

Multiple Feedback Bandpass Filter

$$BW_{(-3\,dB)} = f_{2(-3\,dB)} - f_{1(-3\,dB)}$$

$$BW_{(-3\,dB)} = f_r/Q$$

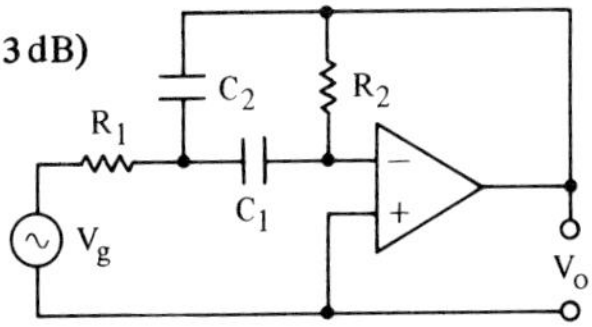

Let
$C_1 = C_2$

$$BW_{(-3\,dB)} = [\pi C_2 R_2]^{-1}$$

$$f_r = \left[2\pi C_2 \sqrt{R_1 R_2}\right]^{-1}$$

$$A_{vo} = R_2/(2R_1)$$

$$Q = .5\sqrt{R_2/R_1}$$

$A_{vol} > 100$ at highest frequency of interest.

Applicable Notes: (1) (3) (10) (11) (12) (15)

d	Damping Factor	Applicable Notes
$d = 1/Q$ *Note:* The general meaning of Q is understood by almost everyone, however an exact definition that is applicable to all circuits is very elusive. The inverse of Q (d or α) is generally used for all high and low pass type circuits when Q is less than 2.		
Low Pass and High Pass Filters $d = 2$ for isolated RC sections response. $d = 1.731$ for best delay, transient or pulse response. (Bessell response.) $d = \sqrt{2}$ for maximally flat and exact 3 dB down cut-off or crossover frequency. (Butterworth response –Recommended for most applications.) $d = 1.045$ for 1 dB peak or hump near the cutoff frequency. (1 dB ripple Chebyshev response.) $d = .895$ for 2 dB peak or hump near the cutoff frequency. (2 dB ripple Chebyshev response.) $d = .767$ for 3 dB peak or hump near the cutoff frequency. (3 dB ripple Chebyshev.) $d > 0$ in all non-oscillating circuits.		(13) (14)

d Damping Factor

Applicable Notes

VCVS Low Pass Filter

$d = 1/Q$

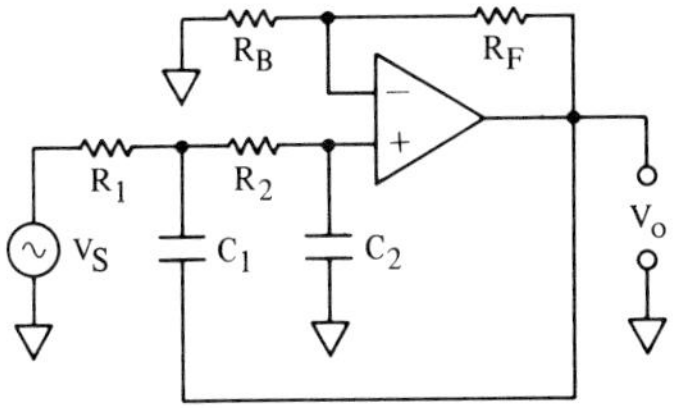

$$d = \frac{\left[C_2(R_1^{-1} + R_2^{-1})\right] - \left[(C_1 R_F)/(R_2 R_B)\right]}{\sqrt{(C_1 C_2)/(R_1 R_2)}}$$

$A_{VO} = (R_F/R_B) + 1$

$f_c = \left[2\pi\sqrt{R_1 R_2 C_1 C_2}\,\right]^{-1}$ when $d = \sqrt{2}$

$$d = \frac{1 + \left[R_2/R_1\right] - \left[R_F/R_B\right]}{\sqrt{R_2/R_1}}$$ when $C_1 = C_2$

$$d = \frac{\left[(2C_2)/C_1\right] - \left[R_F/R_B\right]}{\sqrt{C_2/C_1}}$$ when $R_1 = R_2$

$d = 2\sqrt{C_2/C_1}$ when $R_1 = R_2$ and $A_V = 1$
($R_F = 0$ and/or $R_B = \infty$)

$d = 2 - (R_F/R_B)$ when $R_1 = R_2$ and $C_1 = C_2$

① ③ ⑤ ⑥ ⑦ ⑩ ⑪ ⑫ ⑬ ⑭ ⑮

VCVS High Pass Filter

$d = 1/Q$

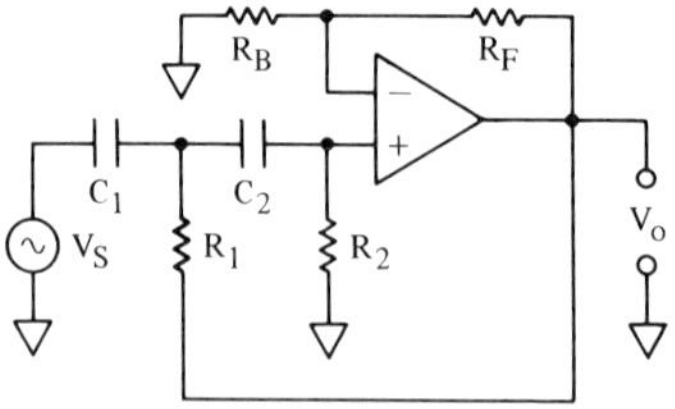

$$d = \frac{\left[R_2^{-1}(C_1 + C_2)\right] - \left[(C_2 R_F)/(R_1 R_B)\right]}{\sqrt{(C_1 C_2)/(R_1 R_2)}}$$

$A_{VO} = (R_F/R_B) + 1$

$f_c = 2\pi\sqrt{R_1 R_2 C_1 C_2}$ when $d = \sqrt{2}$

$$d = \frac{\left[(2R_1)/R_2\right] - \left[R_F/R_B\right]}{\sqrt{R_1/R_2}}$$ when $C_1 = C_2$

$$d = \frac{1 + \left[C_1/C_2\right] - \left[R_F/R_B\right]}{\sqrt{C_1/C_2}}$$ when $R_1 = R_2$

$d = 2\sqrt{R_1/R_2}$ when $C_1 = C_2$ and $A_V = 1$ ($R_F = 0$ and/or $R_B = \infty$)

$d = 2 - [R_F/R_B]$ when $C_1 = C_2$ and $R_1 = R_2$

Applicable Notes: ① ③ ⑤ ⑥ ⑦ ⑩ ⑫ ⑬ ⑭ ⑮

Small Signal	f_C	Cutoff (-3 dB) Frequency	Applicable Notes

$$f_c = \left[BW_{(-3\,dB)}\right] / \left[A_{v(1f)}\right]$$

$$f_c \simeq \left[\sqrt{2}\ BW_{(Av=1)}\right] / \left[A_{v(1f)}\right]$$

$$f_c \simeq \left[\sqrt{2}\ BW_{(Av=1)}\right] \left[R_B/R_F\right]$$

Applicable Notes: ① ③ ⑪ ⑫ ⑮

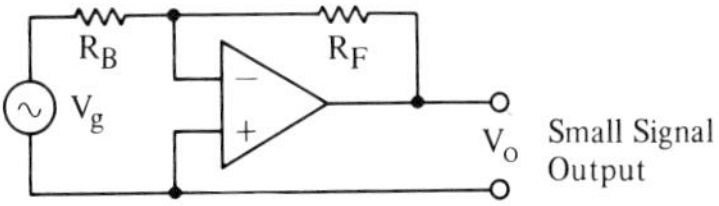

See also—PBW for large signal

$$f_c = \left[BW_{(-3\,dB)}\right] / \left[A_{v(1f)}\right]$$

$$f_c \simeq \left[\sqrt{2}\ BW_{(Av=1)}\right] / \left[A_{v(1f)}\right]$$

$$f_c \simeq \left[\sqrt{2}\ BW_{(Av=1)}\right] \left[(R_B/R_F) - 1\right]$$

Applicable Notes: ① ③ ⑫ ⑮

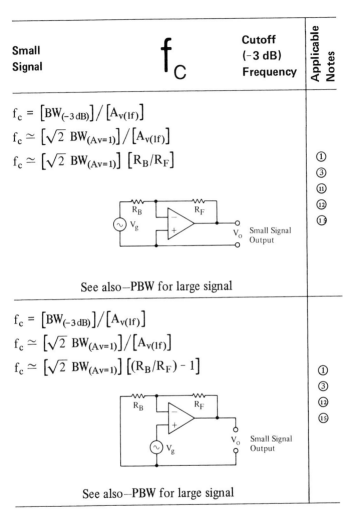

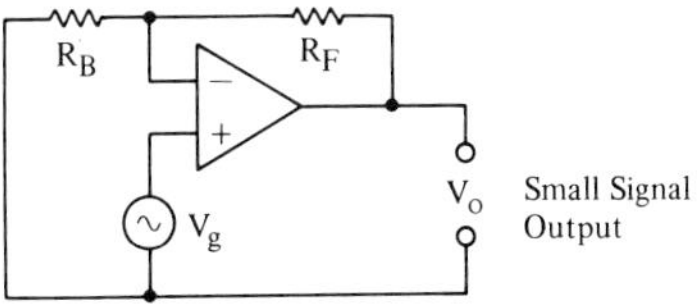

See also—PBW for large signal

Note: To increase f_c:

1. Use opamp with greater unity gain bandwidth.
2. Decrease circuit gain.
3. Use multiple decreased gain stages.

Unity Gain VCVS Low Pass Filter

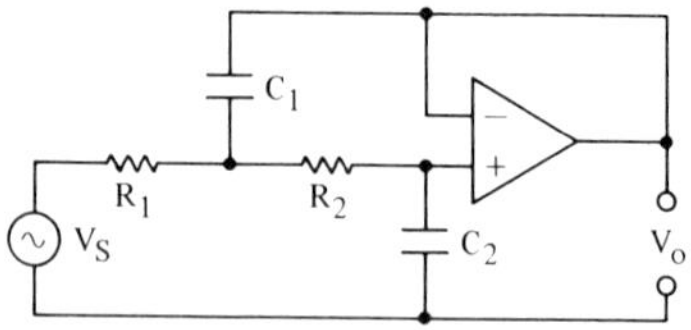

Let $R_1 = R_2$

Let $C_1/C_2 =$

1 for two isolated RC sections response
$1\frac{1}{3}$ for best transient response. (Bessell)
2 for flattest response (Butterworth)
3.3 for 1 dB hump (1 dB ripple Chebyshev)
4.7 for 2 dB hump (2 dB ripple Chebyshev)
6.8 for 3 dB hump (3 dB ripple Chebyshev)

$f_c \simeq \left[2\pi R_1 \sqrt{C_1 C_2}\right]^{-1}$ (exact when $C_1/C_2 = 2$)

Applicable Notes: ① ③ ⑤ ⑥ ⑦ ⑩ ⑪ ⑫ ⑬ ⑮

Unity Gain VCVS High Pass Filter

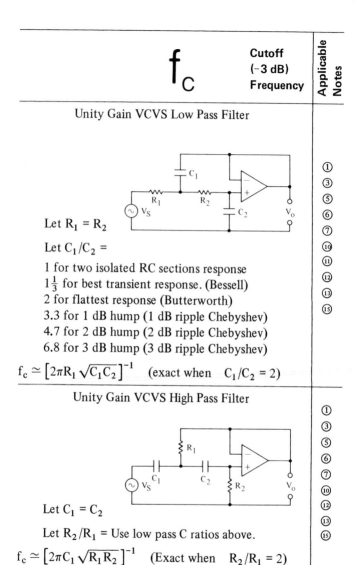

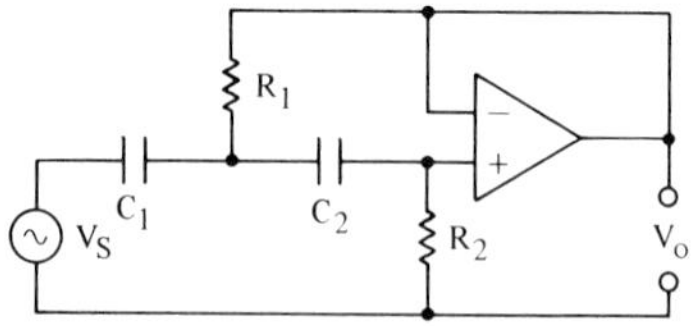

Let $C_1 = C_2$

Let R_2/R_1 = Use low pass C ratios above.

$f_c \simeq \left[2\pi C_1 \sqrt{R_1 R_2}\right]^{-1}$ (Exact when $R_2/R_1 = 2$)

Applicable Notes: ① ③ ⑤ ⑥ ⑦ ⑩ ⑫ ⑬ ⑮

f_C	Cutoff (−3 dB) Frequency	Applicable Notes

Free Gain VCVS Low Pass Filter

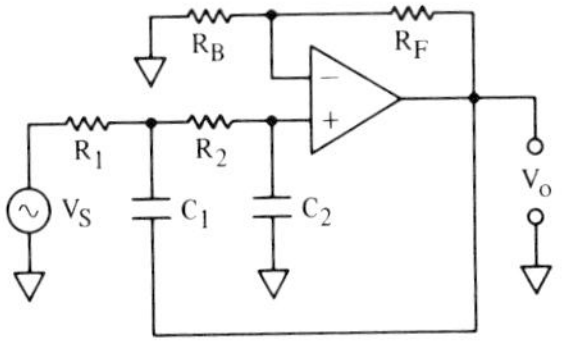

Let $C_1 = C_2$, Let $R_1 = R_2$

Passband gain $(A_{VO}) = (R_F/R_B) + 1$

Let A_{VO} = See below

1.0 for two isolated RC sections response
1.27 for good transient response (Bessell)
1.586 for flattest response (Butterworth)
1.955 for 1 dB hump (1 dB ripple Chebyshev)
2.105 for 2 dB hump (2 dB ripple Chebyshev)
2.233 for 3 dB hump (3 dB ripple Chebyshev)

$f_c \simeq [2\pi R_1 C_1]^{-1}$ (Exact when $A_V = 1.586$)

Applicable Notes: ① ③ ⑤ ⑥ ⑦ ⑩ ⑪ ⑫ ⑬ ⑭ ⑮

Free Gain VCVS High Pass Filter

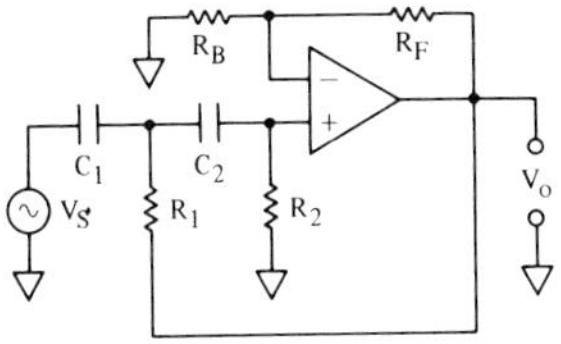

Let $C_1 = C_2$, Let $R_1 = R_2$

Use all low pass formulas above.

Applicable Notes: ① ③ ⑤ ⑥ ⑦ ⑫ ⑬ ⑭ ⑮

f_c Cutoff (-3 dB) Frequency — Applicable Notes

12 dB per Octave Multiple Feedback Butterworth Response Low Pass Filter (Recommended version with gain of 4)

Applicable Notes: ① ③ ⑥ ⑦ ⑩ ⑪ ⑫ ⑮

Let $C_2 = .1\ C_1$

Let $R_2 = 4\ R_1$

Let $R_3 = .8\ R_1$

$f_c = \left[\sqrt{2}\ \pi R_3 C_1\right]^{-1}$

$A_{VO} = 4$

$d = \sqrt{2} \quad (Q = 1/d)$

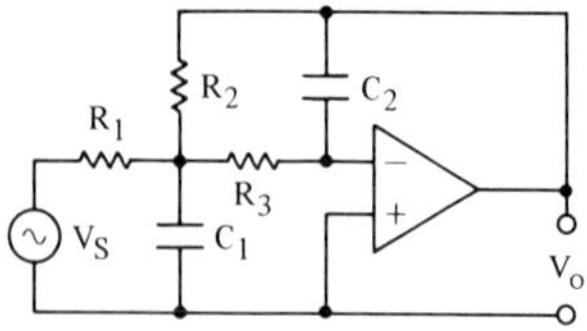

12 dB per Octave Multiple Feedback Butterworth Response High Pass Filter (Recommended unity gain version)

Applicable Notes: ① ③ ⑥ ⑦ ⑫ ⑮

Let $C_2 = C_1$

Let $C_3 = C_1$

Let $R_2 = 4.5\ R_1$

$f_c = \sqrt{2}/(6\pi R_1 C_1)$

$A_{VO} = 1$

$d = \sqrt{2} \quad (Q = 1/d)$

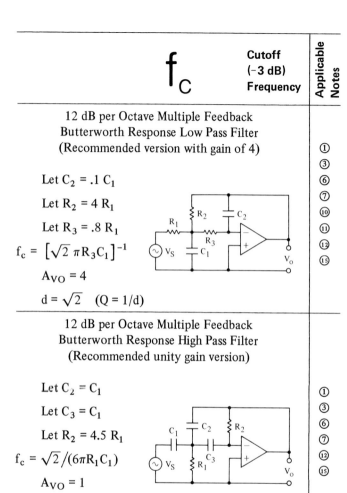

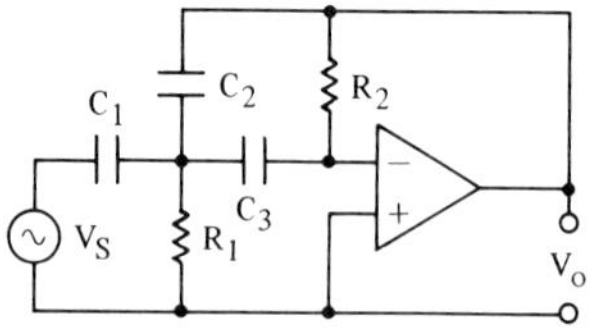

Note that 3 capacitors are required.
VCVS filters with same response require only 2.

f_C Cutoff (−3 dB) Frequency	Applicable Notes

12 dB per Octave Multiple Feedback Butterworth Response Low Pass Filter.

Let $R_3C_1 = \left[\sqrt{2}\,\pi f_c\right]^{-1}$

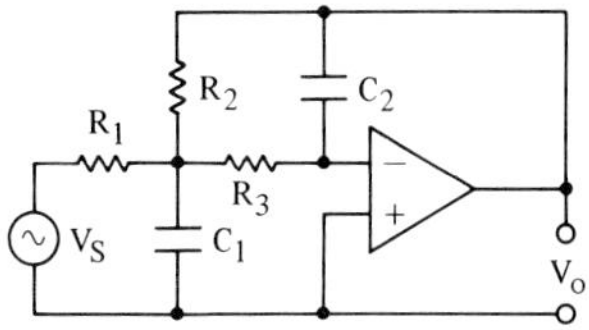

Let $R_2 = R_3(A_{VO} + 1)$

Let $R_1 = R_3(A_{VO}^{-1} + 1)$

Let $C_2 = C_1/\left[2(A_{VO} + 1)\right]$

$$f_c = \left[\sqrt{2}\,\pi R_3 C_1\right]^{-1}$$

$A_{VO} = R_2/R_1, \quad R_3 = (R_1^{-1} + R_2^{-1})^{-1}$

$d = \sqrt{2} \quad (Q = 1/\sqrt{2})$

Applicable Notes: ① ③ ⑥ ⑦ ⑩ ⑪ ⑫ ⑮

12 dB per Octave Multiple Feedback Butterworth Response High Pass Filter

Let $k = 2A_{VO} + 1$

Let $R_1C_3 = \left[\sqrt{2}\,2\pi f_c k\right]^{-1}$

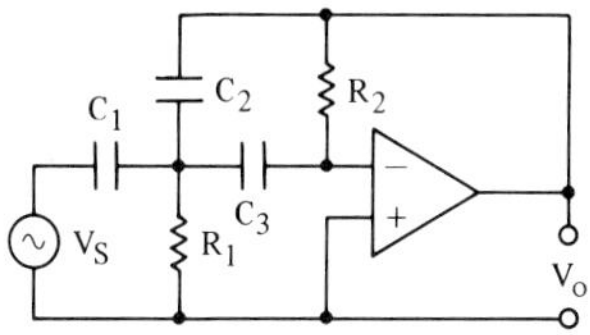

Let $C_1 = C_3$

Let $C_2 = C_1/A_{VO}$

Let $R_2 = .5\,R_1 k^2$

$$f_c = \left[2\pi C_1 \sqrt{R_1 R_2}\,\right]^{-1}$$

$A_{VO} = C_1/C_2, \quad d = \sqrt{2} \quad (Q = 1/\sqrt{2})$

Applicable Notes: ① ③ ⑥ ⑦ ⑩ ⑫ ⑮

18 dB per Octave VCVS Butterworth Response Low Pass Filter with Gain of Two.

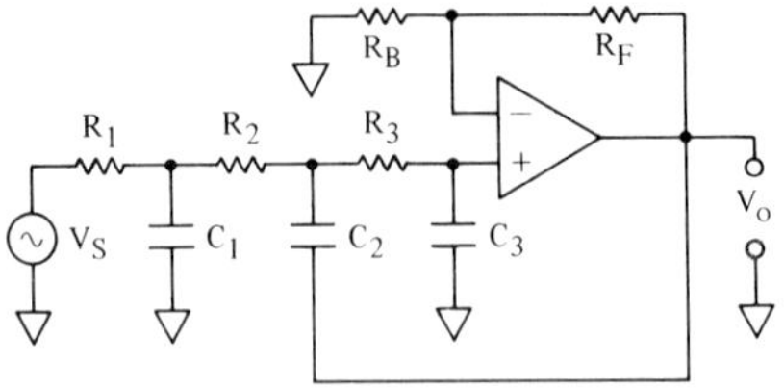

Let $C_1 = C_2 = C_3 = C$

Let $R_F = R_B \approx 7\ R$ ①

Let $R_1 = 1.565\ R$ ③

Let $R_2 = 1.467\ R$ ⑤

Let $R_3 = .435\ R$ ⑥

$f_c = [2\pi RC]^{-1}$ ⑦

$A_{VO} = 2$ ⑩

$d = \sqrt{2} \quad (Q = 1/\sqrt{2})$ ⑪ ⑫ ⑮

Let $R_1 = 3.6\ k$

Let $R_2 = 3.3\ k$

Let $R_3 = 1.0\ k$

Let $R_F = R_B = 15\ k$

$f_c = [14{,}337\ C]^{-1} \quad (C_{(\mu F)} = 69.75/f_c)$

$A_{VO} = 2, \quad d = \sqrt{2}$

f_c Cutoff (−3 dB) Frequency

Applicable Notes

18 dB per Octave VCVS Butterworth Response
Unity Gain High Pass Filter

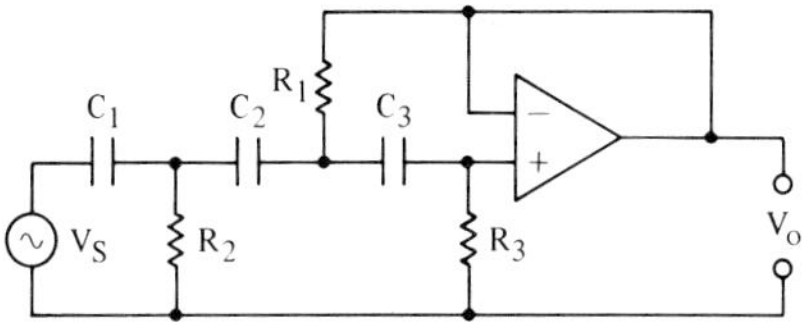

Let $C_1 = C_2 = C_3 = C$

Let $R_3 = 4.94\ R$ ①

Let $R_2 = .282\ R$ ③

Let $R_1 = .718\ R$ ⑤

$f_c = \left[2\pi C(R_1R_2R_3)^{\frac{1}{3}}\right]^{-1}$ ⑥

$f_c = [2\pi RC]^{-1}$ ⑦

$d = \sqrt{2} \quad (Q = 1/\sqrt{2})$ ⑩

$A_{VO} = 1$ ⑪

⑫

⑮

Let $R_3 = 68\ k$

Let $R_2 = 3.9\ k$

Let $R_1 = 10.\ k$

$f_c = [86{,}970\ C]^{-1} \quad (C_{(\mu F)} = 11.5/f_c)$

$d = \sqrt{2} \quad (Q = 1/\sqrt{2})$

$A_{VO} = 1$

Bessell Response VCVS Low Pass Filter
(Best pulse, delay and transient response)

12 dB per Octave (Voltage Gain = 1.269)

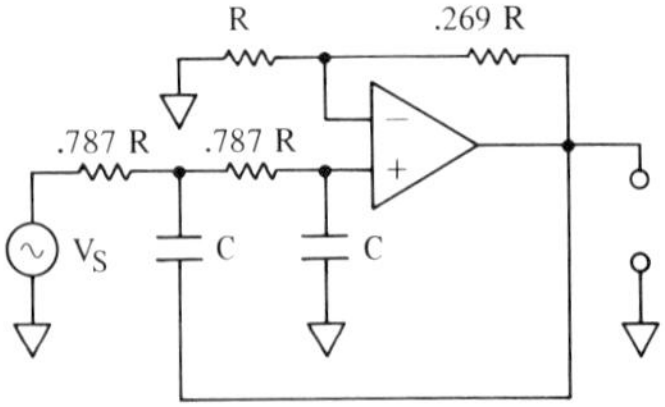

24 dB per Octave (Total Voltage Gain = 1.91)

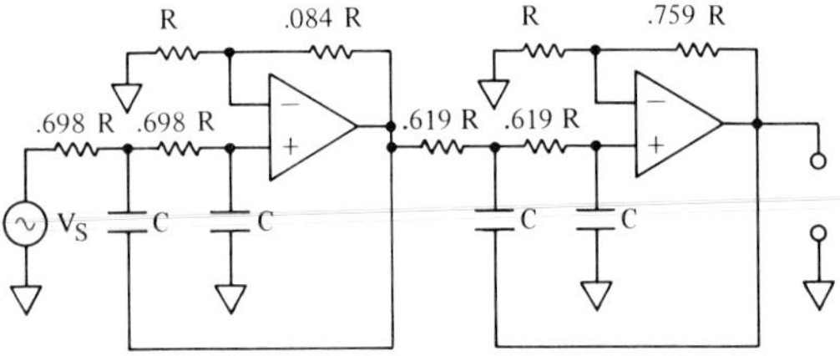

36 dB per Octave (Total Voltage Gain = 2.87)

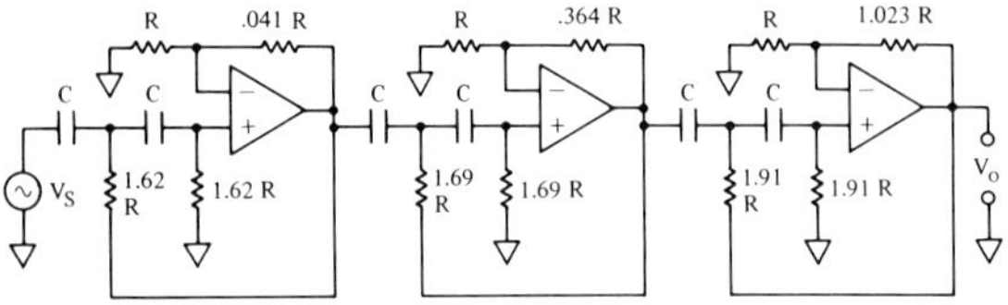

$f_c = [2\pi RC]^{-1}$

Applicable Notes: ① ③ ⑤ ⑥ ⑦ ⑩ ⑪ ⑫ ⑮

Bessell Response VCVS High Pass Filter
(Best pulse, delay and transient response)

12 dB per Octave (Voltage Gain = 1.27)

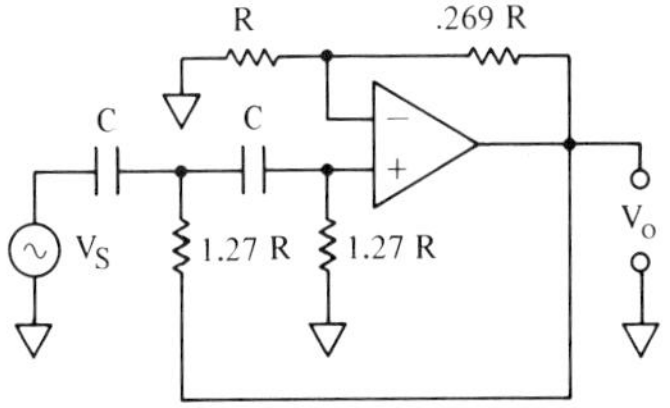

24 dB per Octave (Total Voltage Gain = 1.91)

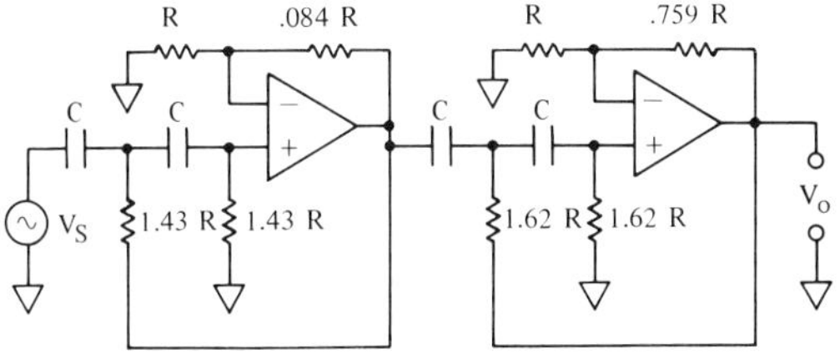

36 dB per Octave (Total Voltage Gain = 2.87)

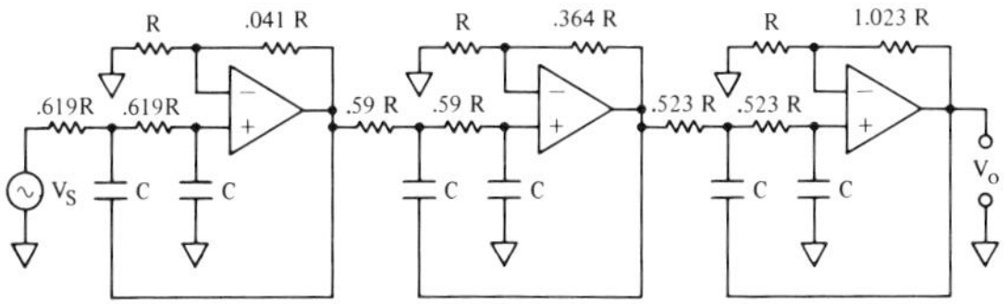

$f_c = [2\pi RC]^{-1}$

① ③ ⑤ ⑥ ⑦ ⑩ ⑫ ⑮

f_o

f_o = 1. Frequency of oscillation. (steady state)
2. The frequency at which oscillation first occurs. (before the signal amplitude reaches a non-linear region of operation)

$f_{o(LC)}$ = The series resonant frequency or the frequency of maximum circulating current regardless of input and output connections. (Not necessarily the frequency of maximum output)

$f_{o(RC)}$ = The frequency where the capacitive reactance equals the resistance.

Passive or Active Bandpass Circuits

f_o = The frequency of maximum output.

f_o = The frequency of resonance (f_r)

Passive or Active Bandstop Circuits

f_o = The frequency of minimum output

f_o = The frequency of antiresonance also known as frequency of resonance (f_r)

Multisection Passive or Active Bandpass Circuits

f_o = Center of pass band

Multisection Passive or Active Bandstop Circuits

f_o = Center of stop band

f_o = Reference frequency

Active Low Pass Circuits

f_o = The frequency of maximum change in output developed by a change of feedback amplitude. (Not necessarily the frequency of maximum output or the cutoff frequency)

f_o = The cutoff (−3 dB) frequency (f_c) only when damping factor (d) equals $\sqrt{2}$ or when Q equals $1/\sqrt{2}$. (when output is Butterworth response)

$f_o = \omega_o/2\pi$

$f_o = f_c \Big/ \left([1 - .5d^2]\,[(1 - .5d^2)^2 + 1]^{\frac{1}{2}}\right)^{\frac{1}{2}}$

$f_o = f_p\sqrt{1 - .5d^2}$ when $(1 - .5d^2) > 0$

Note: f_p = frequency of peak output. Response has no peak when $d \geqq \sqrt{2}$.

$f_o = \left[2\pi\sqrt{R_1R_2C_1C_2}\right]^{-1}$

Active High Pass Circuits

f_o = The frequency of maximum change in output developed by a change of feedback amplitude.

$f_o = f_c$ only when $d = \sqrt{2}$ (Butterworth response)

$f_o = \omega_o/2\pi$

$f_o = f_c\left([1 - .5d^2]\,[(1 - .5d^2)^2 + 1]^{\frac{1}{2}}\right)^{\frac{1}{2}}$

$f_o = f_p/\sqrt{1 - .5d^2}$ when $(1 - .5d^2) > 0$

Notes: 1. f_p = frequency of peak output
2. Response has no peak when $d \geqq \sqrt{2}$

$f_o = \left[2\pi\sqrt{R_1R_2C_1C_2}\right]^{-1}$

Wein Bridge Oscillator

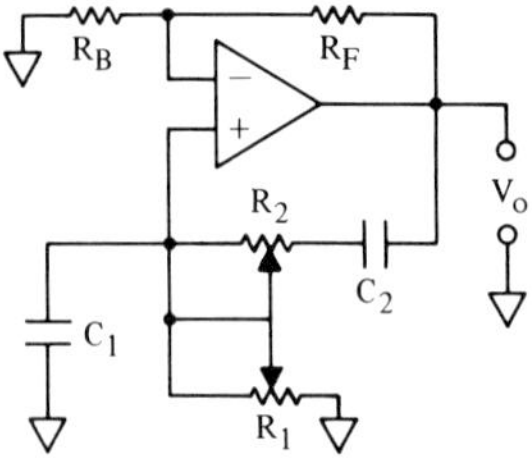

$f_o = [2\pi RC]^{-1}$

when $C_1 = C_2 = C$
$R_1 = R_2 = R,$
$R_F/R_B > 2$

Let $R_1/R_2 = C_1/C_2$

$R_F/R_B = 2R_1/R_2$ minimum

$f_o = [2\pi R_1C_1]^{-1} = [2\pi R_2C_2]^{-1} = \left[2\pi\sqrt{R_1R_2C_1C_2}\right]^{-1}$

Oscillator Using Low Pass VCVS Circuit

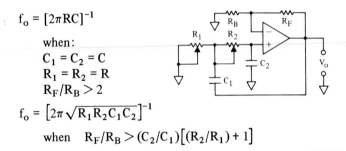

$f_o = [2\pi RC]^{-1}$

when:
$C_1 = C_2 = C$
$R_1 = R_2 = R$
$R_F/R_B > 2$

$f_o = \left[2\pi\sqrt{R_1R_2C_1C_2}\right]^{-1}$

when $R_F/R_B > (C_2/C_1)\left[(R_2/R_1) + 1\right]$

Notes:

1. Loop gain must be only slightly over unity if low distortion sinewave output is desired
2. Potentiometers above should have reverse log tapers if scale is on housing and log tapers if scale is on knob. (reverse audio and audio tapers may be substituted for reverse log and log tapers)

Function Generator

Let $R_3 =$

$[R_1^{-1} + R_2^{-1}]^{-1}$

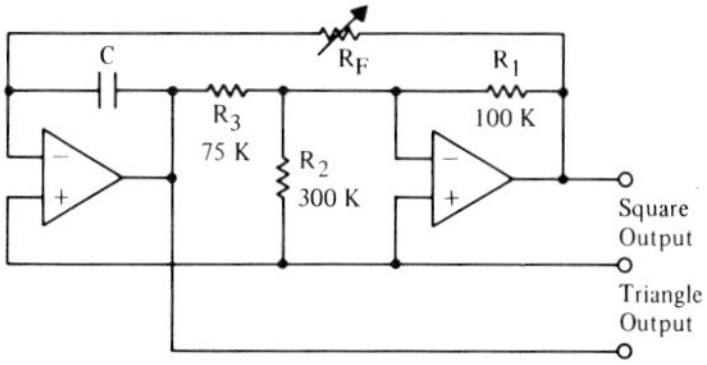

$$f_o = \frac{(R_2/R_1) + 1}{4CR_F}$$

Quadrature Oscillator
(Sinewave generator with 0° and 90° outputs)

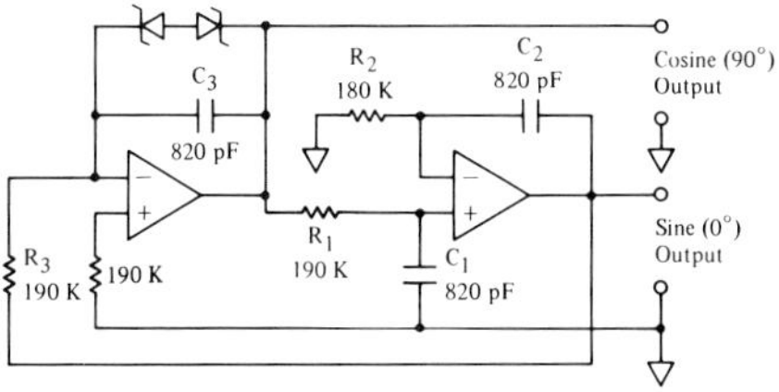

$$f_o = \left[2\pi(R_1R_2R_3C_1C_2C_3)^{\frac{1}{3}}\right]^{-1}$$

Note:

1. Four phase output (0°, 90°, 180°, 270°) can be obtained by the addition of two phase inverters.
2. Zener diode voltage ratings should total <2/3 total supply voltage.

f_o Notch Frequency

Notch Filter
(unity passband gain)

Let $R_2 = R_1$

Let $R_3 = R_1/2$

Let $C_2 = C_1$

Let $C_3 = 2C_1$

$f_o = [2\pi R_1 C_1]^{-1}$

Let $R_1C_1 = R_2C_2$

Let $R_3 = (R_1^{-1} + R_2^{-1})^{-1}$

Let $C_3 = C_1 + C_2$

$f_o = [2\pi R_1 C_1]^{-1}$

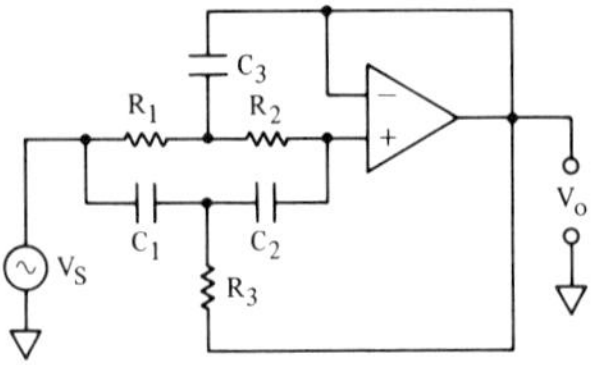

Filter with Notch and Bandpass Outputs

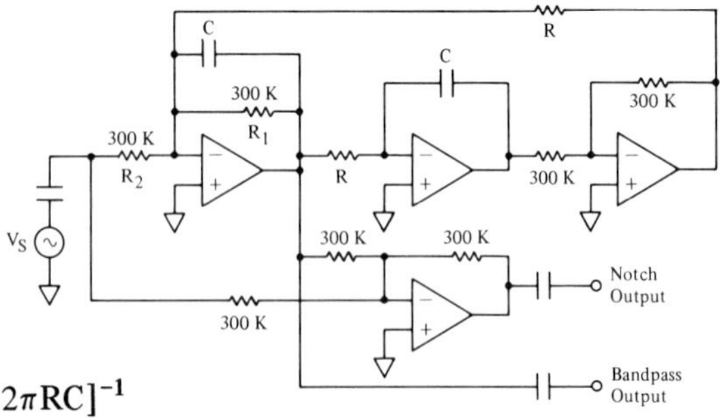

$f_o = [2\pi RC]^{-1}$

$Q = R_1/R, \quad A_{V(BP)} = R_1/R_2$

f_r Resonant Frequency	Applicable Notes
f_r = Resonant frequency f_r = The frequency at which a system will respond with maximum amplitude when driven by an external constant amplitude sinewave signal.	
Multiple Feedback Bandpass Filter Let $C_1 = C_2$ 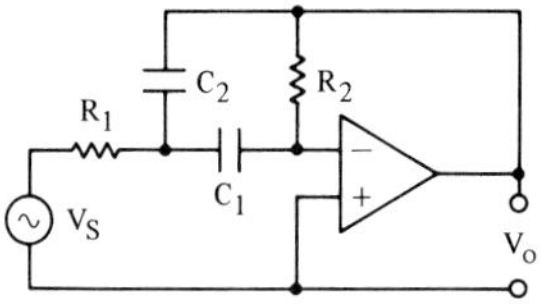 $f_r = \left[2\pi C_1 \sqrt{R_1 R_2}\right]^{-1}$ $A_{VO} = R_2/(2R_1)$ $Q = \sqrt{R_2/4R_1}$	① ③ ⑩ ⑪ ⑫ ⑮
High Z_{in}, Low A_V, High Q Version Let $C_1 = C_2$ 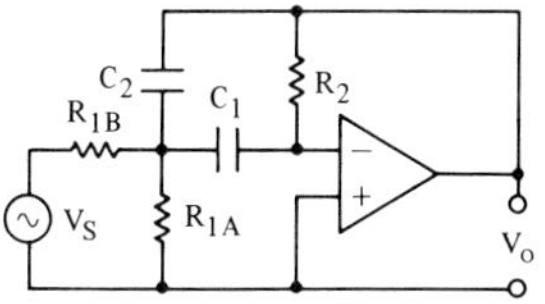 Let $R_1 = (R_{1A}^{-1} + R_{1B}^{-1})^{-1}$ $f_r = [2\pi R_1 C_1]^{-1}$ $A_{VO} = R_2/(2R_{1B})$ $Q = .5\sqrt{R_2(R_{1A}^{-1} + R_{1B}^{-1})}$	① ③ ⑩ ⑪ ⑫ ⑮

PBW

Power Bandwidth

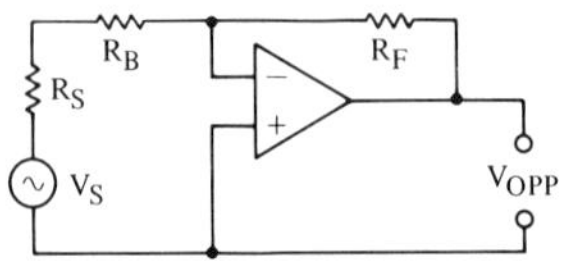

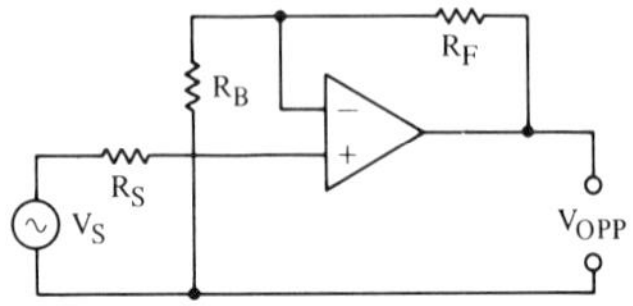

PBW = $BW_P = f_{(max)}$

PBW = In circuits where the low limit bandwidth is zero, the maximum frequency which may be used at a specified peak-to-peak output without the distortion associated with slew rate. (e.g. a sinewave becoming triangular)

PBW = $SR/(\pi V_{OPP})$

$$f_{(max)} = SR/(\pi V_{OPP})$$

$$V_{OPP(max)} = SR/(\pi f_{(max)})$$

$$SR_{(min)} = \pi f_{(max)} V_{OPP}$$

See also – SR, V_{OPP}, V_{OM}, MUF

See also – $BW_{(A_v = 1)}$ for small signal frequency limitations.

Q

Q Factor

Applicable Notes

Multiple Feedback Bandpass Filter

$Q = f_r/(\overline{BW}_{(3\,dB\,DOWN)})$

Let $C_1 = C_2$

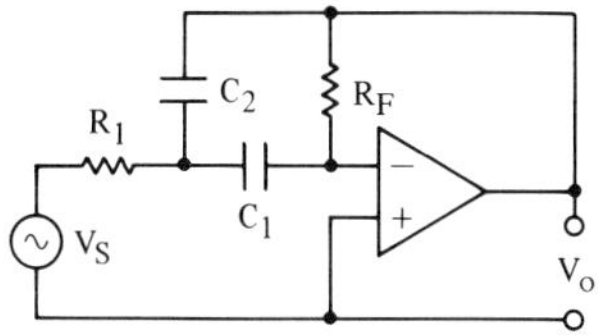

$Q = .5\sqrt{R_F/R_1}$

$Q = .25\sqrt{A_{VO}}$

$A_{VO} = R_F/(2R_1) = 2Q^2$

$f_r = \left[2\pi C_1\sqrt{R_F R_1}\right]^{-1}$

$f_r = 500/\left[\pi C_1\sqrt{R_F R_1}\right] \quad (C_{(\mu F)}, R_{(k\Omega)})$

Applicable Notes: ① ③ ⑩ ⑪ ⑫ ⑮

Higher Z_{in}, Higher Q and Lower A_V Version

Let $C_1 = C_2$

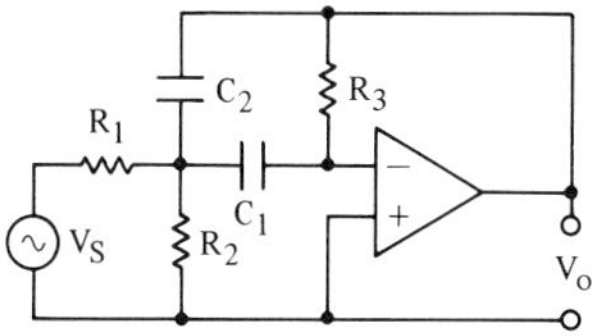

$Q = .5\sqrt{R_3(R_1^{-1} + R_2^{-1})}$

$f_r = \left[2\pi C_1\sqrt{R_3/(R_1^{-1} + R_2^{-1})}\right]^{-1}$

$A_{VO} = R_3/2R_1$

$R_2 = \left[(4Q^2/R_3) - R_1^{-1}\right]^{-1}$

Applicable Notes: ① ③ ⑩ ⑪ ⑫ ⑮

V_{ni} Equivalent Input Total Noise Voltage

$V_{ni} = V_{no}/A_v$

V_{ni} = Total equivalent input rms noise voltage including:

1. Device equivalent input noise voltage (V_n)
2. The product of the device equivalent input noise current (I_n) and the sum of the effective source resistances at both inputs.
3. The thermal noise voltage (V_{nR}) of the effective source resistances at both inputs.

Note: All three noise voltages have components of 1/f noise as well as constant spectral density (white) noise. The device white noise component is shot noise and the white noise of resistance is thermal noise. The 1/f noise of resistances is excess noise or current noise. White noise voltage may be easily calculated from a spot noise voltage by multiplying by the square root of the noise bandwidth but 1/f noise or noise having a significant 1/f noise component must be averaged over the total bandwidth by the rms method. 1/f noise is usually neglected at frequencies above 1 kHz and is often assumed to have straight line response between the 100 Hz and 1 KHz spot noise measurement points.

$V_{ni} = V_{no}$ when $A_v = 1$

$V_{ni} = \sqrt{BW[V_n^2 + I_n^2 R_S^2 + 4K_B T_K R_S]}$

$K_B = 1.38 \cdot 10^{-23}$

$T_K = °C + 273.15$

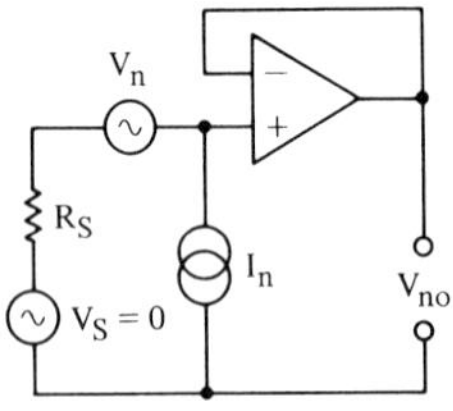

V_{ni} Equivalent Input Total Noise Voltage

$V_{ni} = V_{no}/A_v$

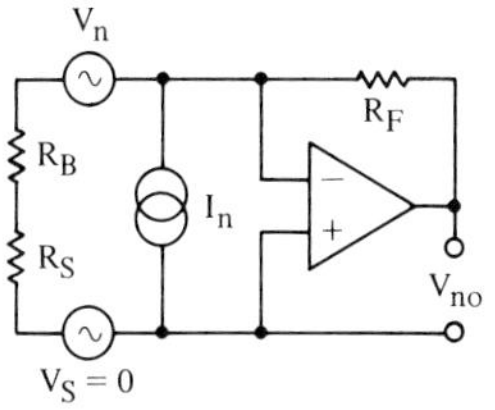

$V_{ni} = \sqrt{BW[V_1^2 + V_2^2 + V_3^2]}$

where $V_1 = V_n$

$V_2 = I_n R_X$

$R_X = \left[R_F^{-1} + (R_S + R_B)^{-1}\right]^{-1}$

$V_3 = \sqrt{4k_B T_K R_X}$

k_B = Boltzmann constant ($1.38 \cdot 10^{-23}$ J/°K)

T_K = Kelvin Temperature. (°C + 273.15)

$V_{ni} = V_{no}/A_v$

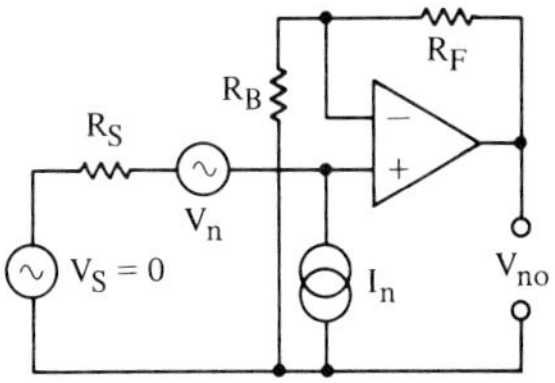

$V_{ni} = \sqrt{BW[V_1^2 + V_2^2 + V_3^2]}$

where $V_1 = V_n$, $V_2 = I_n R_X$

$R_X = R_S + (R_F^{-1} + R_B^{-1})^{-1}$

$V_3 = \sqrt{4k_B T_K R_X}$

k_B = Boltzmann constant ($1.38 \cdot 10^{-23}$)

T_K = Kelvin temperature. (°C + 273.15)

V_{no}

Noise Voltage Output

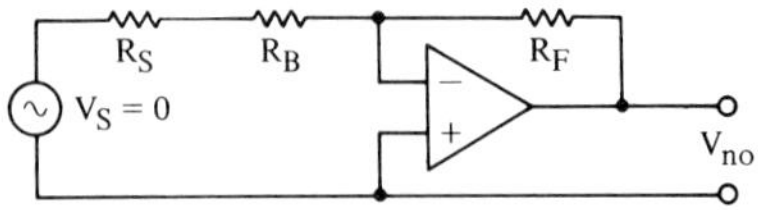

$$V_{no} = A_v \sqrt{BW} \cdot \sqrt{V_n^2 + (R_S + R_B)^2 (I_n^2 + 4kT_K R_F^{-1}) + 4k_B T_K (R_S + R_B)}$$

V_n = Equivalent input spot noise voltage of opamp at a given frequency. (usually given in $nV/\sqrt{Hz}$ at 1 kHz)

I_n = Equivalent input spot noise current of opamp at a given frequency. (usually given in $pA/\sqrt{Hz}$ at 1 kHz)

k_B = Boltzmann constant

$k_B = 1.38 \cdot 10^{-23}$ J/°K

T_K = Temperature in Kelvin

T_K = °C + 273.15

A_v = Closed loop circuit voltage amplification (A_{vcl} or V_o/V_S)

$A_v = R_F/(R_S + R_B)$

R_S = Source resistance

Notes:

1. Formula does not include opamp 1/f noise. (opamp 1/f noise usually is insignificant above 1 kHz)
2. Formula includes thermal noise of all external resistances, but does not include resistor excess noise (current noise or 1/f resistor noise)
3. Noise measurements require a bandwidth correction factor for all except rectangular response curves. See—BW_{NOISE}

V_{no}

Noise Voltage Output

Let BW = 10 kHz

Let $T \approx 27^\circ C$

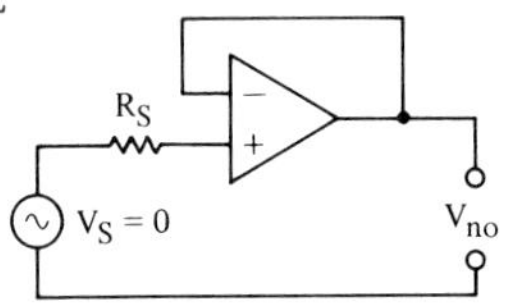

$$V_{no} = 100\sqrt{V_n^2 + I_n^2 R_S^2 + 1.656 \cdot 10^{-20} R_S}$$

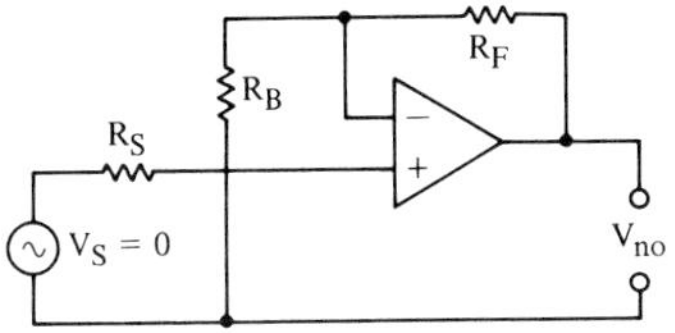

$$V_{no} = A_v \sqrt{\overline{BW}}\sqrt{V_n^2 + I_n^2(R_S + R_X)^2 + 4k_B T_K R_S}$$

V_n = Equivalent input spot noise voltage of opamp at a given frequency. (usually given in $nV/\sqrt{Hz}$ at 1 kHz)

I_n = Equivalent input spot noise current of opamp at a given frequency. (usually given in $pA/\sqrt{Hz}$ at 1 kHz)

T_K = Kelvin temperature ($^\circ C + 273.15$)

k_B = Boltzmann's constant ($1.38 \cdot 10^{-23}$)

$R_X = R_F \| R_B = (R_F^{-1} + R_B^{-1})^{-1}$

$A_v = (R_F/R_B) + 1$

See–Preceding page notes. See also–BW_{NOISE}

V_O	**Output Voltage**	**Applicable Notes**

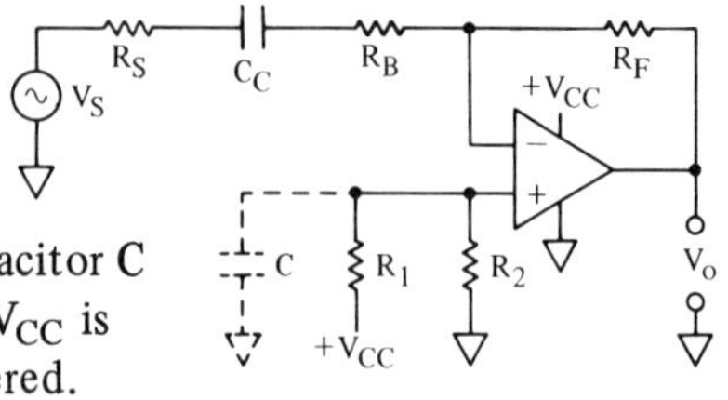

Add capacitor **C** unless $+V_{CC}$ is well filtered.

$$V_{O(DC)} = V_{CC}/2 \quad \text{when } R_1 = R_2$$

$$V_{O(DC)} = V_{CC}/\left[(R_1/R_2) + 1\right]$$

$$V_{O(AC)} = -A_V V_S$$

$$V_{O(AC)} = -(V_S R_F)/(R_B + R_S)$$

Applicable notes: ① ③ ④ ⑮

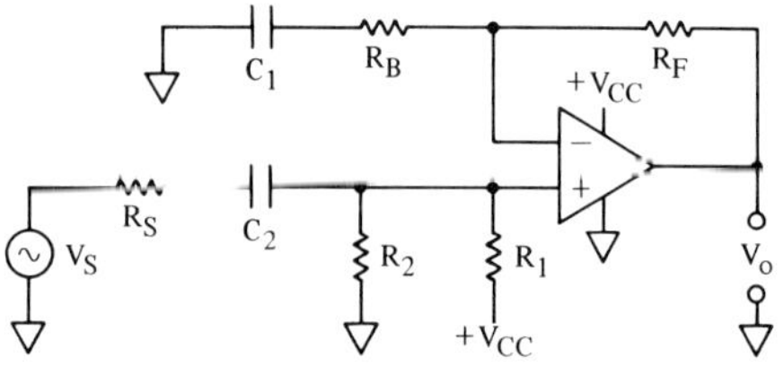

$$V_{O(DC)} = V_{CC}/2 \quad \text{when } R_1 = R_2$$

$$V_{O(DC)} = V_{CC}/\left[(R_1/R_2) + 1\right]$$

$$V_{O(AC)} \approx A_V V_S$$

$$V_{O(AC)} = V_S\left[(R_F/R_B) + 1\right]/\left[R_S(R_1^{-1} + R_2^{-1}) + 1\right]$$

Note: $+V_{CC}$ to R_1 must be well filtered. If R_S is low, such as the output impedance of another opamp stage, a large coupling capacitor (C_2) may provide proper filtering.

Applicable notes: ① ③ ⑮

V_O Output Voltage

Applicable Notes

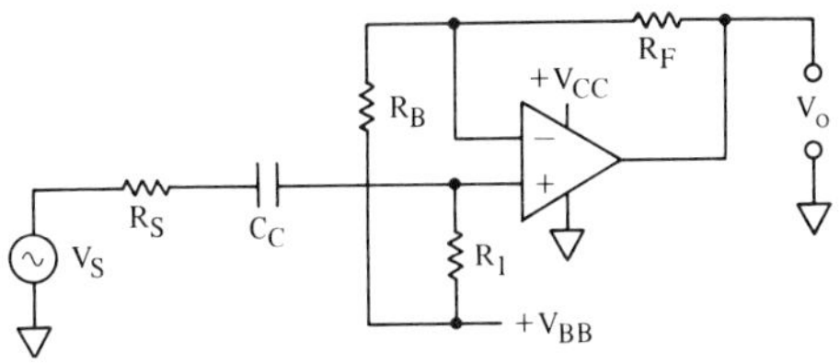

① ③ ⑮

$$V_{O(DC)} = +V_{BB} \pm V_{OO} \quad \text{See}-V_{OO}$$

$$V_{O(AC)} = A_V V_S$$

$$V_{O(AC)} = V_S\left[(R_F/R_B) + 1\right]/\left[(R_S/R_1) + 1\right]$$

Note: Maximum undistorted output is obtained when $V_{O(DC)} = V_{CC}/2$. Typically, V_{BB} must be between +2V and (V_{CC} − 2V) to prevent saturation.

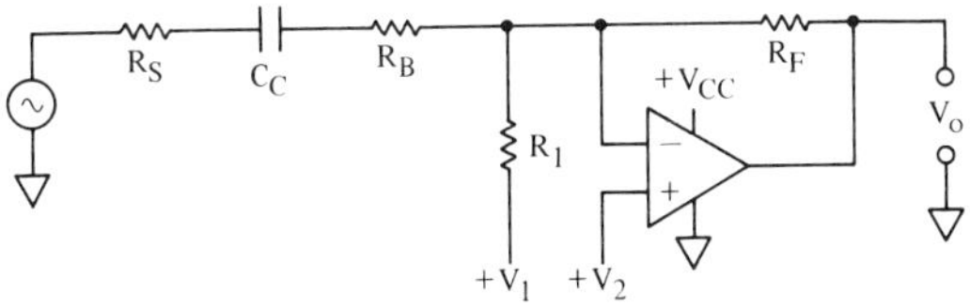

① ③ ⑮

$$V_{O(DC)} = \left(V_2\left[(R_F/R_1) + 1\right]\right) - \left[V_1(R_F/R_1) \pm V_{OO}\right]$$

$$V_{O(AC)} = A_V V_S$$

$$V_{O(AC)} = V_S\left[R_F/(R_B + R_S)\right]$$

Note: Maximum undistorted output is obtained when $V_{O(DC)} = V_{CC}/2$. Typically, V_1 and V_2 each must be between +2V and (V_{CC} − 2) to prevent saturation.

V_{OO} Output Offset Voltage

Applicable Notes

Output Offset Voltage
(input voltage(s) = 0)

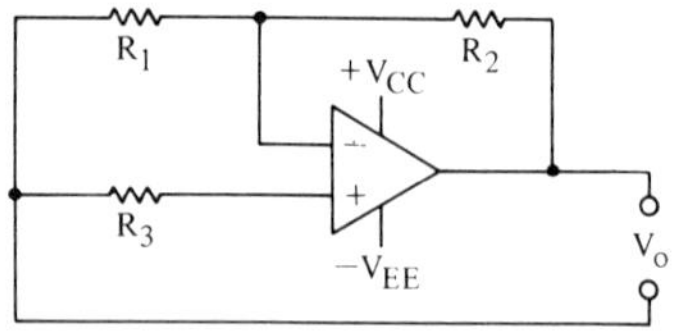

Output from input offset voltage (V_{IO}) only
($I_{IO} = 0, I_{IB} = 0$)

$$V_{OO} = V_{IO}(A_V + 1)$$

$$V_{OO} = V_{IO}\left[(R_2/R_1) + 1\right]$$

① ⑮

Output from input offset current (I_{IO}) only
($V_{IO} = 0, R_3 = [R_1^{-1} + R_2^{-1}]^{-1}$)

$$V_{OO} = I_{IO}R_3(A_V + 1)$$

$$V_{OO} = I_{IO}R_3\left[(R_2/R_1) + 1\right]$$

Output from bias current (I_{IB}) only
($I_{IO} = 0, V_{IO} = 0$)

$$V_{OO} = I_{IB}\left[R_3(A_V + 1) - R_2\right]$$

$$V_{OO} = I_{IB}\left[R_3(R_2/R_1 + 1) - R_2\right]$$

Total Output Offset Voltage

$$V_{OO} = \left[V_{IO}(A_V + 1)\right] + I_{IB}\left[R_3(A_V + 1) - R_2\right] \pm \left[I_{IO}R_3(A_V + 1)\right]$$

V_{OPP} **Maximum Peak to Peak Output Voltage**

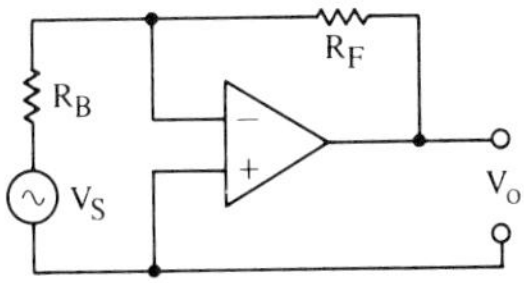

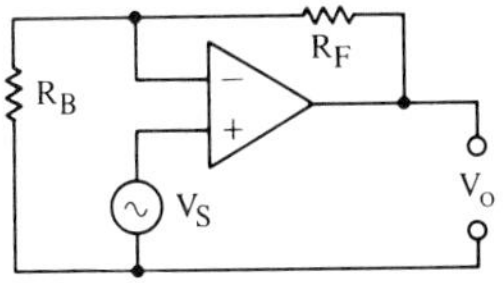

$V_{OPP} = SR/(2\pi f)$
when V_O is limited only by SR

$V_{OPP} <$ Total supply voltage
Typically $V_{OPP} > 2/3$ total supply voltage when output load resistance is 2 kΩ or higher.

Note: A sinewave input signal is transformed into a triangular wave output signal by the effects of SR at outputs above $V_{O(MAX\ P\text{-}P)}$.

Resistor R_F is effectively in parallel with the output load resistance R_L.

Opamp *Notes:*

④ A negative resultant for A_V or V_O indicates that a phase inversion has taken place. (output 180° out of phase with the input)

⑤ VCVS is the abbreviation for voltage controlled voltage source. VCVS highpass and lowpass filters are characterized by high input impedance, low output impedance and by non-inverted passband output.

Low Frequency Input Impedance — Z_i — **Input Impedance** — **Applicable Notes**

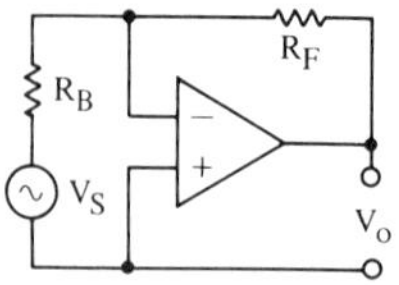

$Z_i \simeq R_B$

$Z_i = R_B + \left[R_F/(A_{VOL} + 1)\right]$

$Z_i = R_B + \left[R_F\Big/\left(\left[\log^{-1}(A_{VOL(dB)}/20)\right] + 1\right)\right]$

Applicable Notes: ① ③ ⑨ ⑪ ⑫ ⑮

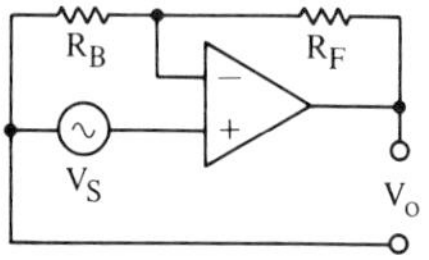

$Z_i \simeq [r_i A_{VOL}]/A_{VCL}$

$Z_i = \dfrac{r_i(A_{VOL} + 1)}{(R_F/R_B) + 1}$

$Z_i = r_i\left(\log^{-1}\left[(A_{VOL(dB)} - A_{VCL(dB)})/20\right] + 1\right)$

$A_{VCL(dB)} = 20\left(\log\left[(R_F/R_B) + 1\right]\right)$

Applicable Notes: ① ③ ⑨ ⑫ ⑮

Opamp *Notes:*

⑥ 12 dB per octave filters are also known as second order filters.

⑦ 6 dB per octave equals 20 dB per decade, 12 dB per octave equals 40 dB per decade, 18 dB per octave equals 60 dB per octave etc.

⑧ $|x|$ = The magnitude or the absolute value of x

⑨ $\log x = \log_{10} x$, $\log^{-1} x = \text{antilog}_{10}\, x = 10^x$

⑩ $x^{-1} = 1/x$, $x^{\frac{1}{2}} = \sqrt{x}$

⑪ Source resistance, if significant, must be considered as an additional resistance in series with the input.

Z_o Output Impedance

Applicable Notes: ① ③ ⑨ ⑫ ⑮

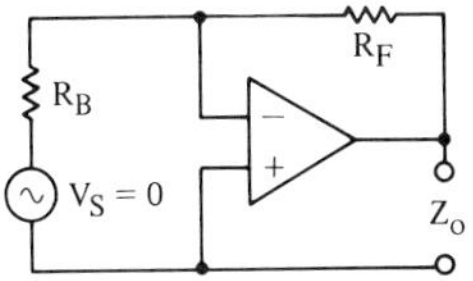

$$Z_o \simeq (r_o A_{VCL})/A_{VOL}$$

$$Z_o = \frac{r_o}{(A_{VOL}/A_{VCL}) + 1}$$

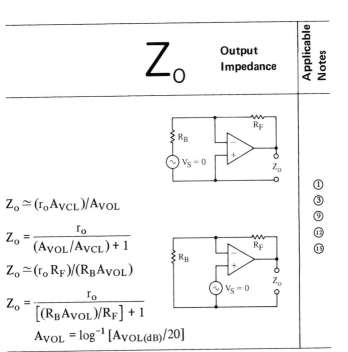

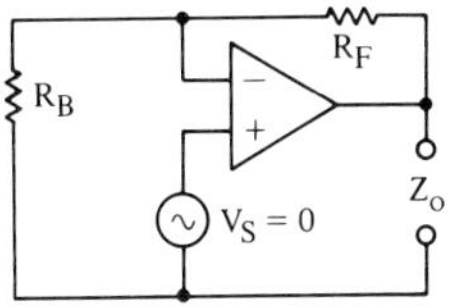

$$Z_o \simeq (r_o R_F)/(R_B A_{VOL})$$

$$Z_o = \frac{r_o}{[(R_B A_{VOL})/R_F] + 1}$$

$$A_{VOL} = \log^{-1} [A_{VOL(dB)}/20]$$

Opamp *Notes:*

⑫ When supply voltage connections are not shown, a split supply is assumed with V_{CC} positive with respect to common (ground) and V_{EE} negative with respect to common (ground).

⑬ Low gain Butterworth or Bessel response VCVS filters are relatively insensitive to value changes except for the cutoff frequency. 5% tolerance resistors and 10% tolerance capacitors are recommended for most applications.

⑭ Free gain Butterworth VCVS filters may be empirically determined by simultaneously adjusting the gain just under that which produces a slight hump while adjusting R_1, R_2, C_1 and/or C_2 for the proper 3 dB down frequency.

⑮ See–Appendix A for table of ratios possible with standard 5% value components.

APPENDIX A
RATIOS AVAILABLE FROM 5% COMPONENT VALUES

5% Component Values

**10	**33
11	36
*12	*39
13	43
**15	**47
16	51
*18	*56
20	62
**22	**68
24	75
*27	*82
30	91

** are also 10% and 20% values
* are also 10% values

Above values are available over the range of .1 ohm to 10 megohms in resistors.

5% capacitor values are not as available and demand a much greater premium than resistors and are not recommended. Resistor values may be changed to accept 20% value (not 20% tolerance) capacitors in almost all RC circuits. 10% tolerance capacitors and 5% tolerance resistors (7.5% overall) are recommended for most applications.

RATIOS 5% Component Value Ratios

Ratio	Values	Ratio	Values	Ratio	Values
9.38	150/16	8.24	750/91	6.96	390/56
9.23	120/13	8.23	510/62	6.94	430/62
9.23	360/39	8.20	82/10	6.92	270/39
9.22	470/51	8.18	180/22	6.91	470/68
9.17	110/12	8.18	270/33	6.88	110/16
9.17	220/24	8.15	220/27	6.83	82/12
9.17	330/36	8.13	130/16	6.83	560/82
9.15	430/47	8.00	120/15	6.82	75/11
9.15	750/82	8.00	160/20	6.82	150/22
9.12	620/68	8.00	240/30	6.81	620/91
9.11	510/56	7.69	100/13	6.80	68/10
9.10	91/10	7.69	300/39	6.80	510/75
9.09	100/11	7.68	430/56	6.67	100/15
9.09	200/22	7.67	330/43	6.67	120/18
9.09	300/33	7.66	360/47	6.67	160/24
9.07	390/43	7.65	390/51	6.67	180/27
9.07	680/75	7.58	91/12	6.67	200/30
9.03	560/62	7.58	470/62	6.67	220/33
9.01	820/91	7.56	620/82	6.67	240/36
9.00	180/20	7.50	75/10	6.50	130/20
9.00	270/30	7.50	120/16	6.47	330/51
8.89	160/18	7.50	150/20	6.43	360/56
8.89	240/27	7.50	180/24	6.38	300/47
8.67	130/15	7.50	270/36	6.32	430/68
8.46	110/13	7.50	510/68	6.31	82/13
8.46	330/39	7.47	560/75	6.29	390/62
8.43	430/51	7.47	680/91	6.28	270/43
8.39	470/56	7.45	82/11	6.27	470/75
8.37	360/43	7.41	200/27	6.25	75/12
8.33	100/12	7.33	110/15	6.25	100/16
8.33	150/18	7.33	220/30	6.25	150/24
8.33	200/24	7.27	160/22	6.22	510/82
8.33	300/36	7.27	240/33	6.20	62/10
8.30	390/47	7.22	130/18	6.18	68/11
8.29	680/82	7.06	360/51	6.15	240/39
8.27	91/11	7.02	330/47	6.15	560/91
8.27	620/75	7.00	91/13	6.11	110/18
8.24	560/68	6.98	300/43	6.11	220/36

RATIOS

5% Component Value Ratios

Ratio	Values	Ratio	Values	Ratio	Values
6.07	91/15	5.16	470/91	4.41	300/68
6.06	200/33	5.13	200/39	4.40	330/75
6.00	120/20	5.13	82/16	4.39	360/82
6.00	180/30	5.12	220/43	4.35	270/62
5.93	160/27	5.11	240/47	4.33	130/30
5.91	130/22	5.10	51/10	4.31	56/13
5.89	330/56	5.09	56/11	4.31	220/51
5.88	300/51	5.06	91/18	4.30	43/10
5.81	360/62	5.00	75/15	4.29	240/56
5.77	75/13	5.00	100/20	4.29	390/91
5.74	270/47	5.00	110/22	4.27	47/11
5.74	390/68	5.00	120/24	4.26	200/47
5.73	430/75	5.00	150/30	4.25	51/12
5.73	470/82	5.00	180/36	4.25	68/16
5.69	91/16	4.85	160/33	4.19	180/43
5.67	68/12	4.85	330/68	4.17	75/18
5.64	62/11	4.84	300/62	4.17	100/24
5.64	220/39	4.82	270/56	4.17	150/36
5.60	56/10	4.81	130/27	4.14	91/22
5.60	510/91	4.80	360/75	4.13	62/15
5.58	240/43	4.77	62/13	4.10	82/20
5.56	100/18	4.76	390/82	4.10	160/39
5.56	150/27	4.71	240/51	4.07	110/27
5.56	200/36	4.70	47/10	4.02	330/82
5.50	110/20	4.69	75/16	4.00	120/30
5.47	82/15	4.68	220/47	4.00	300/75
5.45	120/22	4.67	56/12	3.97	270/68
5.45	180/33	4.65	200/43	3.96	360/91
5.42	130/24	4.64	51/11	3.94	130/33
5.36	300/56	4.62	180/39	3.93	220/56
5.33	160/30	4.58	110/24	3.92	47/12
5.32	330/62	4.56	82/18	3.92	51/13
5.29	270/51	4.55	91/20	3.92	200/51
5.29	360/68	4.55	100/22	3.91	43/11
5.24	430/82	4.55	150/33	3.90	39/10
5.23	68/13	4.53	68/15	3.88	62/16
5.20	390/75	4.44	120/27	3.87	240/62
5.17	62/12	4.44	160/36	3.85	150/39

RATIOS 5% Component Value Ratios

Ratio	Values	Ratio	Values	Ratio	Values
3.83	180/47	3.29	270/82	2.80	56/20
3.79	91/24	3.27	36/11	2.79	120/43
3.78	68/18	3.25	39/12	2.78	75/27
3.75	75/20	3.24	220/68	2.78	100/36
3.73	56/15	3.23	200/62	2.77	36/13
3.73	82/22	3.21	180/56	2.77	130/47
3.72	160/43	3.20	240/75	2.76	91/33
3.70	100/27	3.19	150/47	2.75	33/12
3.67	110/30	3.18	51/16	2.73	30/11
3.66	300/82	3.14	160/51	2.73	82/30
3.64	120/33	3.13	47/15	2.70	27/10
3.63	330/91	3.13	75/24	2.69	43/16
3.62	47/13	3.11	56/18	2.68	150/56
3.61	130/36	3.10	62/20	2.68	220/82
3.60	36/10	3.09	68/22	2.67	200/75
3.60	270/75	3.08	120/39	2.65	180/68
3.58	43/12	3.06	110/36	2.64	240/91
3.57	200/56	3.04	82/27	2.61	47/18
3.55	39/11	3.03	91/30	2.60	39/15
3.55	220/62	3.03	100/33	2.58	62/24
3.53	180/51	3.02	130/43	2.58	160/62
3.53	240/68	3.00	30/10	2.56	100/39
3.50	56/16	3.00	33/11	2.56	110/43
3.49	150/43	3.00	36/12	2.55	51/20
3.44	62/18	3.00	39/13	2.55	56/22
3.42	82/24	2.97	270/91	2.55	120/47
3.41	75/22	2.94	47/16	2.55	130/51
3.40	51/15	2.94	150/51	2.54	33/13
3.40	68/20	2.94	200/68	2.53	91/36
3.40	160/47	2.93	220/75	2.52	68/27
3.37	91/27	2.93	240/82	2.50	30/12
3.33	100/30	2.90	180/62	2.50	75/30
3.33	110/33	2.87	43/15	2.48	82/33
3.33	120/36	2.86	160/56	2.45	27/11
3.33	130/39	2.83	51/18	2.44	39/16
3.31	43/13	2.83	68/24	2.44	200/82
3.30	33/10	2.82	62/22	2.42	150/62
3.30	300/91	2.82	110/39	2.42	220/91

RATIOS

5% Component Value Ratios

Ratio	Values	Ratio	Values	Ratio	Values
2.40	24/10	2.08	75/36	1.78	91/51
2.40	36/15	2.07	56/27	1.77	39/22
2.40	180/75	2.07	62/30	1.77	110/62
2.39	43/18	2.06	33/16	1.76	120/68
2.35	47/20	2.06	68/33	1.76	160/91
2.35	120/51	2.00	20/10	1.74	47/27
2.35	160/68	2.00	22/11	1.74	68/39
2.34	110/47	2.00	24/12	1.74	75/43
2.33	56/24	2.00	30/15	1.74	82/47
2.33	91/39	2.00	36/18	1.73	130/75
2.33	100/43	2.00	150/75	1.72	62/36
2.32	51/22	1.98	180/91	1.70	51/30
2.32	130/56	1.96	47/24	1.70	56/33
2.31	30/13	1.96	100/51	1.69	22/13
2.30	62/27	1.96	110/56	1.69	27/16
2.28	82/36	1.95	39/20	1.67	20/12
2.27	68/30	1.95	43/22	1.67	30/18
2.27	75/33	1.95	160/82	1.65	33/20
2.25	27/12	1.94	91/47	1.65	150/91
2.25	36/16	1.94	120/62	1.64	18/11
2.21	150/68	1.92	75/39	1.64	36/22
2.20	22/10	1.91	130/68	1.63	39/24
2.20	33/15	1.91	82/43	1.63	91/56
2.20	180/82	1.89	51/27	1.62	110/68
2.20	200/91	1.89	68/36	1.61	82/51
2.18	24/11	1.88	30/16	1.61	100/62
2.17	39/18	1.88	62/33	1.60	16/10
2.16	110/51	1.87	56/30	1.60	24/15
2.15	43/20	1.85	24/13	1.60	75/47
2.14	47/22	1.83	22/12	1.60	120/75
2.14	120/56	1.83	33/18	1.59	43/27
2.13	51/24	1.83	150/82	1.59	62/39
2.13	100/47	1.82	20/11	1.59	130/82
2.13	160/75	1.80	18/10	1.58	68/43
2.12	91/43	1.80	27/15	1.57	47/30
2.10	82/39	1.80	36/20	1.56	56/36
2.10	130/62	1.79	43/24	1.55	51/33
2.08	27/13	1.79	100/56	1.54	20/13

RATIOS 5% Component Value Ratios

Ratio	Values	Ratio	Values	Ratio	Values
1.50	15/10	1.32	62/47	1.11	62/56
1.50	18/12	1.32	82/62	1.11	91/82
1.50	24/16	1.32	120/91	1.10	11/10
1.50	27/18	1.31	47/36	1.10	22/20
1.50	30/20	1.31	51/39	1.10	33/30
1.50	33/22	1.30	13/10	1.10	43/39
1.50	36/24	1.30	39/30	1.10	56/51
1.47	22/15	1.30	43/33	1.10	68/62
1.47	75/51	1.30	56/43	1.10	75/68
1.47	91/62	1.25	15/12	1.10	100/91
1.47	100/68	1.25	20/16	1.09	12/11
1.47	110/75	1.25	30/24	1.09	24/22
1.46	82/56	1.23	16/13	1.09	36/33
1.46	120/82	1.23	27/22	1.09	47/43
1.45	16/11	1.22	22/18	1.09	51/47
1.45	68/47	1.22	33/27	1.09	82/75
1.44	39/27	1.22	62/51	1.08	13/12
1.44	56/39	1.22	100/82	1.08	39/36
1.44	62/43	1.21	47/39	1.07	16/15
1.43	43/30	1.21	68/56		
1.43	130/91	1.21	75/62	1.00	ALL
1.42	47/33	1.21	82/68		
1.42	51/36	1.21	91/75		
1.38	18/13	1.21	110/91		
1.38	22/16	1.20	12/10		
1.38	33/24	1.20	18/15		
1.36	15/11	1.20	24/20		
1.36	30/22	1.20	36/30		
1.35	27/20	1.19	43/36		
1.34	75/56	1.19	51/43		
1.34	91/68	1.19	56/47		
1.34	110/82	1.18	13/11		
1.33	16/12	1.18	39/33		
1.33	20/15	1.15	15/13		
1.33	24/18	1.13	18/16		
1.33	36/27	1.13	27/24		
1.33	68/51	1.11	20/18		
1.33	100/75	1.11	30/27		

This listing of all possible ratios between 10 and 1 may also be used for all other possible ratios by moving the proper decimal points.

APPENDIX B

ELECTRONIC TERMS AND THEIR SYMBOLS

This is an alphabetical listing of passive, bipolar-transistor and operational-amplifier (opamp) linear-circuit electronics terms with their corresponding symbols. Included also, are selected electronic, magnetic, acoustic, electrical, mechanical, mathematical and physical terms with their corresponding symbol, abbreviation, sign or acronym.

An attempt has been made to include present common usage (USA), traditional and recognized standard symbols, however, the preferred symbol (listed last) is often the author's projection of present trend, personal preference or arbitrary selection and does not necessarily represent an accepted industry standard.

This listing is intended as a reference source of electronic symbols, but may also be used to locate formulas having unfamiliar resultant symbols. It should be noted, however, that several different symbols may be shown for a given term and that the last-listed symbol is not always the one used in the formula and definition sections, since the last-listed symbol may be the author's projection of present trend.

Textbooks and scientific journals conventionally use italic (slanted) type for quantity symbols, however, this handbook follows the example of almost all technical manuals where roman (upright) type is used for both quantity and unit symbols. Unit symbols are clearly indicated as such in this appendix.

Common electronic abbreviations should be written without periods and generally in lower case letters as listed, however, certain abbreviations are capitalized and certain others are capitalized when used as a noun.

An asterisk is used to indicate schematic letter symbols.

No attempt has been made to include terms or symbols associated with computing systems, control systems, digital systems, digital devices, non-linear circuits, non-linear devices, vacuum tubes or field effect transistors.

a

about equal to	$\approx$
absolute temperature	
(quantity)	T, T_K
(unit)	K
absolute value (of x)	$\lvert x \rvert$
See also–magnitude	
absolute zero temperature	T_o
acceleration, angular	α
acceleration, linear	a
acoustic	
angular frequency	ω
angular velocity	ω
attenuation coefficient	α
damping coefficient	δ
frequency	f
impedance	Z_a
loudness level	L_N
mechanical impedance	Z_m
period	T
resonant frequency	f_r
reverberation time	T, T_{60}
sound power	P
sound power level	PWL, L_P
sound pressure	P
sound pressure level	SPL, L_p
sound velocity	c, v
specific impedance	Z_s
wavelength	λ
admittance	Y
input	Y_{in}, Y_i
magnitude	$\lvert Y \rvert, Y$
output	Y_o
vector	$\vec{Y}, \mathbf{Y}$
admittance, transistor (hybrid parameters)	
output	
common base	h_{ob}
common collector	h_{oc}
common emitter	h_{oe}
admittance, transistor, (y parameters)	
common base	
forward transfer	y_{fb}
input	y_{ib}
output	y_{ob}
reverse transfer	y_{re}
common emitter	
forward transfer	y_{fe}
input	y_{ie}
output	y_{oe}
reverse transfer	y_{re}
alpha (greek letter)	α
alpha, transistor	
small signal	α, h_{fb}
static (dc)	$\bar{\alpha}, h_{FB}$
alpha cutoff frequency	$f_{\alpha b}$
alternating current	AC, ac
ambient temperature	t_A, T_A
American wire gage	AWG
ampere (unit)	A
ampere-hour (unit)	$A \cdot h$, Ah

ampere-squared-seconds (unit)	I^2t
ampere-turn (unit of magnetomotive force)	A · t, A, At
ampere per meter (unit of magnetic field strength)	At/m, A/m
amplification	
(quantity)	A
(unit)	dg, (numeric), dB
See also–gain	
amplification,	
dc or large signal	
current	A_I
power (gain)	G_P
voltage	A_V
small signal	
current	A_i
power (gain)	G_p
voltage	A_v
amplification factor (vacuum tube)	μ
amplitude modulation	AM
angle, loss	δ
angle, phase	ϕ, θ
admittance	θ_Y
current	θ_I
impedance	θ_Z
voltage	$\phi_E, \phi_V, \theta_E, \theta_V$
angle, phase margin	ϕ_m, θ_m
angle, plane	ϕ, θ
angle, solid	Ω
angular frequency	ω
angular velocity	ω
antilogarithm (of x)	
base 10	lg^{-1}, 10^x, log^{-1}
base ϵ	e^x, ϵ^x, ln^{-1}
common	lg^{-1}, 10^x, log^{-1}
natural	e^x, ϵ^x, ln^{-1}
antiresonant frequency	f_o, f_r
apparent power	
(quantity)	S, P_s, VA
(unit)	VA
approximately equal to	$\approx$
arc cosine	arccos, cos^{-1}
hyperbolic	arcosh, $cosh^{-1}$
arc cosecant	arcsec, sec^{-1}
hyperbolic	arsech, $sech^{-1}$
arc cotangent	arccot, cot^{-1}
hyperbolic	arcoth, $coth^{-1}$
arc secant	arcsec, sec^{-1}
hyperbolic	arsech, $sech^{-1}$
arc sine	arcsin, sin^{-1}
hyperbolic	arsinh, $sinh^{-1}$
arc tangent	arctan, tan^{-1}
hyperbolic	artanh, $tanh^{-1}$
area	A
area, cross-sectional	S, A
atmosphere	atm
attenuation coefficient	α
atto (unit prefix for 10^{-18})	a
audio frequency	a-f
automatic frequency control	AFC

automatic gain control — AGC
average
- current — I_{av}
- noise current — $i_N, \overline{i_n}, i_n, I_N, \overline{I_n}, I_n$
- noise voltage — $E_N, \overline{E_n}, e_N, \overline{e_n}, \overline{V_n}, V_n$
- power (long term average) — $\overline{P}, P_{av}$
- power (short term or one cycle average) — P
- voltage — E_{av}, V_{av}

b

bandwidth
- 3dB down — $(f_2 - f_1)$, B, B_3, BW, $BW_{-3\,dB}$
- half power — $(f_2 - f_1)$, B, B_3, BW, $BW_{-3\,dB}$
- noise — B, BW, $\overline{B}$, B_n, $\overline{BW}$, BW_n
- unity gain — B_1, $BW_{(A_v = 1)}$

*base (transistor) — B
base 10 logarithm — lg, $\log_{10}$, log
base of natural logarithms — ε, e, ϵ
base ϵ logarithm — $\log_\epsilon$, ln
*base capacitor (transistor) — C_B
base current (transistor)
- small signal — I_b
- static (dc) — I_B

base resistance (transistor)
- external — R_B
- internal
  - small signal — r_b
  - static (dc) — r_B

base spreading resistance (transistor) — $r_{bb'}$, r_{bb}
base supply voltage (transistor) — V_{BB}
base-to-emitter voltage (transistor) — V_{BE}
- active — $V_{BE(ON)}$
- saturated — $V_{BE(SAT)}$

base voltage (transistor) — V_B
bel See—decibel
beta (greek letter) — β
beta, transistor
- small signal — β, h_{fe}
- static (dc) — $\overline{\beta}$, h_{FE}

bias current, input (op amp) — I_{IB}
Boltzmann constant — k, k_B
*bootstrap capacitor — C_B
breakdown, second (transistor)
- current — $I_{S/b}$
- energy — $E_{S/b}$

- breakdown voltage (transistor)
  - collector-to-base
    - emitter open BV_{CBO}, $V_{(BR)CBO}$
  - collector-to-emitter
    - base-emitter
      - circuit BV_{CEX}, $V_{CEX(SUS)}$
      - resistance BV_{CER}, $V_{CER(SUS)}$
      - shorted BV_{CES}, $V_{CES(SUS)}$
      - voltage BV_{CEV}, $V_{CEV(SUS)}$
    - base open BV_{CEO}, $V_{CEO(SUS)}$
  - emitter-to-base, collector open BV_{EBO}, $V_{(BR)EBO}$
- breadth (width) b
- British thermal unit Btu
- broadband
  - noise current $\overline{i_n}$, $\overline{I_n}$
  - noise voltage E_N, $\overline{E_n}$, $\overline{e_n}$, $\overline{V_n}$
  - voltage gain, common emitter transistor G_{VE}
- Brown and Sharpe wire gauge (American wire gage) AWG
- *bypass capacitor C_B

c

- capacitance C
  - parallel C_P, C_p
  - resonant C_o, C_r
  - series C_S, C_s
- capacitance, transistor
  - collector-to-base C_{cb}
  - collector-to-case C_c
  - emitter-to-base C_{eb}
  - feedback C_{FB}, $C_{b'c}$
  - input, common base C_{ib}
  - output, common base C_{ob}
    - open circuit C_{obo}
- capacitive
  - current $-I_X$, I_C
  - reactance $-X$, X_C
  - susceptance $-B$, B_C
  - voltage $-E_X$, $-V_X$, E_C, V_C
- *capacitor C
  - bootstrap C_B
  - bypass C_B
  - coupling C_C
  - feedback C_{FB}, C_F
- carrier frequency f_c
- case temperature t_C, T_C
- cathode-ray tube CRT
- Celsius temperature
  - (quantity) $t_{°C}$, t, T_C
  - (unit) °C
- cent ¢
- centi (unit prefix for 10^{-2}) c
- centigrade See—Celsius

centimeter (unit) cm
  cubic (unit) cm^3
  square (unit) cm^2
centimeter-gram-second
  (unit system) cgs, CGS
characteristic impedance Z_O
charge, electric Q
charge, elementary
  (charge of electron) e, q
closed-loop voltage
  amplification (op amp) A_{VCL}, A_V
coefficient,
  attenuation α
  coupling k
  damping δ
  temperature α, TC
*collector (transistor) C
collector current (transistor)
  small signal I_c
  static (dc) I_C
collector cutoff current
  (transistor)
  base open I_{CEO}
  base-to-emitter
    circuit I_{CEX}
    resistance I_{CER}
    short I_{CES}
    voltage I_{CEV}
collector dissipation
  (transistor) P_C
collector efficiency
  (transistor) η, η_C
collector resistance
  (transistor)
  external R_C
  internal, T equiv. r_c
collector supply voltage
  (transistor, op amp) V_{CC}
collector voltage
  (transistor)
  small signal e_c, E_c, V_c
  static (dc) V_C
common base (transistor)
  forward current
    transfer ratio h_{fb}
  input impedance h_{ib}
  output admittance h_{ob}
  reverse voltage
    transfer ratio h_{rb}
common collector (transistor)
  forward current
    transfer ratio h_{fc}
  input impedance h_{ic}
  output admittance h_{oc}
  reverse voltage
    transfer ratio h_{rc}
common emitter (transistor)
  forward current ratio
    small signal h_{fe}
    static (dc) h_{FE}
  input impedance h_{ie}
  output impedance h_{oe}
  reverse voltage
    transfer ratio h_{re}

common emitter (transistor)
  voltage gain G_{ve}
    broadband G_{VE}
common logarithm
  lg, $\log_{10}$, log
common mode (op amp)
  input voltage V_{ICM}
    range V_{ICR}
  rejection ratio CMRR
  voltage V_{CM}
complex quantity
  (phasor quantity)
  admittance $\vec{Y}$, **Y**
  current $\vec{I}$, **I**
  impedance $\vec{Z}$, **Z**
  voltage $\vec{E}$, $\vec{V}$, **E**, **V**
conductance G
conductance, mutual g_m
  See also—transconductance
  large signal G_m
conductance, transistor
  (real part of y
  parameters)
  common base
    forward transfer g_{fb}
    input g_{ib}
    output g_{ob}
    reverse transfer g_{rb}
  common emitter
    forward transfer g_{fe}
    input g_{ie}
    output g_{oe}
    reverse transfer g_{re}

constant,
  acceleration of free fall g
  Boltzmann k, k_B
  dielectric k, k_d
  gravitational G
  Planck h
  time τ, T
  See also—coefficient and
  factor
conversion gain G_c
conversion
  transconductance
  g_c, g_{mc}
cosecant cosec
  hyperbolic cosech
cosine cos
  hyperbolic cosh
cotangent cot
  hyperbolic coth
coulomb (unit) Q
*coupling capacitor C_C
coupling coefficient k
  critical k_c
critical
  angular frequency ω_c
  angular velocity ω_c
  coupling coefficient k_c
  frequency f_c
  wavelength λ_c
crossover
  angular frequency ω_c
  angular velocity ω_c

- crossover
  - frequency — f_c
  - wavelength — λ_c
- cubic units
  - centimeter — cm^3
  - foot — cu ft, ft^3
  - inch — cu in, in^3
  - meter — m^3
  - yard — cu yd, yd^3
- current — I
  - alternating — I_{AC}, I_{ac}, I
  - average — I_{av}
  - capacitive — $+jI_X$, I_C
  - direct — I_{DC}, I_{dc}, I
  - effective — I_{eff}, I_{rms}, I
  - generator — I_g
  - inductive — $-jI_X$, I_L
  - input — I_{in}, I_i
  - instantaneous — i
  - lagging — $-jI_X$
  - leading — $+jI_X$
  - magnitude — I
  - noise — i_N, I_N, i_n, I_n
  - output
    - small signal — I_o
    - large signal — I_O
  - peak — I_{pk}, i_p, I_p
  - peak-to-peak — $I_{p\text{-}p}$
  - phasor — $\vec{I}$, $\mathbf{I}$
  - polar form — $\mathbf{I}_{POLAR}$
  - rectangular form — $\mathbf{I}_{RECT}$
  - root-mean-square — I_{rms}, I
- current
  - second breakdown — $I_{S/b}$
  - vector (phasor) — $\vec{I}$, $\mathbf{I}$
- current, opamp
  - bias — I_B
  - device
    - negative supply — $I-$, I_{D-}, I_{EE}
    - non-inverting input
      - grounded — I_{DG}
      - open — I_{DO}
    - positive supply — $I+$, I_{D+}, I_{CC}
  - input
    - bias — I_{IB}
    - offset — I_{IO}
    - signal — I_{IN}, I_{in}
  - noise, equivalent input
    - 1/f — I_{nf}
    - device — I_n
    - shot — I_{ns}
  - noise, thermal noise of
    - input resistance — I_{nR}
  - output
    - large-signal — I_O
    - maximum — I_{OM}
    - negative swing — I_{O-}
    - peak-to-peak — I_{OPP}
    - positive swing — I_{O+}
    - shorted — I_{OS}
    - small-signal — I_o

- current, transistor
  - base, small-signal I_b
  - base, static (dc) I_B
  - collector, small-signal I_c
  - collector, static (dc) I_C
  - emitter, small-signal I_e
  - emitter, static (dc) I_E
- current, transistor
  - collector cutoff
  - base-emitter
    - circuit I_{CEX}
    - resistance I_{CER}
    - shorted I_{CES}
    - voltage I_{CEV}
- current, transistor
  - emitter cutoff
  - collector open I_{EBO}
- customary temperature t
- cutoff
  - angular frequency ω_c
  - angular velocity ω_c
  - frequency f_c
  - wavelength λ_c
- cycle, duty
  - See–duty factor
- cycles per second cps, c/s, Hz
  - See also–hertz

d

- damping coefficient δ
- damping factor α, δ, d
- deci (unit prefix for 10^{-1}) d
- decibel (ratio unit for power, voltage and current) dB
- decibel level See–level
- decilog dg
- decimal point .
- degree °
- deka (unit prefix for 10) (rare in USA) da
- delay time t_d
- delta (greek letter)
  - capital Δ
  - script δ
- depth d
- device under test DUT
- diameter d
  - inside d_i, d_{in}, ID
  - outside d_o, d_{out}, OD
- dielectric constant k, k_d
- dissipation
  - collector (transistor) P_C
  - device P_D
  - power P_D
  - total P_t, P_T
- dissipation factor D
- distance d
- distortion
  - intermodulation IM, IMD
  - total harmonic THD
- direct current DC, dc
- double pole (switch)
  - double throw DPDT
  - single throw DPST
- drain See–FET literature

duty cycle
See—duty factor
duty factor F_D, D, df
dynamic resistance r
See—vaccum tube literature
See also—internal small-signal resistance (transistor and opamp)
dyne (CGS unit) dyn

e

effective
bandwidth
B, BW, BW_{NOISE}, BW_{eff}, $\overline{B}$, $\overline{BW}$, B_n, BW_n
current (ac)
I_{eff}, I_{rms}, I
power P
radiated power ERP
voltage (ac)
E_{eff}, E_{rms}, V_{rms}, E, V
See also—equivalent and total
efficiency η
electric charge Q
electromotive force
emf, E, V
See also—voltage
elementary charge (charge of electron) e, q
*emitter (transistor) E
breakdown voltage
BV_{EBO}, $V_{(BR)EBO}$
emitter (transistor)
*capacitor C_E
resistance, external R_E
resistance, internal
small signal r_e
static (dc) r_E
*emitter resistor R_E
energy e, E, W
second breakdown $E_{S/b}$
epsilon (greek letter) ε, ϵ
equal =
approximately $\approx$
identically $\equiv$
not ‡, $\neq$
very nearly $\cong$, $\simeq$
equivalent (of x)
x_{equiv}, x_T, x_t, x =
Note: The resultant of formulas is the equivalent quantity.
equivalent series resistance ESR
erg (CGS unit) erg
eta (greek letter) η
exa (unit prefix for 10^{18}) E
excess noise voltage
E_{EX}, $E_{N(EX)}$, $V_{nR(EX)}$

f

factor
damping α, δ, d
dissipation D
energy See—quality
flare (flaring) m

- factor
  - magnification See—quality
  - merit See—quality
  - noise (transistor) (noise figure) — F, NF, F_n
  - power — $\cos\theta$, PF, pf, F_P
  - Q — Q
  - quality — Q
  - storage See—quality
- Fahrenheit temperature
  - (quantity) — t, $t_{°F}$, T_F
  - (unit) — °F
- fall time — t_f
- farad (unit) — F
- feedback
  - * capacitor — C_{FB}, C_F
  - * resistor — R_{FB}, R_F
  - transfer ratio — β
- femto (unit prefix for 10^{-15}) — f
- field effect transistor — FET
- field strength, electric
  - (quantity) — E
  - (unit) — V/m
- field strength, magnetic
  - (quantity) — $\mathbf{H}$, H
  - (unit) — Oe, At/m, A/m
- figure, noise (transistor) (noise factor) — F, NF, F_n
- flare (acoustic horn)
  - cutoff frequency — f_{FC}
  - factor — F_F, m
- flux density, magnetic
  - (quantity) — B
  - (unit) — G, T
- flux, total magnetic
  - (quantity) — Φ, ϕ
  - (unit) — Mx, Wb
- foot (unit) — ′, ft
  - cubic (unit) — cu ft, ft^3
  - square (unit) — sq ft, ft^2
- force
  - electromotive — emf, E, V
  - magnetizing See—magnetic field strength
  - magnetomotive
    - (quantity) — F, $\mathcal{F}$, F_m
    - (unit) — A · t, A, At
  - mechanical
    - (quantity) — F
    - (unit) — kgf, lbf, N
- forward current (semiconductor) — I_F
- forward current transfer ratio
  - common base — h_{fb}
  - common collector — h_{fc}
  - common emitter
    - small-signal — h_{fe}
    - static (dc) — h_{FE}
- forward transfer admittance
  - common base — y_{fb}
  - common emitter — y_{fe}

frequency	f
angular	ω
carrier	f_c
critical	f_c
critical angular	ω_c
crossover	f_c
crossover angular	ω_c
cutoff	f_c
cutoff angular	ω_c
deviation	f_d
Doppler shift	f_D
extremely high	ehf
flare cutoff	f_o, f_c, f_{FC}
high	hf
input	f_i, f_{in}
intermediate	i-f
low	lf
lowest satisfactory horn loading	f'
maximum usable	MUF
midband	f_o
modulation	f_m
oscillation	f_{osc}, f_o
pulse repetition	f_p
reference	f_{ref}, t_o
resonant	f_o, f_r
resonant, angular	ω_0, ω_r
superhigh	shf
transition, transistor	f_T, f_t
ultra high	uhf
very high	vhf
very low	vlf
frequency modulation	FM
function	F, f

g

gain (amplification)	
current	
large-signal	A_I
small-signal	A_i
margin	ϕ_m, θ_m
voltage	
large signal	A_V
small signal	A_v
transistor	G_{ve}
broadband	G_{VE}
gain (power)	
large signal	G_P
small-signal	G_p
transistor	
common base, large signal	G_{PB}
common base small signal	G_{pb}
common emitter large-signal	G_{PE}
common emitter small-signal	G_{pe}
gain-bandwidth	
product	GBW
opamp (unity gain frequency)	B_1, $BW_{(A_V = 1)}$
transistor (transition frequency)	f_t, f_T

gamma (greek letter) γ
gate See–FET literature
gauss (CGS unit) G
generator current i_g, I_g
generator voltage e_g, E_g, V_g
giga (unit prefix for 10^9) (pronouced jiga) G
gilbert (CGS unit) Gb
gram (CGS unit) g
gravitational
- acceleration g
- acceleration, standard g_n
- constant G

greater than (x) $>x$
- not $\ngtr x$
- or equal to $\geqslant x$

grid See–vacuum tube literature

h

harmonic distortion, total THD
heater See–vacuum tube literature
heatsink temperature t_S, T_S
hecto (unit prefix for 10^2) (rare USA) h
height h
henry (unit) H
hertz (unit) Hz
high frequency (3-30 MHz) hf
high frequency
- extremely (30-300 GHz) ehf
- super (3-30 GHz) shf
- ultra (300 MHz-3 GHz) uhf
- very (30-300 MHz) vhf

horn, acoustic
- flare cutoff
  - frequency f_o, f_c, f_{FC}
- flaring factor F_F, m
- lowest frequency for
  - satisfactory loading f'

horsepower (unit) hp
hour (unit) h
hour, ampere (unit) A · h, Ah
hybrid parameter (transistor)
- forward current ratio
- small signal
  - common base h_{fb}
  - common collector h_{fc}
  - common emitter h_{fe}
- static (dc)
  - common emitter h_{FE}

hybrid parameter (transistor)
- input impedance
  - common base h_{ib}
  - common collector h_{ic}
  - common emitter h_{ie}

hybrid parameter	
(transistor)	
output admittance	
— common base	h_{ob}
— common collector	h_{oc}
— common emitter	h_{oe}
reverse voltage ratio	
— common base	h_{rb}
— common collector	h_{rc}
— common emitter	h_{re}

i

idling current	I_i, I_q
idling current drift	ΔI_i, ΔI_q
imaginary number	i, j
imaginary part of (x)	Im x
imaginary part of transistor	
y parameters	
common base	
— forward transfer admittance	$\pm b_{fb}$
— input admittance	$\pm b_{ib}$
— output admittance	$\pm b_{ob}$
— reverse transfer admittance	$\pm b_{rb}$
common emitter	
— forward transfer admittance	$\pm b_{fe}$
— input admittance	$\pm b_{ie}$
— output admittance	$\pm b_{oe}$
— reverse transfer admittance	$\pm b_{re}$

impedance	Z
— characteristic	Z_O
— input	Z_{in}, Z_i
— magnitude	Z
— mechanical	Z_m
— output	Z_o
— parallel	Z_P, Z_p
— phasor	$\vec{Z}$, $\mathbf{Z}$
— polar form	$\mathbf{Z}_{POLAR}$
— primary	Z_p
— rectangular form	$\mathbf{Z}_{RECT}$
— scalar	Z
— secondary	Z_s
— series	Z_S, Z_s
— vector (phasor)	$\mathbf{Z}$
impedance, opamp	
— small signal	
— input	
— — high frequency	z_i
— — — common mode	z_{ic}
— — low frequency	R_i, r_i
— output	z_o
impedance, transistor	
— small signal	
— input, high frequency	
— — common base	z_{ib}
— — common emitter	z_{ie}
— input, low frequency	
— — common base	h_{ib}
— — common collector	h_{ic}
— — common emitter	h_{ie}
— output	z_o
— — See also–admittance	

inch (unit) in
  cubic (unit) cu in, in^3
  square (unit) sq in, in^2
increment Δ
indefinite number n
index, noise NI
inductance L
  mutual M
  parallel L_P, L_p
  primary L_p
  resonant L_r
  secondary L_s
  series L_S, L_s
induction, magnetic
  See—magnetic field
    strength
inductive
  current $-jI_X$, $+I_X$, I_L
  reactance $+X$, X_L
  susceptance $-jB$, $+B$, B_L
  voltage
    $+E_X$, $+V_X$, E_L, V_L
*inductor L
*inductor, mutual L_M
infinity ∞
infra-red IR
input admittance Y_{in}, Y_i
  transistor
    common base y_{ib}
    common emitter y_{ie}
input capacitance C_{in}, C_i
  transistor
    common base C_{ib}, C_{ibo}
    common emitter
      C_{ie}, C_{ieo}
input equivalent noise
  (opamp and transistor)
  current i_n, I_n
  total e_{ni}, V_{ni}
  voltage e_n, V_n
input frequency f_i, f_{in}
input impedance
  opamp z_{in}, z_i
    common mode z_{ic}
input impedance, transistor
  common base h_{ib}
  common collector h_{ic}
  common emitter h_{ie}
    high frequency z_{ie}
    low frequency r_{ie}
input offset current
  (opamp) I_{IO}
input offset voltage
  (opamp) V_{IO}
input power P_{in}, P_i
input resistance R_{in}, R_i
  opamp R_i, r_i
    differential r_{id}
  transistor
    common base
      R_{ib}, $Re(h_{ib})$, r_{ib}
    common emitter
      R_{ie}, $Re(h_{ie})$, r_{ie}

instantaneous
current i
peak current i_{pk}, i_p
peak power p_{pk}, P_{pk}
peak voltage e_{pk}, e_p, v_{pk}, v_p
power P
voltage e, v
*integrated circuit IC
intermediate frequency i-f
intermodulation IM
intermodulation distortion IM, IMD
internal resistance, opamp
input R_i, r_i
output R_o, r_o
internal resistance, transistor (T equivalent)
base r_b
collector r_c
emitter r_e
intrinsic standoff ratio (unijunction transistor) η
inverse See–arc, negative reciprocal or reverse

j

joint army-navy specification JAN
joule (unit) W · s, Ws, J

k

kelvin (unit) K
kelvin temperature (thermodynamic temperature) T, T_K
kilo (unit prefix for 10^3) K, k
knot(unit) kn

l

lambda (greek letter)
capital Λ
script λ
lead temperature t_L, T_L
leakage coefficient σ
leakage current I_L
transistor See–cutoff current
leakage inductance L'_s, l_s
primary L'_p, l_p
secondary L'_s, l_s
length ℓ
less than (x) $<x$
or equal to $\leqq x$
not $\nless x$
level (in decibels)
current
ref. 1 pA $L_{I/pA}$
power
ref. 1 mW dBm, $L_{P/mW}$
ref. 1 fW $L_{P/fW}$
sound power
ref. 1 pW PWL, $L_{P/pW}$
sound pressure
ref. 20 μPa/m^2 SPL, $L_{p/20\,\mu Pa}$

level (in decibels)	
voltage	
ref. 1V	dBV, $L_{V/V}$
ref. $1V_{p-p}$	dBv, $L_{V/V_{p-p}}$
light amplification by stimulated emission of radiation	LASER, laser
light dependent resistor	LDR
light emitting diode	LED
line (of magnetic flux) (unit) See–Maxwell	
liter (unit)	l, L
load admittance	Y_L
load impedance	Z_L
load resistance	R_L
*load resistor	R_L
loaded Q	Q_L
logarithm	
base 10	lg, $\log_{10}$, log
base ϵ	$\log_{\epsilon}$, ln
common	lg, $\log_{10}$, log
natural	$\log_{\epsilon}$, ln
loss angle	δ
lot tolerance percent defective	LTPD
low frequency	lf
very	vlf

m

magnetic field strength	
(quantity)	**H**, H
(unit)	Gb/cm, Oe, At/m, A/m
magnetic flux	
(quantity)	Φ, ϕ
(unit)	Mx, Wb
magnetic flux density	
(quantity)	B
(unit)	G, T
(magnetic) permeability	
(quantity)	μ
(unit)	G/Oe, (numeric)
(magnetic) reluctance	
(quantity)	R, $\mathcal{R}$
(unit)	A/Wb, At/Wb
magnetizing force See–magnetic field strength	
magnetomotive force	
(quantity)	$\mathcal{F}$, F, F_m
(unit)	A · t, A, At
magnitude (of x)	$\|x\|$
magnitude of	
admittance	$\|Y\|$, Y
capacitive reactance	X_C
capacitive susceptance	B_C
current	$\|I\|$, I
impedance	$\|Z\|$, Z
inductive reactance	X_L
inductive susceptance	B_L
input offset current (opamp)	$\|I_{IO}\|$, I_{IO}
input offset voltage (opamp)	$\|V_{IO}\|$, V_{IO}
reactance	X
susceptance	B

magnitude of	
voltage	$\lvert E \rvert$, $\lvert V \rvert$, E, V
magnification factor (Q factor or quality factor)	Q
margin, gain	A_m
margin, phase	ϕ_m, θ_m
mark See—sign	
mass	m
maximum (device)	
available gain	MAG
output current	I_{OM}
peak-to-peak	I_{OPP}
output swing bandwidth	B_{OM}
output voltage	V_{OM}
usable frequency	MUF
maxwell (CGS unit)	Mx
mean-time-between-failures	MTBF
mean-time-to-failure	MTTF
mean-time-to-first-failure	MTTFF
mechanical	
efficiency	η
energy	E, W
force	F
impedance	Z_m
power	P
pressure	p
torque	T
work	W

medium frequency (300 kHz–3 MHz)	mf
mega (unit prefix for 10^6)	M
merit factor	Q
See also—quality factor	
meter (unit)	m
cubic (unit)	m^3
square (unit)	m^2
mho (unit)	mho, S, ℧, Ω^{-1}
See also—seimens	
micro (unit prefix for 10^{-6})	μ
mile (unit)	mi
square (unit)	mi^2
mile per hour (unit)	mph, mi/h
milli (unit prefix for 10^{-3})	m
milli-inch (unit)	mil
mode, common	
rejection	CMR
rejection ratio	CMRR
mouth area (acoustic horn)	S_M, A_M
mu (greek letter)	μ
*mutual capacitor	C_M
mutual conductance (transconductance)	g_m
transistor	
common emitter	g_{me}
large-signal	G_{me}, g_{ME}
mutual inductance	M
*mutual inductor	L_M
mutual impedance	Z_M

n

nano (unit prefix for 10^{-9}) n
naperian logarithm $\log_\epsilon$, ln
natural logarithm $\log_\epsilon$, ln
natural resonant
  frequency f_n
negative –
negative quantity
  See–specific quantity
"negative reactance" $-X$, X_C
negative supply (opamp
  or npn transistor)
  current I_{EE}
  voltage V_{EE}
neper (power ratio unit) Np
net parallel susceptance
  $(B_L - B_C)$, $\pm B$
net series reactance
  $(X_L - X_C)$, $\pm X$
neutralizing capacitor C_N
newton (unit) N
*no connection NC
noise current
  average
    (broadband) $\overline{i_n}$, $\overline{I_n}$
  spot (1 Hz BW)
    i_n, I_n, $I_{n/\sqrt{Hz}}$
noise current, device
  equivalent input
  average (broadband) $\overline{i_n}$, $\overline{I_n}$
  spot (1 Hz BW)
    i_n, I_n, $I_{n/\sqrt{Hz}}$

noise, excess
  (quantity)
    E_{EX}, $E_{N(EX)}$, $V_{nR(EX)}$
  (unit) $\mu V/V_{dc}$
noise factor
  (quantity) F, NF, F_n
  (unit) dB
noise figure
  See–noise factor
noise index
  (quantity) NI
  (unit) dB
noise power N, P_n
noise, resistance
  See–thermal noise
noise temperature T_N
noise, thermal
  current i_N, $I_{n(th)}$, I_{nR}
  power N_{th}, $P_{n(th)}$, P_{nR}
  voltage e_N, $E_{n(th)}$, V_{nR}
noise voltage
  average (broadband)
    e_n, $\overline{e_n}$, $\overline{E_n}$, $\overline{V_n}$
  spot (1 Hz BW)
    e_n, V_n, $e_{n/\sqrt{Hz}}$, $V_{n/\sqrt{Hz}}$
noise voltage, device
  equivalent input
  1/f E_{nf}, e_{nf}, V_{nf}
  average (broadband)
    e_n, V_n, $\overline{E_n}$, $\overline{e_n}$, $\overline{V_n}$
  shot e_s, e_{ns}, V_{ns}
  spot (1 Hz BW)
    e_n, V_n, $e_{n/\sqrt{Hz}}$, $V_{n/\sqrt{Hz}}$

noise voltage, device equivalent input total (V_n, $I_n R_S$ and V_{nR_S}) — E_{ni}, e_{ni}, V_{ni}
noise voltage output — E_{no}, e_{no}, V_{no}
*non-polar (capacitor) — NP
*normally closed (contact) — NC
*normally open (contact) — NO
number — n, N
  definite — N
  imaginary — i, j
  indefinite — n
  pairs of poles — N_{pp}
  poles — N_p
  primary turns — N_p
  secondary turns — N_s
  turns — N, N_t
  turns ratio — n, $N_{p/s}$

O

oersted (CGS unit) — Oe
ohm (unit) — Ω
omega (greek letter)
  capital — Ω
  script — ω
on-off ratio
  See—duty factor
open loop voltage amplification (opamp) — A_{VOL}
operating temperature — t_{opr}, T_{OPR}
operational amplifier — op amp, opamp
operational transconductance amplifier — OTA
optimum resistance — R_{opt}
oscillation frequency — f_{osc}, f_o
output admittance — Y_o
output admittance, transistor
  h parameters
    common base — h_{ob}
    common collector — h_{oc}
    common emitter — h_{oe}
  y parameters
    common base — y_{ob}
    common emitter — y_{oe}
output capacitance — C_{out}, C_o
output capacitance, transistor — C_{out}, C_o
  common base — C_{ob}
    open circuit — C_{obo}
  common emitter — C_{oe}
    open circuit — C_{oeo}
output current — I_o
output current (opamp)
  maximum — I_{OM}
  peak-to-peak — I_{OPP}
  shorted output — I_{OS}
output frequency — f_{out}, f_o

output impedance	
circuit	Z_o
opamp	z_o
transistor	z_o
See also—output admittance	
output power	P_o
output resistance	
circuit	R_o
opamp	R_o, r_o
transistor	r_o
See also—output conductance	
output voltage	E_o, V_o
output voltage, opamp	
maximum (peak)	V_{OM}
peak-to-peak	V_{OPP}
overshoot	OS, os

p

parallel	
capacitance	C_P, C_p
impedance	Z_P, Z_p
inductance	L_P, L_p
resistance	R_P, R_p
parameters, hybrid	
See—hybrid parameters	
passband voltage amplification	A_{vo}
peak current	I_{pk}, i_p, I_p
peak inverse voltage	PIV
peak reverse voltage	PRV
peak power	p, P_{pk}, P_p
peak voltage	$e_{pk}, e_p, E_{pk}, E_p, V_{pk}, V_p$
peak-to-peak	
current	I_{p-p}
voltage	E_{p-p}, V_{p-p}
peak-to-peak (opamp)	
current	I_{OPP}
voltage	V_{OPP}
percent	%
period	.
period, time	T
permeability (magnetic)	μ
permeance (magnetic)	$\mathscr{P}$
permittivity (dielectric constant)	k, k_d
peta (unit prefix for 10^{15})	P
phase angle	ϕ, θ
admittance	ϕ_Y, θ_Y
current	ϕ_I, θ_I
impedance	ϕ_Z, θ_Z
voltage	$\phi_E, \phi_V, \theta_E, \theta_V$
phase margin	ϕ_m, θ_m
phasor quantities	
admittance	$\mathbf{Y}$
polar	$\mathbf{Y}_{POLAR}$
rectangular	$\mathbf{Y}_{RECT}$
current	$\mathbf{I}$
polar	$\mathbf{I}_{POLAR}$
rectangular	$\mathbf{I}_{RECT}$
impedance	$\mathbf{Z}$
polar	$\mathbf{Z}_{POLAR}$
rectangular	$\mathbf{Z}_{RECT}$

phasor quantities	
voltage	$\mathbf{E}$, $\mathbf{V}$
polar	$\mathbf{E}_{POLAR}$
rectangular	$\mathbf{E}_{RECT}$
phi (greek letter)	ϕ
pi (greek letter)	π
pico (unit prefix for 10^{-12}) (pronounced "peeko")	p
Planck constant	h
plate See—vacuum tube literature	
*plug (male connector)	P
polar	
admittance	$Y\angle\theta_Y$, $\mathbf{Y}_{POLAR}$
current	$I\angle\theta_I$, $\mathbf{I}_{POLAR}$
impedance	$Z\angle\theta_Z$, $\mathbf{Z}_{POLAR}$
voltage	$E\angle\theta_E$, $\mathbf{E}_{POLAR}$
	$V\angle\theta_V$, $\mathbf{V}_{POLAR}$
pole frequency (poles and zeros)	f_p
positive	+
positive quantities See—specific quantity	
positive supply, opamp	
current	I_{D+}, I_{CC}
voltage	V_{D+}, V_{CC}
positive supply, transistor	
npn	
current	I_{CC}
voltage	V_{CC}
pnp	
current	I_{EE}
voltage	V_{EE}

potential See—voltage	
pound (unit)	lb
pound per square inch	psi
power	P
power amplifier	PA
power factor	$\cos\theta$, PF, pf, F_P
power gain	G_P
transistor, large-signal	
common base	G_{PB}
common emitter	G_{PE}
transistor, small signal	
common base	G_{pb}
common emitter	G_{pe}
power, device	P_D
power dissipation	P_D
power, effective radiated	ERP
power input	P_{in}, P_i
power level (quantity)	
reference 1 fW	$L_{P/fW}$
reference 1 mW	$L_{P/mW}$
power level (unit)	
reference 1 fW	dBf
reference 1 mW	dBm
power level, acoustic	
reference 1 pW	PWL, $L_{P/pW}$
power output	P_{out}, P_o
power, radiated	P_R
power ratio (unit)	dB
power, signal	S, P_s
power, total	P_T, P_t

prefix, unit multiplier		
atto	(10^{-18})	a
centi	(10^{-2})	c
deci	(10^{-1})	d
deka	(10)	da
exa	(10^{18})	E
femto	(10^{-15})	f
giga	(10^{9})	G
(pronounced jiga)		
hecto	(10^{2})	h
kilo	(10^{3})	k
mega	(10^{6})	M
micro	(10^{-6})	μ
milli	(10^{-3})	m
nano	(10^{-9})	n
peta	(10^{15})	P
pico	(10^{-12})	P
(pronounced peeko)		
tera	(10^{12})	T

primary	
current	I_p
impedance	Z_p
voltage	E_p, V_p
printed circuit	PC
printed circuit board	PCB
printed wiring board	PWB
programable unijunction transistor	PUT
psi (greek letter)	ψ
public address (system)	PA
pulse energy test	PET

q

Q factor	Q
quality assurance	QA
quality control	QC
quality factor	Q
quantity of charge	
(quantity)	Q
(unit)	C
quench frequency	f_q
quiescent current	I_q
quiescent voltage	E_q, V_q

r

radian (unit)	rad
radius	r
radiated power	P_R
effective	ERP
radiation efficiency	η, η_R
radiation resistance	R_R
radio detection and ranging	RADAR, radar
radio frequency	rf, r-f
radio frequency choke	RFC
radio frequency interference	RFI
random noise	
See–thermal noise	
rate, repetition (frequency)	f
ratio (of x to y)	x/y, x : y

- ratio (unit)
  - current, voltage
  - or power — (numeric), dB
  - other — (numeric)
- ratio, power supply rejection
  - (opamp) — PSRR
- ratio, transistor
  - forward current transfer
  - small signal
    - common base — h_{fb}
    - common collector — h_{fc}
    - common emitter — h_{fe}
  - static (dc)
    - common emitter — h_{FE}
- ratio, transistor
  - reverse voltage transfer
    - common base — h_{rb}
    - common collector — h_{rc}
    - common emitter — h_{re}
- ratio, turns — n, $N_{p/s}$
- reactance — X
  - capacitive — -X, X_C
  - inductive — +X, X_L
  - parallel — $\pm X_P$, X_p
  - series — $\pm X_S$, X_s
- reactive
  - current — $\pm I_X$, I_X
  - power — P_q, var
  - voltage — $\pm E_X$, $\pm V_X$, E_X, V_X
- real part of (x) — Re (x)
- real part of transistor
  - admittance
  - common base
    - forward transfer — Re (y_{fb}), g_{fb}
    - input — Re (y_{ib}), g_{ib}
    - output — Re (h_{ob}), Re (y_{ob}), g_{ob}
    - reverse transfer — Re (y_{rb}), g_{rb}
  - common emitter
    - forward transfer — Re (y_{fe}), g_{fe}
    - input — Re (y_{ie}), g_{ie}
    - output — Re (h_{oe}), Re (y_{oe}), g_{oe}
    - reverse transfer — Re (y_{re}), g_{re}
- rectangular form
  - admittance — $\mathbf{Y}_{RECT}$
  - current — $\mathbf{I}_{RECT}$
  - impedance — $\mathbf{Z}_{RECT}$
  - voltage — $\mathbf{E}_{RECT}$, $\mathbf{V}_{RECT}$
- reference — re, ref
  - angular frequency — ω_o
  - angular velocity — ω_o
  - current — I_{ref}
  - frequency — f_o
  - voltage — E_{ref}, V_{ref}
- reluctance (magnetic) — $\mathcal{R}$
- reluctivity (magnetic) — v, μ^{-1}

repetition rate	
(frequency)	f
resistance	R
device input	r_i
device output	r_o
generator	R_g
input	R_{in}, R_i
output	R_{out}, R_o
parallel	R_P, R_p
series	R_S, R_s
source	R_S
resistance, opamp	
input	R_i, r_i
output	R_o, r_o
resistance, transistor	
input	
common base	h_{ib}, Re (h_{ib}), r_{ib}
common emitter	h_{ie}, Re (h_{ie}), r_{ie}
output See also– output conductance	r_o
resistance, transistor,	
saturation	$r_{CE(SAT)}$
resistive current	I_R
resistive voltage	E_R, V_R
resistivity	ρ
*resistor	R
* base (transistor)	R_B
* collector (transistor)	R_C
* emittor (transistor)	R_E
* feedback	R_F

resonant	
angular frequency	ω_o, ω_r
angular velocity	ω_o, ω_r
capacitance	C_o, C_r
frequency	f_o, f_r
inductance	L_o, L_r
wavelength	λ_o, λ_r
reverberation time	T_{RVB}, T, T_{60}
reverse current	I_R
reverse transfer	
admittance (transistor)	
common base	y_{rb}
common emitter	y_{re}
reverse voltage	V_R
reverse voltage, peak	PRV
reverse voltage transfer	
ratio (transistor)	
common base	h_{rb}
common collector	h_{rc}
common emitter	h_{re}
revolutions per	
minute (unit)	r/min, rpm
second (unit)	rps, r/s
rho (greek letter)	ρ
rise time	t_r
root-mean-square	rms

S

saturation	SAT
saturation resistance	
(transistor)	$r_{ce(SAT)}$

scalar See—magnitude	
screen See—vacuum tube literature	
second (angle unit)	″
second (time unit)	s
second breakdown (transistor)	
current	$I_{S/b}$
energy	$E_{S/b}$
secondary	
current	I_s
impedance	Z_s
turns	N_s
voltage	E_s, V_s
sectional area	S, A
sensitivity	S
sensitivity, power supply (opamp)	PSS
series	
aiding inductance	L_{SA}
capacitance	C_S, C_s
impedance	Z_S, Z_s
inductance	L_S, L_s
opposing inductance	L_{SO}
reactance	X_S, $\pm X_s$, X_s
resistance	R_S, R_s
short-circuit output current (opamp)	I_{OS}
shot noise See—noise	
siemens (unit)	S
See also—mho	

sigma (greek letter)	
capital	Σ
script	s, σ
signs and marks	
absolute value	| |
addition	+
approaches	≐
ampersand	&
and	&
angle	∠
apostrophe	’
asterisk	*
at	@
because	∵
braces	{ }
brackets	[ ]
breve	˘
caret	^
cent	¢
circumflex	ˆ
colon	:
comma	,
congruent	≅
dagger	†
decimal point	.
degree	°
difference	~
directly proportional	∝
division	÷
dollar	$
double dagger	‡
em dash	—
en dash	–

signs and marks	
equal to	=
approximately	≈
congruently	≅
identically	≡
not	≢
nearly	≃
not	ǂ, ≠
very nearly	≑, ≅
equivalent	⇳
exclamation mark	!
factorial	!
greater than	>
not	≯
or equal to	≥, ≧
hyphen	-
inch	″
infinity	∞
integral	∫
less than	<
not	≮
or equal to	≤, ≦
macron	—
mean value	—
minute	′
minus	–
multiplication	×, ·
negative	–
not	
equal to	ǂ, ≠
greater than	≯
identical	≢
less than	≮

signs and marks	
number	#
paragraph	¶
parallel	‖
parentheses	( )
partial differential	∂
percent	%
period	.
plus	+
plus or minus	±, ±
positive	+
positive or negative	±, ±
pound	#
prime	′
double (second)	″
triple (third)	‴
proportion	::
proportional, directly	∝
question mark	?
quotation marks	″ ″, " "
radical sign	√
ratio	:
second	″
sectional symbol	§
semicolon	;
solidus	/
subtraction	–
therefore	∴
varies as	∝
viculum	—
virgule	/
signal	S, sig

signal generator
- current I_g
- impedance Z_g
- resistance R_g
- voltage E_g, V_g

signal, large
- See–specific quantity

signal level
- See–level

signal power P_s

signal, small
- See–specific quantity

signal source
- current I_S
- voltage E_S, V_S

signal-to-noise ratio S/N

silicon controlled rectifier SCR

silicon controlled switch SCS

silicon unilateral switch SUS

sine sin
- hyperbolic sinh

sinewave power P_{sine}

single pole (switch)
- double throw SPDT
- single throw SPST

single sideband SSB

sink temperature (heatsink) t_S, T_S

small-signal
- See-specific quantity

*socket (receptacle or female connector) S

sound navigation and ranging SONAR, sonar

sound power P

sound power level, ref. 1 pW PWL, $L_{P/pW}$

sound pressure p

sound pressure level, ref. 20 μPa/m^2 SPL, $L_{p/20\,\mu Pa}$

source
- current I_S
- impedance Z_S
- resistance R_S
- voltage E_S, V_S

source (field effect transistor)
- See–FET literature

spacing s

speed
- See also–velocity
- light c
- sound c, v

spot noise See–noise

square units
- centimeter cm^2
- foot sq ft, ft^2
- inch sq in, in^2
- meter m^2
- mile sq mi, mi^2
- yard sq yd, yd^2

square wave power P_{sqr}

standing wave ratio
- power SWR
- voltage S, VSWR

static transistor parameter
See–specific parameter
storage factor
See–quality factor
sum Σ
summation Σ
super high frequency shf
supply voltage sensitivity
(opamp) PSS, k_{SVS}
susceptance B
capacitive B_C
inductive B_L
susceptance, transistor
(imaginary part of
y parameters)
common base
forward transfer
$\pm jb_{fb}, \pm b_{fb}, b_{fb}$
input $\pm jb_{ib}, \pm b_{ib}, b_{ib}$
output $\pm jb_{ob}, \pm b_{ob}, b_{ob}$
reverse transfer
$\pm jb_{rb}, \pm b_{rb}, b_{rb}$
common emitter
forward transfer
$\pm jb_{fe}, \pm b_{fe}, b_{fe}$
input $\pm jb_{ie}, \pm b_{ie}, b_{ie}$
output $\pm jb_{oe}, \pm b_{oe}, b_{oe}$
reverse transfer
$\pm jb_{re}, \pm b_{re}, b_{re}$
sustaining voltage
See–voltage
*switch S, SW

t

tangent tan
hyperbolic tanh
tau (greek letter) τ
television TV
temperature
ambient t_A, T_A
case t_C, T_C
Celsius $t_{°C}$, t, T_C, T
centigrade
See–Celsius
coefficient α, TC
Fahrenheit t, t_F, T_F
junction t_J, T_J
Kelvin T, T_K
lead t_L, T_L
noise T_n, T_N
sink (heat) t_S, T_S
tab t_T, T_T
tera (unit prefix for 10^{12}) T
tesla (magnetic unit) T
thermal conductance G_θ
thermal conductivity λ
thermal noise
See–noise
thermal resistance θ, R_θ
theta (greek letter)
capital Θ
script θ
threshold current I_{TH}
throat area
(acoustic horn) S_o, A_o
time t

time constant	τ, T
time, delay	t_d
time, fall	t_f
time of one cycle	T
time,	
periodic	T
phase propagation	t_ϕ
pulse duration	t_p
rise	t_r
reverberation	T_{RVB}, T, T_{60}
storage	T_S, t_S, t_{STG}
total	t_{TOT}
torque	T
total (also meaning effective or equivalent)	
admittance	Y_T, Y_t
capacitance	C_T, C_t
conductance	G_T, G_t
current	I_T, I_t
dissipation	P_t, P_T
harmonic distortion	THD
impedance	Z_T, Z_t
inductance	L_T, L_t
power	P_t, P_T
resistance	R_T, R_t
susceptance	B_T, B_t
time	t_{TOT}
voltage	E_T, V_T, E_t, V_t
transadmittance See–admittance	
transconductance	g_m
See also–mutual conductance	

*transformer	T
*transistor	Q, TR
transistor parameters See–specific parameter	
transistor-under-test	TUT
transmission loss (attenuation)	
(quantity)	α
(unit)	(numeric), dB
*tube, vacuum	V
turn(s)	n, N
ampere (magnetic unit)	A · t, A, At
primary	N_p
ratio	n, $N_{p/s}$
secondary	N_s

u

ultra-high-frequency	uhf
ultra-violet	UV
unijunction (transistor)	UJT
unipolar transistor (field effect transistor)	FET
unknown	
capacitance	C_x
current	I_x
impedance	Z_x
inductance	L_x
resistance	R_x
voltage	E_x, V_x
unloaded Q	Q_u

V

*vacuum tube V
vacuum tube voltmeter VTVM
variable frequency oscillator VFO
vector See also–phasor
  admittance $\mathbf{Y}$
  current $\mathbf{I}$
  impedance $\mathbf{Z}$
  voltage $\mathbf{E}$, $\mathbf{V}$
velocity See also–speed
  (quantity) v
  (unit) ft/s, m/s
velocity of light See–speed
velocity of sound c, v
very high frequency (30-300 MHz) vhf
very low frequency (3-30 kHz) vlf
very nearly equal to $\cong$
volt (unit) V
  ac VAC, V AC, V ac
  average V av
  dc VDC, V DC, V dc
  peak V pk
  peak-to-peak V p-p
  root-mean-square V rms
voltage (quantity)
  ac E_{ac}, V_{ac}
  amplification A_V, A_v
voltage (quantity)
  average E_{av}, V_{av}
  capacitive E_C, V_C
  dc E_{dc}, V_{dc}
  effective E_{rms}, V_{rms}, E, V
  gain (amplification) A_V, A_v
  generator E_g, V_g
  inductive E_L, V_L
  input E_{in}, V_{in}, E_i, V_i
  instantaneous e, v
    peak e_p, v_p
  level L_V
  output E_o, V_o
  peak E_{pk}, V_{pk}, E_p. V_p
  peak-to-peak E_{p-p}, V_{p-p}
  polar $E\angle\theta_E$, $\mathbf{E}_{POLAR}$, $V\angle\theta_V$, $\mathbf{V}_{POLAR}$
  power supply E_{PS}, V_{PS}
  primary E_p, V_p
  rectangular $\mathbf{E}_{RECT}$, $\mathbf{V}_{RECT}$
  resistive E_R, V_R
  root-mean-square E_{rms}. V_{rms}, E, V
  source E_S, V_S
voltage controlled oscillator VCO
voltage controlled resistor VCR

voltage, transistor
(general)
base V_B
base supply V_{BB}
base-to-emitter V_{BE}
active $V_{BE(ON)}$
saturated $V_{BE(SAT)}$
collector V_C
collector supply V_{CC}
collector-to-base V_{CB}
emitter open V_{CBO}
collector-to-emitter V_{CE}
base-emitter
circuit V_{CEX}
resistance V_{CER}
short V_{CES}
voltage V_{CEV}
base open V_{CEO}
emitter V_E
emitter supply V_{EE}
emitter-to-base V_{EB}
open collector V_{EBO}

voltage, transistor,
breakdown
collector-to-base
emitter open
BV_{CBO}, $V_{(BR)CBO}$
collector-to-emitter
base-emitter
circuit
BV_{CEX}, $V_{CEX(SUS)}$

voltage, transistor
breakdown
base-emitter
resistance
BV_{CER}, $V_{CER(SUS)}$
base-emitter
short
BV_{CES}, $V_{CES(SUS)}$
base open
BV_{CEO}, $V_{CEO(SUS)}$
emitter-to-base
collector open
BV_{EBO}, $V_{(BR)EBO}$

voltage, transistor,
sustaining LV, $V_{(SUS)}$
collector-to-emitter
base-emitter resistance
LV_{CER}, $V_{CER(SUS)}$
base-emitter short
LV_{CES}, $V_{CES(SUS)}$
base-emitter voltage
LV_{CEV}, $V_{CEV(SUS)}$
base open
LV_{CEO}, $V_{CEO(SUS)}$

voltage, working WV

voltampere
(apparant power)
(quantity) S, P_s, VA
(unit) VA

volt-ohm meter VOM

volume (cubic content) V

volume unit (similar to dBm)
vu, VU

W

watt (unit)	W
watthour (unit)	W · h, Wh
wattsecond (unit)	W · s, Ws
(joule)	J
wavelength	λ
weber (magnetic unit)	Wb
weight	W
See also–mass	
white noise See–noise	
width (breadth)	b
wire gage (gauge)	
American	AWG
British standard	SWG
steel	Stl WG
wirewound (resistor)	WW
work	
(quantity)	W
(unit)	KWh, W · s, J
working voltage	WV
wye connection	Y

X Y Z

xi (greek letter)	
capital	Ξ
script	ξ
zener (semiconductor)	
current	I_Z
impedance	Z_Z
voltage	V_Z
zeta (greek letter)	ζ

LIBRARY
ST. LOUIS COMMUNITY COLLEGE
AT FLORISSANT VALLEY

SPRING '83